4th Regional Process Technology Conference 2012

League City, Texas, USA
4-5 October 2012

ISBN: 978-1-5108-1595-7

Printed from e-media with permission by:

Curran Associates, Inc.
57 Morehouse Lane
Red Hook, NY 12571

Some format issues inherent in the e-media version may also appear in this print version.

Printed by Curran Associates, Inc. (2015)

For permission requests, please contact AIChE
at the address below.

AIChE
120 Wall Street, FL 23
New York, NY 10005-4020

Phone: (800) 242-4363
Fax: (203) 775-5177

www.aiche.org

Additional copies of this publication are available from:

Curran Associates, Inc.
57 Morehouse Lane
Red Hook, NY 12571 USA
Phone: 845-758-0400
Fax: 845-758-2634
Email: curran@proceedings.com
Web: www.proceedings.com

TABLE OF CONTENTS

PETROLEUM REFINING TECHNOLOGY 2

PROCESS DEVELOPMENT & CHEMICAL MANUFACTURING

TECHNICAL SESSIONS FOR YOUNG PROFESSIONALS

Jim Griffin

Plant Manager, Dianal America

A Community's Perspective

1981 Events

- Federal Spending = $678 billion
- Federal Debt = $995 billion
- The First Launch of the Space Shuttle (*Columbia*)
- Academy Award for Engineering went to Eastman Kodak for the Prostar Microfilm Processer
- First DeLorean sports car rolled off of the assembly line

Henry, IL.

A Nation Growing in Population

Total US Population (All Ages) - 56 Year Graph
USA Population Growth Graph
The below graph shows monthly Total USA Population Growth (All Ages). Measurement is in Millions. Updated Monday, March 26, 2012.

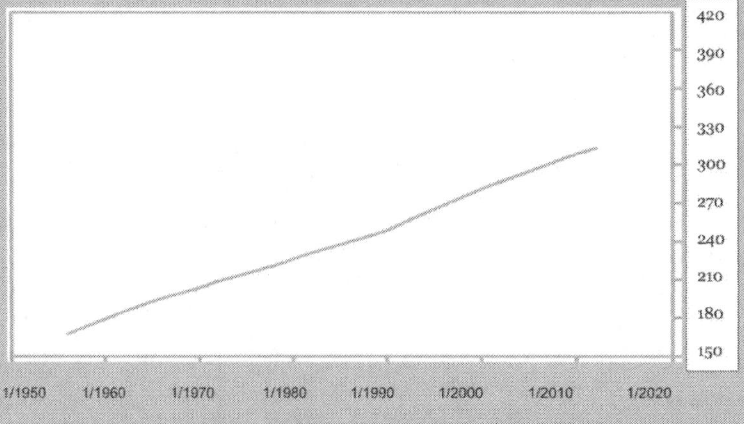

Population Growth Chart: Harris County 2005-2040

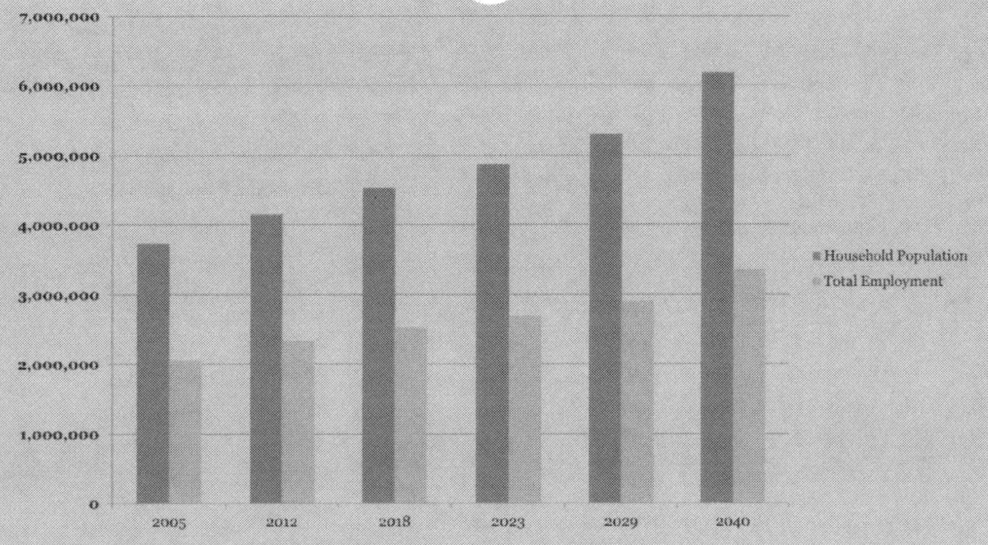

U.S. Petrochemical Industry

- We are one of America's top exporters, with $171 billion in annual exports, accounting for more than 10 cents of every dollar in total U.S. merchandise exports.
- We have reduced our fuel and power energy consumed (per unit of output) by more than half since 1974.
- We invest $55 billion in research and development annually.
- 1/5 of U.S. Patents are chemistry or chemistry-related.

Source - ACC

Our Product Line

- The science and materials of chemistry make the lives of Americans and others throughout the world healthier, safer, and more sustainable and more productive.
- Our food, safe water supply, clothing, shelter, health care, computer technology, clean energy sources, transportation and every other facet of modern life, all depend upon the business of chemistry.

Source: ACC

Economic Growth

- 2% Growth in the past 10 years
    - Potential
    - Barriers
- U.S. Trade Deficit equals $600 billion
    - 4% of GDP
    - Move Up the Value Added Ladder
- Energy, Energy, Energy
    - US Imports of Petroleum have declined
    - Natural Gas Boom
    - Next Generation Technologies
    - Uneconomical Options
- U.S. Energy Independence: Is it possible?

Profit Maximization

"Provducing the products earning the highest returns, and producing these products at the lowest possible cost."

Source: Dr. Paul Rubin

- Which products produce the highest returns?
  - More of these products equal more profits
- What if business does not maximize profits?
  - Consumers are harmed
- How to address the potential tension between local impact & social benefit

Capital Expenditures

- ACC: $25 billion annually
  - Past 10 years
- EHCMA: $1.7 billion spent annually on operating and capital improvement related to pollution control and reduction
- This is a very capital-intense business that requires planning & predictability

Regulations

- Code of Federal Regulations growing
- Predictability decreasing
- Lack of cost-benefit analysis
- Absence of peer review

East Harris County Manufactures Association

- 30,000 Direct Jobs
- Members bring an estimated $14 billion annually to the local economy
- 11 Community Advisory Panels
- 300,000 volunteer hours annually
- Channel Industries Mutual Aid (CIMA) is the largest of its kind in the world.
- 1,000 Employees devote more than 50% of their job function to site security
- ~240,000 jobs are indirectly tied to EHCMA operations

East Harris County Air Partners

- Promotes measurable improvement in air quality
- Implements reasonable and practical initiative to reduce air emissions
- Fosters effective education and communication among stakeholders in the community

Concerns of NGO

- Transparency
- Data Collection
- Influence
- Lack of Relationship
- Limited Use of Practices

"A Day in the Life"
Bay Cap

"A Day in the Life"

"A Day in the Life"

What I've Learned at CAP'S

- Behavior changes from dialog, not information.
- The Value of an Ongoing Relationship
- Target the "Opinion Leaders"
- Realistic Goals for the Relationship
    - Trust
    - Candor

"Here is something I have done different because of our discussion."

Success in Our Community

- Trust
- Buy-In
- Momentum

Community Trust Components

- Character
  - Reliability
  - Participation
  - Reasonable
- Competence
  - Incident Severity
  - Plant Incident Frequency
  - Involvement
- Follow-up Actions
  - Timely
  - Easily Understandable Language
  - Focus on people

Community Buy-In

- You cannot separate the leader from the cause he promotes.
- Time is needed for Buy-In.
- People want to go along with people they get along with.

All It Takes Is A Little Momentum

- P = MV
- M = 270,000 People
- V = Leadership
- Momentum is easier to steer than start.
- Leaders always find a way to make things happen.

Water for Manufacturing in Texas

Dan Hardin, Director of Water Resource Planning

Texas Water Development Board

Water Planning: Legislative Response to Drought

- Late 1950s Drought of Record
  - 1957: Creation of TWDB
  - $200 million Water Development Fund
  - 9 State Water Plans, 1961-2012
- Late 1990s: Potential New Drought of Record
  - ~$6 billion economic losses in '96 (mostly agriculture)
  - ~300 entities with threat to water supplies
  - 1997 & 2001: Passage of SB 1 & 2 which created & refined regional water planning

2

Regional Water Planning: By the Numbers

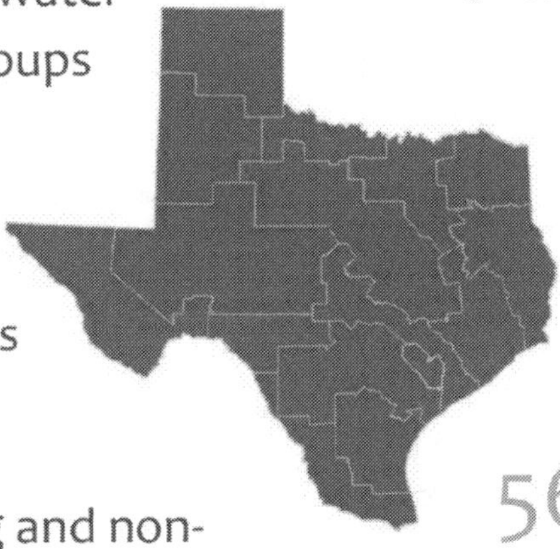

16 regional water planning groups

12 interest groups

450 voting and non-voting planning group members

6 water use categories

2,900 water user groups

562 recommended water management strategies to meet needs

3

Regional water planning process

- Project future population and water demand
- Quantify existing and future water supplies available during repeat of drought of record
- Identify surpluses and needs
- Evaluate and recommend water management strategies
- Make policy recommendations
- Adopt the plan

Existing Water Supplies − Projected Water Demand = Surplus (+) or Need (-)

Projected Texas Population

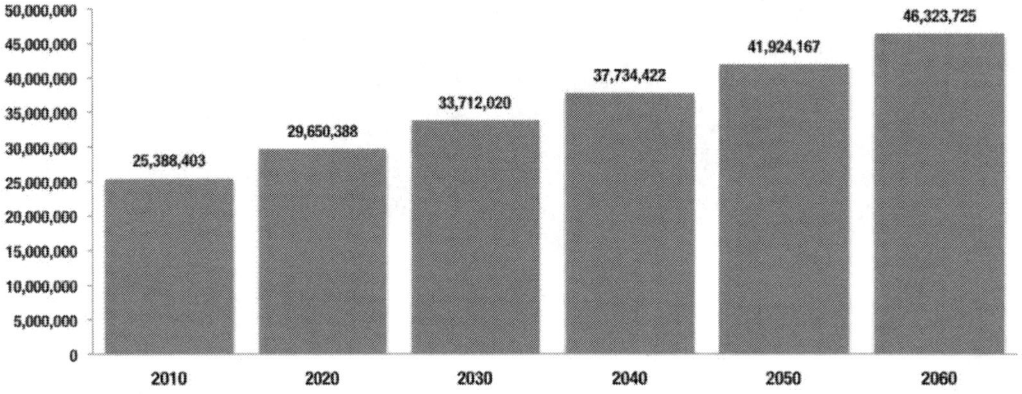

5

Projected Population Growth in Texas Counties

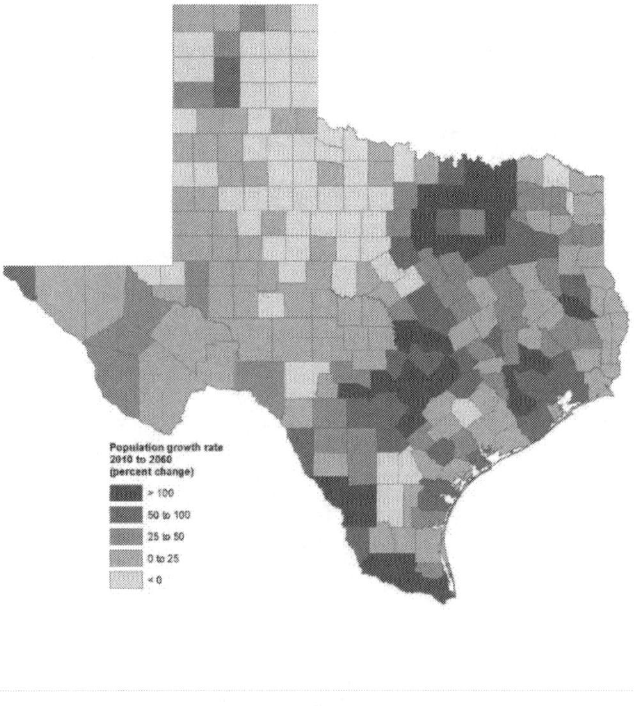

6

Water Demand Projections

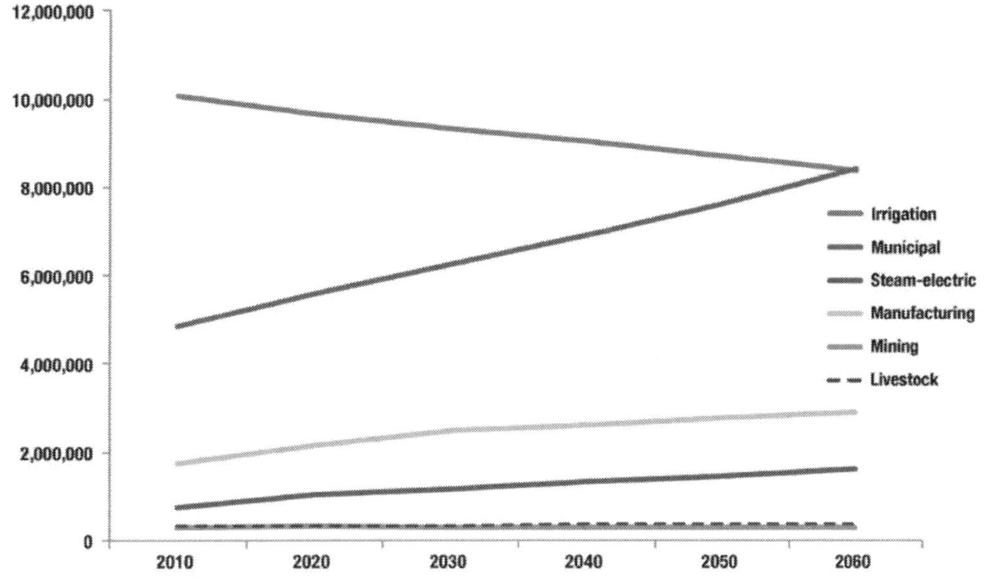

Projected Water Demands and Existing Supplies

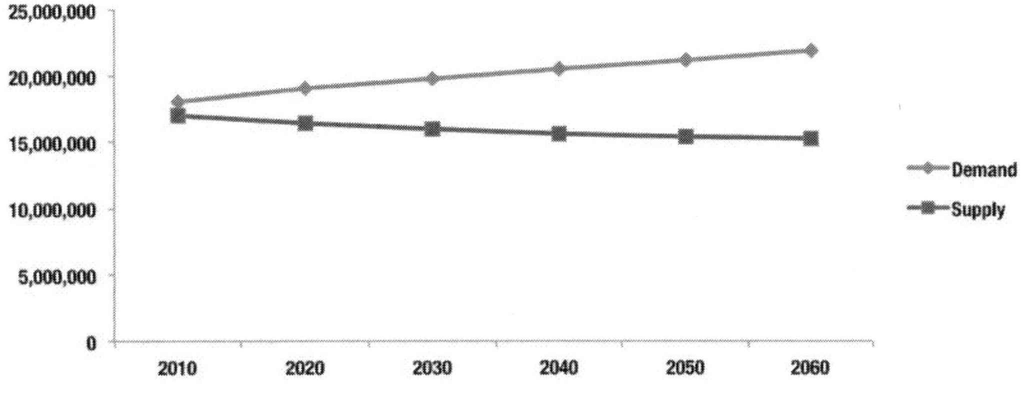

8

Projected Need for Additional Water in Times of Drought

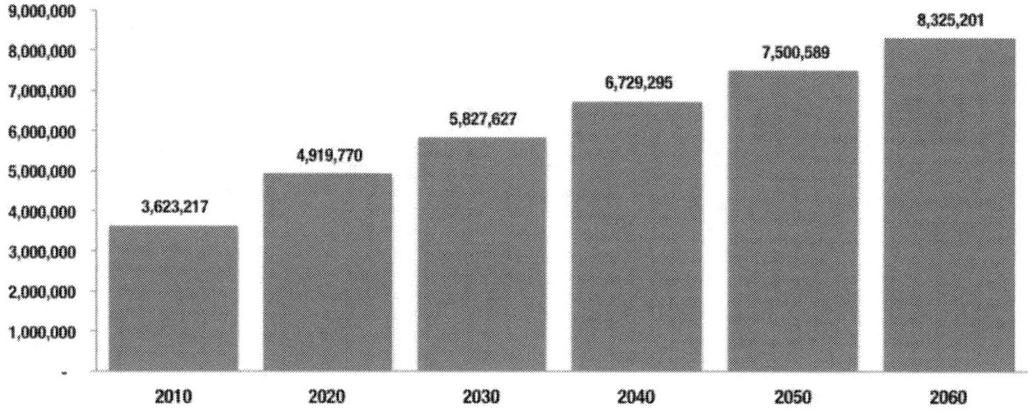

9

Water Supplies from Water Management Strategies

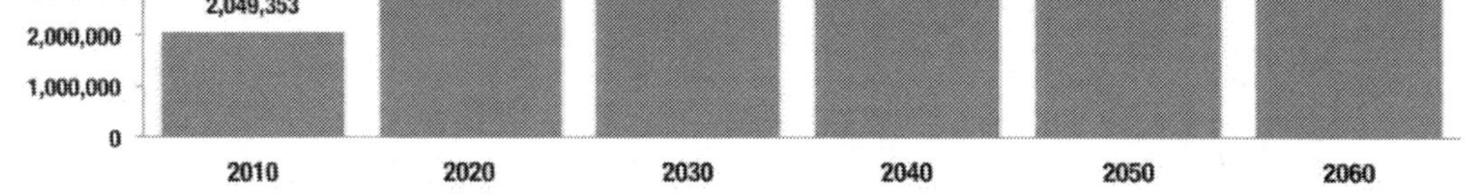

10

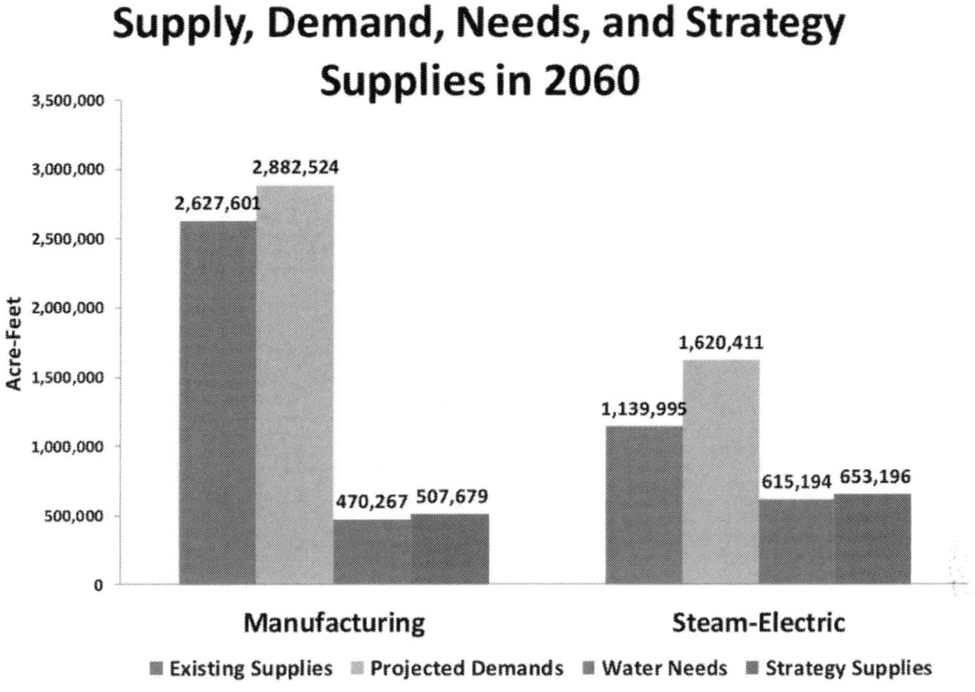

Relative Volumes of Recommended Strategies (2060)

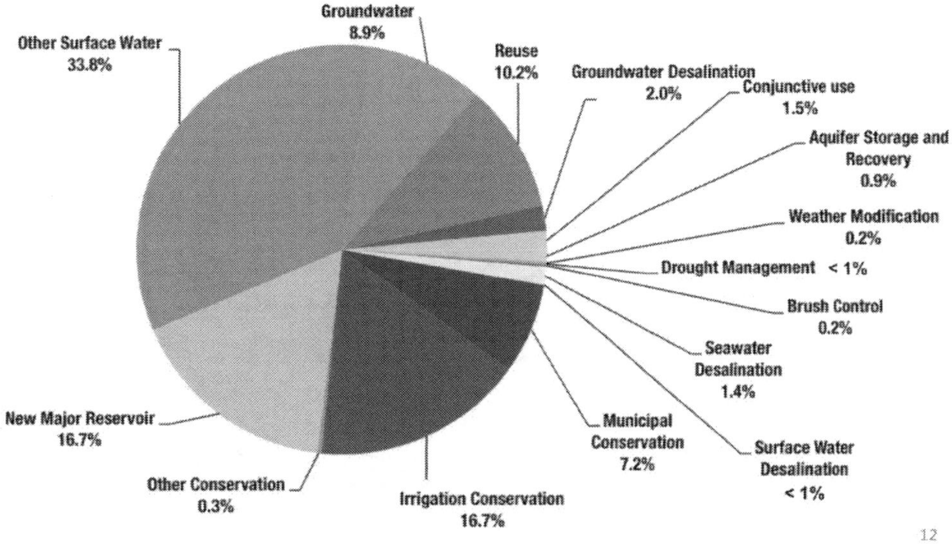

12

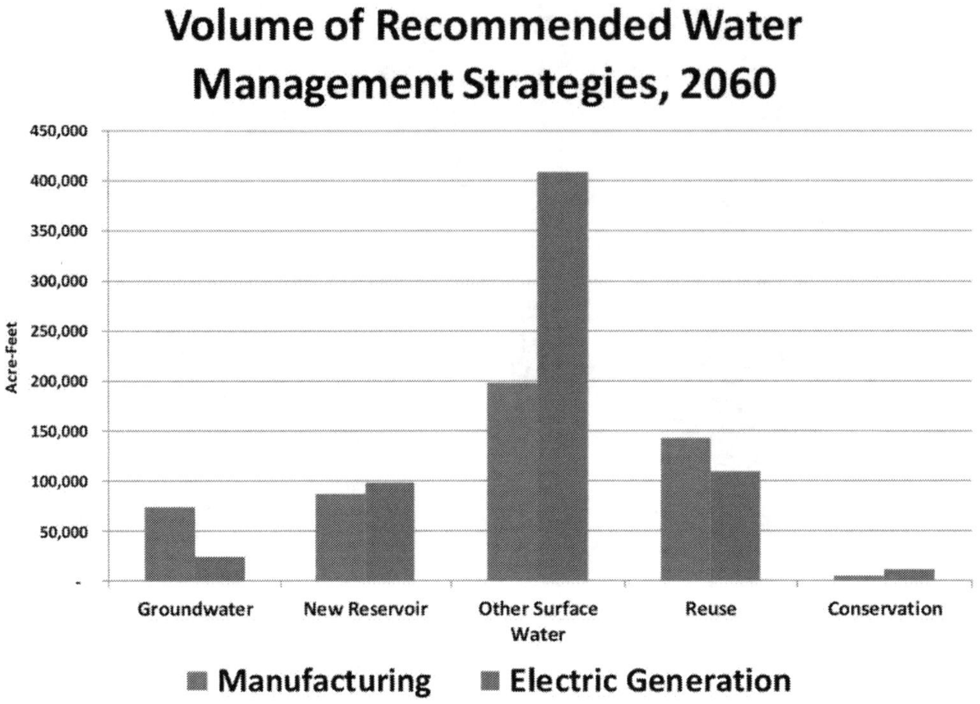

Cost of 2012 State Water Plan

- $53.1 billion to implement
- Project sponsors need access to $26.9 billion of project capital costs through state assistance

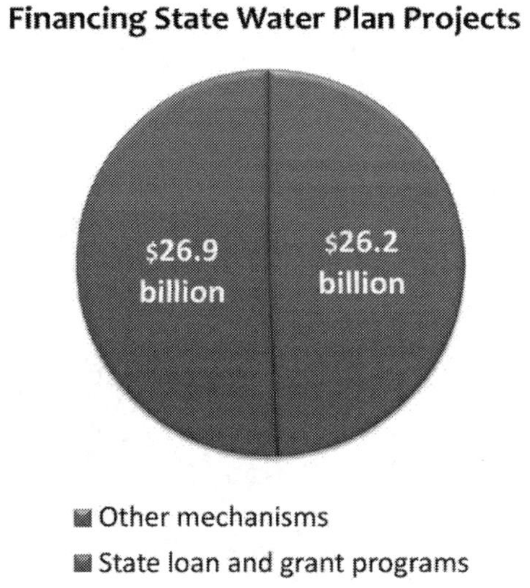

14

Cost of Not Implementing Plan Recommendations

- $12 billion lost income - 2010
- $116 billion lost income – 2060

- State/local business taxes lost: $1 billion – 2010
- State/local business taxes lost: $10 billion – 2060

- Lost jobs : 115,000 – 2010
- Lost jobs: 1 million – 2060

- Lost population growth: 1.4 million - 2060

15

Cost of Not Implementing Manufacturing and Electric Generation Strategies

- $8.7 billion lost income - 2010
- $95.2 billion lost income – 2060

- State/local business taxes lost: $819 thousand– 2010
- State/local business taxes lost: $7.9 billion – 2060

- Lost jobs : 80 thousand – 2010
- Lost jobs: 769 thousand– 2060

- **Total Capital Cost of Implementing Strategies to meet electric generation needs - $5.7 billion**

16

Barriers to Implementation

- State has no authority to require implementation of projects
- State has devoted limited financial resources to encourage implementation
- Local entities reluctant to impose costs of projects and associated debt on constituents
- Lack of recognition that there is a problem or of the "big picture" benefits of taking action

17

Implementing the State Water Plan

"What's in it for me?"

18

For More Information

2012 State Water Plan:
http://www.twdb.texas.gov/wrpi/swp/swp.asp

Dan.Hardin@twdb.texas.gov

Texas Water
Development Board

Tools to Assess Risk and Identify Opportunities

Davé, S.,

Originally presented on: 10/4/2012 12:25:00 - 12:50:00

The demand on water resources continues to grow with growth in world population, industrialization, and energy demand. Companies are increasingly competing with other users such as agricultural and domestic users for the same water resources, putting their operations at risk from the interruption of supplies. Oil & Gas, refining and petrochemical companies are some of the largest water users and need a systematic approach to assess their water risk and develop management plans to mitigate water risk to their operations.

The Global Environmental Management Initiative (GEMI) Local Water Tool (LWT) is a free spreadsheet tool developed by a group of 40+ companies to help identify the external impacts, business risks, and opportunities related to water use and discharge at a specific site or operation. Companies may use the information generated in the LWT for developing management plans and communicating outcomes at their discretion. Two versions of the LWT were created: General LWT (multi-sector); and Oil and Gas LWT.

The LWT helps companies with assessment and management of water-related issues via a common and consistent visualization platform for internal and external communications. It provides connectivity between global and local water risk assessments and a uniform approach between site assessments. It provides a central repository of information to create reports for internal and external stakeholders. The data can be updated as the internal and external circumstances change to continuously update water risk and management plans.

Texas Gulf Coast Water Risk Assessment

Beth Jones
LyondellBasell
October 4, 2012

lyondellbasell

LyondellBasell Fast Facts

- One of the world's largest plastics, chemicals and refining companies with revenues of $51billion (2011)
- The global leader in polyolefins technology, production and marketing
- A pioneer in propylene oxide and derivatives
- A producer of fuels and refined products, including biofuels
- Listed on the New York Stock Exchange (NYSE) as a publicly traded company. Ticker symbol: LYB

Tarragona, Spain

lyondellbasell.com lyondellbasell 2

Global Reach

58 plants in 18 countries

More than 14,000 employees worldwide

Sales in more than 100 countries

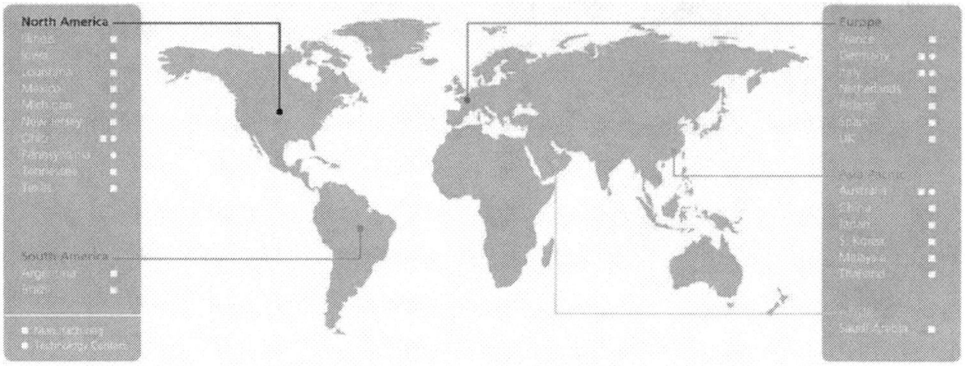

Owned and operated by LyondellBasell, its subsidiaries and/or joint ventures

Diversified and Vertically Integrated Portfolio

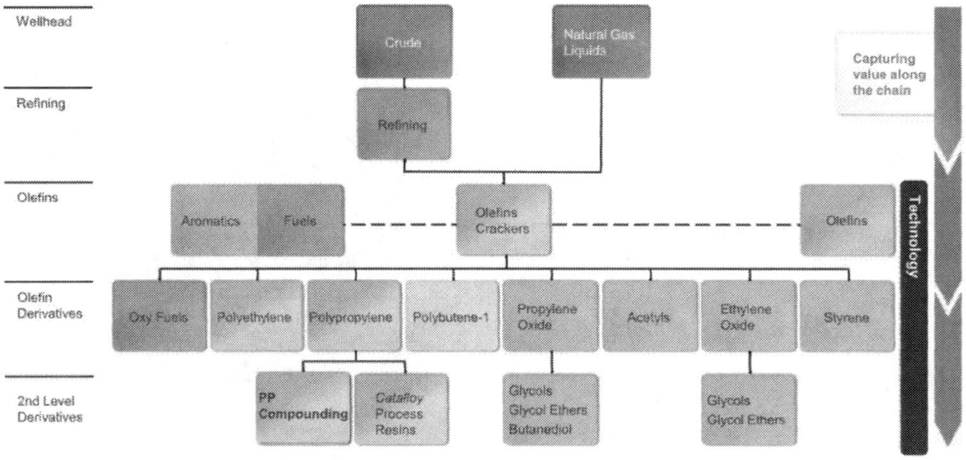

- Olefins & Polyolefins Americas
- Olefins & Polyolefins Europe Asia & International
- Intermediates & Derivatives
- Refining & OxyFuels
- Technology

Disclaimer And Trademarks

- *Catalloy, Spheripol, Spherizone, Hostalen* and *Lupotech* are trademarks owned and/or used by the LyondellBasell family of companies. *Spheripol, Spherizone* and *Hostalen* are registered in the United States Patent and Trademark Office.
- All information ("Information") contained herein is provided without compensation and is intended to be general in nature. You should not rely on it in making any decision. LyondellBasell accepts no responsibility for results obtained by the application of this Information, and disclaims liability for all damages, including without limitation, direct, indirect, incidental, consequential, special, exemplary or punitive damages, alleged to have been caused by or in connection with the use of this Information. LyondellBasell disclaims all warranties, including, but not limited to, the implied warranties of merchantability and fitness for a particular purpose, that might arise in connection with this information.
- LyondellBasell does not sell PB-1 for use in pipe applications intended for use in North America, and requires its customers not to sell products made from PB-1 into pipe applications for North America.

Drought Progression

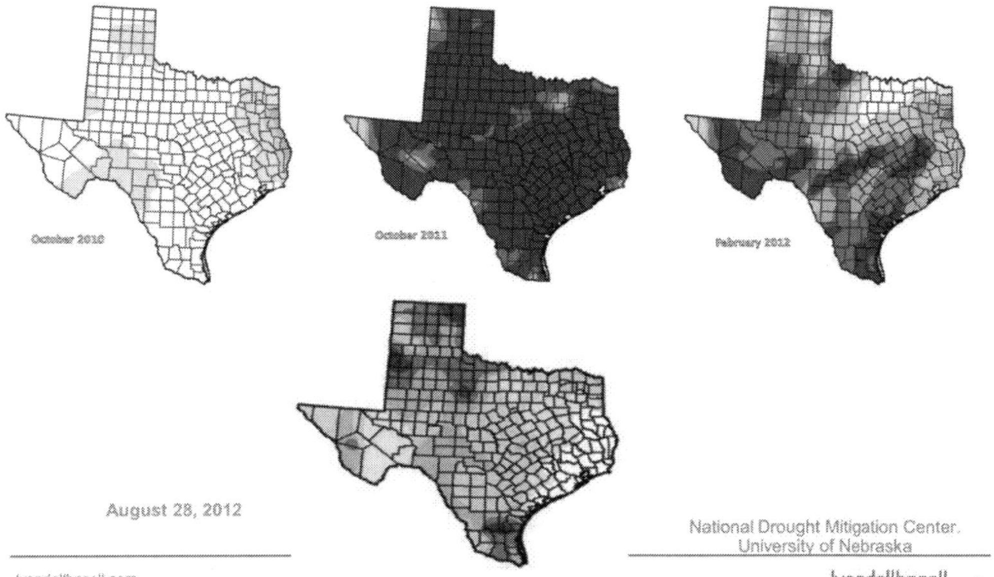

August 28, 2012

National Drought Mitigation Center, University of Nebraska

Background

After the 2011 drought, NA Plant Managers requested assistance:

To assess vulnerabilities associated with water availability at LYB's Texas Gulf Coast plants and develop a strategic plan to avoid business interruption

In A Way That:

- Determines applicable laws and governing bodies
- Considers each watershed and all sites dependent on it separately
  - Determines current drought contingency plans and initiating events
- Works with each site to develop a strategic water plan:
  - Considers current contracts, alternative water sources, and appropriate conservation measures
  - Develops cost estimates and risk/benefit analyses as needed.
  - Coordinates with hosts of tenant plants
- Identifies opportunities to influence future water legislation and decisions

So That LYB facilities experience no business interruption due to water supply.

Team Agenda

- Understand current and future water demand
- Identify current risks and mitigation opportunities
- Identify future risks and mitigation opportunities

Water Strategy Team

- Global HSE
- Energy
- Utilities Specialist
- Consulting Engineer, Water
- Procurement
- Site Operations

Texas Gulf Coast Water Use

2011 Internal Data

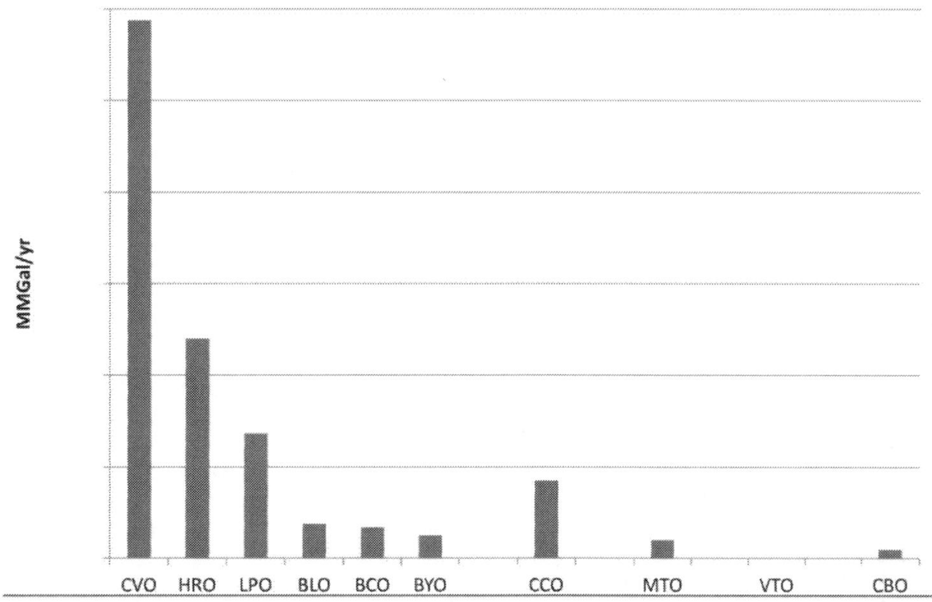

Water Usage Relative to Other Industrial Users

State Water Region	Site	Region Rank Per State Data	% Region's Industrial Water
H	CVO	7**	1.8%
	HRO	14	0.7%
	LPO	36	0.2%
	BLO	48	0.3% combined
	BCO	50	
	BYO	58	
	CBO	*	0.0%
K	MTO	13	0.7%
L	VTO	*	0.0%
N	CCO	6	8.4%

* Included in host site's data
** CHO & CXO Combined Equivalent

Risk Assessment

- Methodology
  - Engaged CH2MHill
  - Performed studies and evaluations with Local Water Tool
  - Considered local plant and regional / watershed influences
  - Ranked current and future risks in 17 categories
- Validation and buy-in
  - Sought input and validated results in meetings with each site
  - Solicited ideas
    - Past studies or plans
    - Priority / interruptible uses
    - Potential future opportunities

Current Threats to Supply

- Right to supply:
  - Several sites buy from municipal systems
  - Others are supplied from "host" sites or groundwater
- Drought
  - Continued drought impacts around Corpus Christi and Matagorda
  - Proposed City of Houston drought response ordinance

City of Houston Response to 2011 Drought

- Drafted ordinance defining 4 levels of drought stress
  - Included force majeure on contract users at levels 3 & 4
    - Last year's drought fell 1-2 months short of Level 3
    - Stage 4 requires non-residential customers to cut 15% immediately
  - Based reductions on previous 8 months' use
    - No summer use; no T/A offsets
  - Intended for quick passage
- East Harris County Manufacturers Association became involved
- Mayor's Task Force on Water Conservation was created
  - Best outcome would be:
    - Reasonable baseline calculation
    - Recognition of strategic importance of industry to the economy
    - Treatment multi-site consumers as a single entity
      - Let us decide how best to reduce net water use across all our sites
    - Recognition of prior conservation efforts

Threats to Future Availability

- Population growth (expected to be >50% over 50 years)
  - Municipal water rights will continue to supercede industry's
- Industrial growth: Primarily power plants
- Regulation:
  - Receiving body quality
  - Surface subsidence from ground water depletion
  - Ecological harm to habitats
    - Aransas lawsuit
    - LCRA upstream restrictions
    - New minimum flow rate regulations on several basins

Water: Another Resource to Manage

- Measure what you use
  - Know where it goes and why it's going there
  - Review billing information for changes in usage
- Don't waste it
  - Find and fix leaks
  - Look for pumping system short-circuiting
- Use it efficiently
  - Cascade usage when possible (ex: cooling tower blowdown for fire water)
  - Cycle up cooling towers
  - Recover condensate
  - Treat and recycle "easy" streams
- Know what usage can be "designed out" economically

completing the energy sustainability puzzle
ENERGY and WATER
Emerging Cooling Water Issues and Directions
AICHE Regional Process Technology Conference
Mike Hightower
Sandia National Laboratories
mmhight@sandia.gov, 505-844-5499

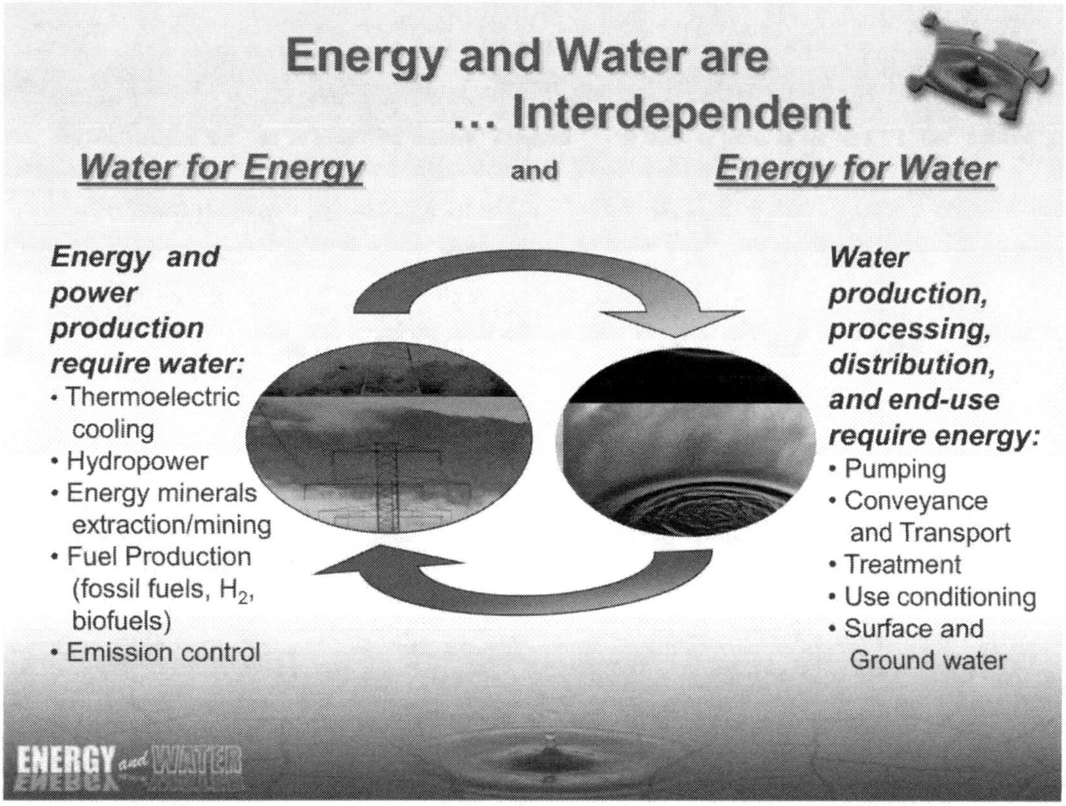
Energy and Water are
... Interdependent
Water for Energy
and
Energy for Water
Energy and power production require water:
• Thermoelectric cooling
• Hydropower
• Energy minerals extraction/mining
• Fuel Production (fossil fuels, H2, biofuels)
• Emission control
Water production, processing, distribution, and end-use require energy:
• Pumping
• Conveyance and Transport
• Treatment
• Use conditioning
• Surface and Ground water
ENERGY and WATER

Water Provider Assessment of Global and Regional Water Stress

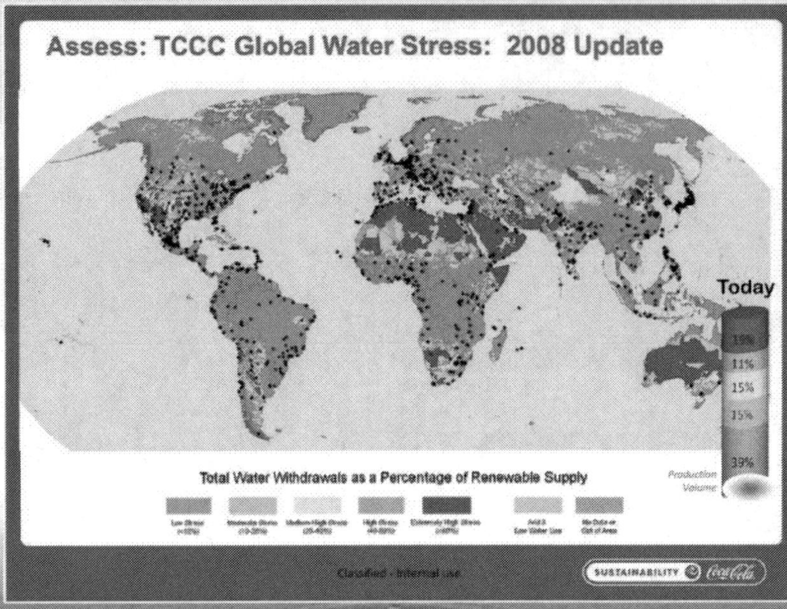

Growing Limitations on Fresh Surface and Ground Water Availability

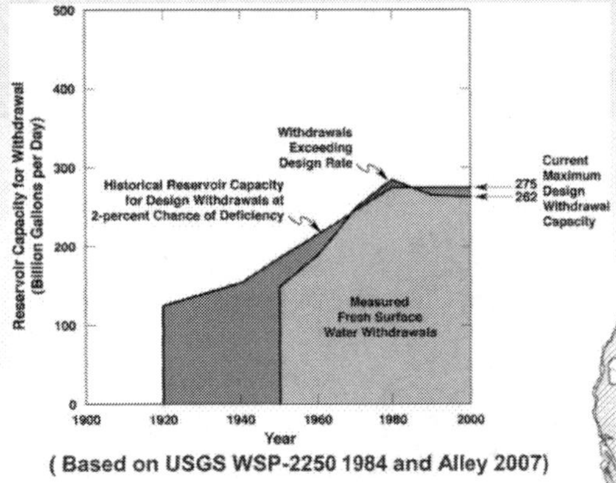

(Based on USGS WSP-2250 1984 and Alley 2007)

- Many major ground water aquifers seeing reductions in water quality and yield
- Little increase in surface water storage capacity since 1980
- Concerns over climate impacts on surface water supplies

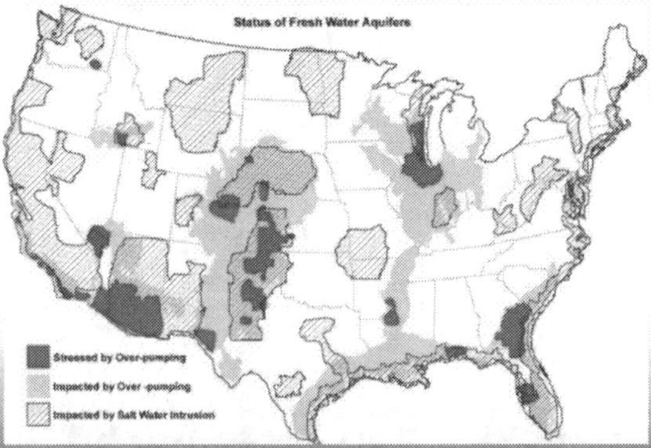

(Shannon 2007)

ENERGY and WATER

Most State Water Managers Expect Shortages Over The Next Decade Under Average Conditions

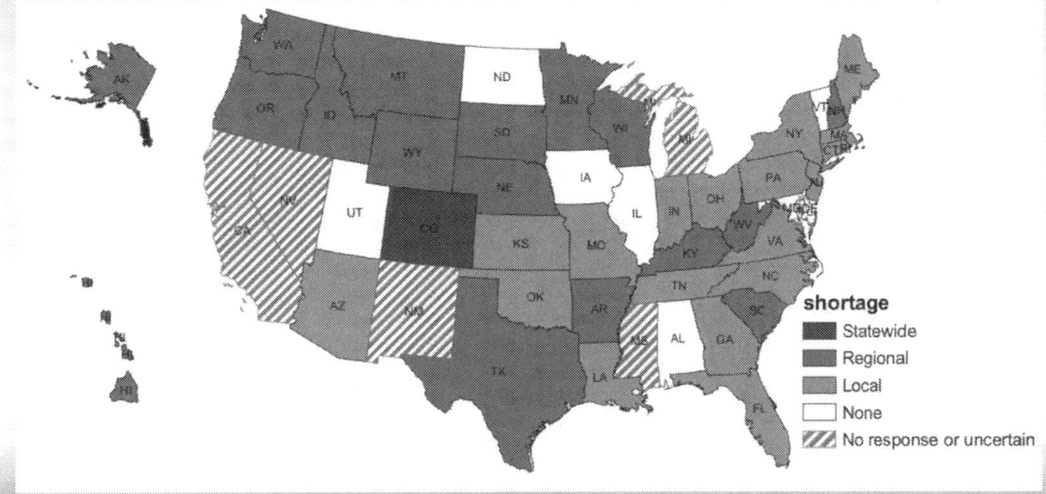

Source: GAO 2003

Water Use and Consumption for Electric Power Generation

Plant-type	Cooling Process	Water Use Intensity (gal/MWh$_e$)		
		Steam Condensing		Other Uses
		Withdrawal	Consumption	Consumption
Fossil/ biomass steam turbine	Open-loop	20,000–50,000	~200-300	~30
	Closed-loop	300–600	300–480	
Nuclear steam turbine	Open-loop	25,000–60,000	~400	~30
	Closed-loop	500–1,100	400–720	
Natural Gas Combined-Cycle	Open-loop	7,500–20,000	100	7–10
	Closed-loop	230	180	
Integrated Gasification Combined-Cycle	Closed-loop	200	180	150
Carbon sequestration for fossil energy generation	~80% increase in water withdrawal and consumption			
Geothermal Steam	Closed-loop	2000	1350	50
Concentrating Solar	Closed-loop	750	740	10
Wind and Solar Photovoltaic	N/A	0	0	1-2

Cooling Systems at Utility Steam Electric Plants – Source: EEI, 1996

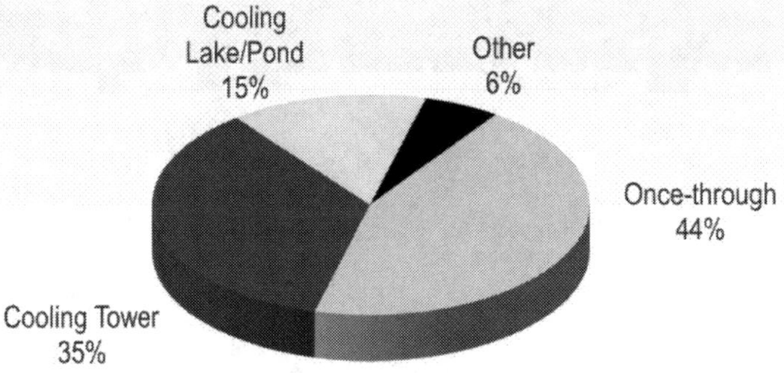

Total capacity = 569,000 MW

ENERGY and WATER

2003 Heat Wave Impact on French Electric Power Generation

- Loss of 7 to 15% of nuclear generation capacity for 5 weeks
- Loss of 20% of hydro generation capacity
- Large-scale load shedding and shut off transmission to Italy
- Sharp increase of spot-market prices: 1000 to 1500 $ / MWh for most critical days

Bort-les-Orgues Réservoir

Normal conditions in August

August 27, 2003

ENERGY and WATER

Summary of Major National Needs and Issues Identified in Regional Workshops

Better resources planning and management

- Improved water supply and demand characterization, monitoring, and modeling
- Integrated regional energy and water resource planning and decision support tools
- Framework for incorporating infrastructure, regulatory, and policy considerations for improved energy/water efficiency planning

Improved water and energy use efficiency

- Improved water efficiency in thermoelectric power generation
- Improved biofuels/biomass water use efficiency
- Reduced water intensity for emerging energy resources

Development of alternative water resources and supplies

- Non-traditional and oil and gas produced water use and reuse
- Improved energy efficiency for non-traditional water treatment and use

ENERGY and WATER

Research Program for Electric Power Sector

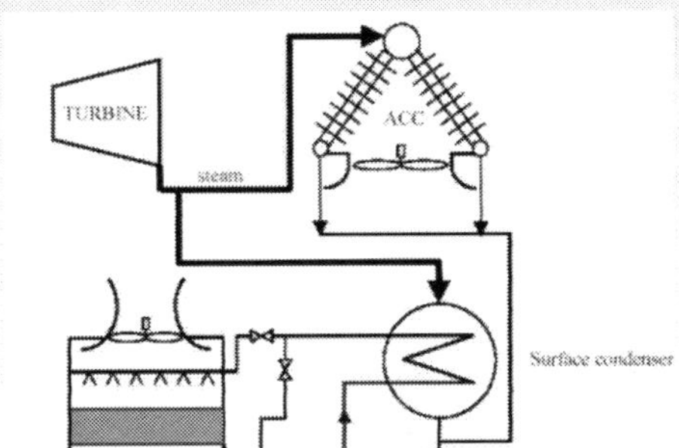

Hybrid Wet-Dry Cooling System

- Improve dry and hybrid cooling system performance
- Improve ecological performance of intake structures for hydro and once-through cooling
- Improve materials and cooling approaches compatible with use of degraded water
- Electric grid infrastructure upgrades to improve low water use renewable technology integration

ENERGY and WATER

Growing Use of Non-traditional Water Resources

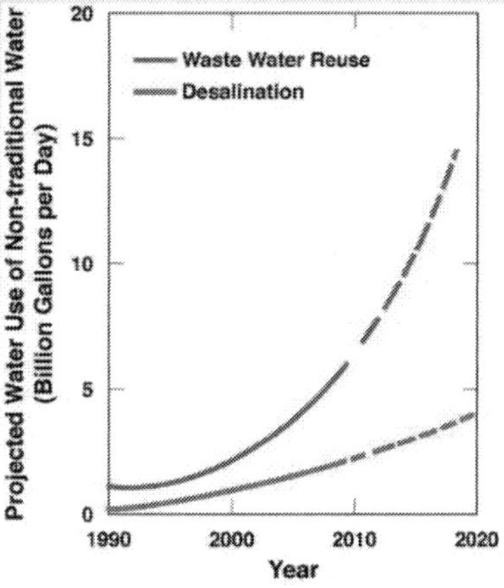

(Modified from Water Reuse 2007, EPA 2004, Mickley 2003)

Power Requirements For Treating

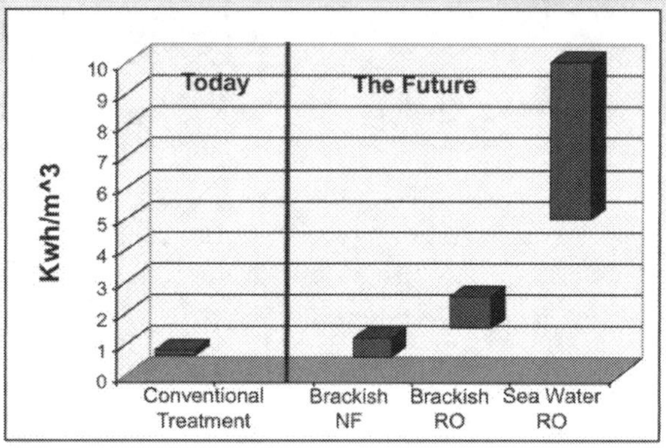

(Einfeld 2007)

- Desal growing at 10% per year, waste water reuse at 15% per year
- Reuse not accounted for in USGS assessments
- Non-traditional water use is energy intensive

ENERGY and WATER

Alternative Cooling Issues and Challenges

Pilot Spray-enhanced Dry Cooling

Dry-Cooled Power Plant

- High capital cost and large footprint
- Hot weather penalties
  - Power capacity reduced during peak summer demand
  - Grid stability concerns with many regional air cooled plants during summer months
  - Increased GHG emissions for fossil thermoelectric plants
- Wind effects reduce cooling efficiency
- Not currently licensed for nuclear power plants

ENERGY and WATER

Dry and Hybrid Cooling Issues and Technology Concerns

- Up to 90% less water consumption with hybrid cooling
- Total 6 % loss in production
- As much as 20% reduced capacity at hottest hours
- 10% increase in capital costs
- 1-2 ¢/kWh increase in cost of power

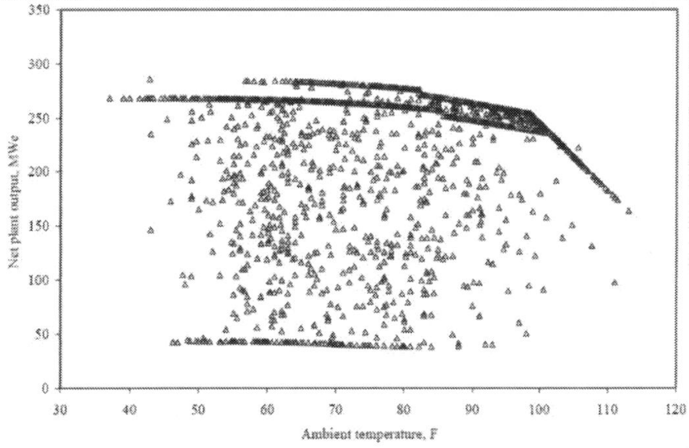

Figure 5 Net Plant Output as a Function of Ambient Temperature; Dry Heat Rejection

*ALL VALUES ARE SITE SPECIFIC

ENERGY and WATER

Results Of Argonne Study on Energy Penalties (% reduction in output)

Convert from Once-Through to:	Summer Peak Conditions	Annual Average Conditions
Wet Tower	**2.4 – 4.0**	**0.8 – 1.5**
Dry Tower - 20° approach	**8.9 – 12.1**	**4.2 – 5.2**
Dry Tower - 40° approach	**12.7 – 15.9**	**7.9 – 8.8**

ENERGY and WATER

EPRI Cooling Technology Research Program

Drivers

- Majority of freshwater use (~90%) in plants for cooling
- Current water-conserving technologies, such as air-cooled condensers and hybrid cooling with significant capital outlays, lower efficiency, and operational and maintenance issues.

Water Use Reduction Opportunities Identified for Power Plants

Illustration of a 500 MW Power Plant Steam and Cooling Water Flow System.

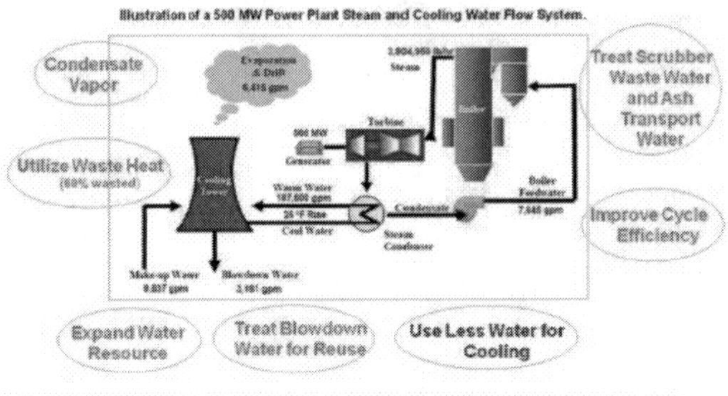

Top Priority in 2011

Develop Water Conserving Cooling Technologies

ENERGY and WATER

EPRI Cooling Research Project

Schematic Illustration of a Typical Adsorption Chiller

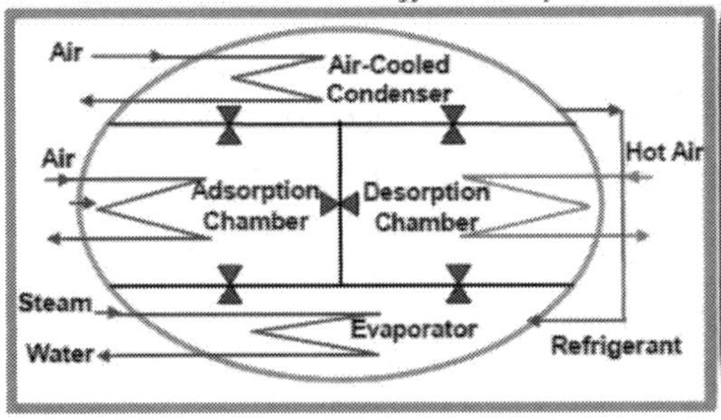

Key Potential Benefits

- Evaporative cooling
  - Significantly higher heat removal rate than convective water cooling
- Closed loop cooling system
  - **Near Zero** water use and consumption
- Reduced condensation temperature
  - As low as **10 °C**
  - Potential for increased power production by 10%
- Waste Heat Utilization

Project Scope

- Explore best power plant system level approaches to utilize waste heat for desorption
- Perform system integration energy and mass flow balance analysis for a 500 MW coal-fired power plant
- Develop and build a 5 kW Green Adsorption Chiller Prototype
- Perform prototype testing
- Perform technical and economic feasibility study

ENERGY and WATER

EPRI Cooling Research Project

As ambient conditions permit, the thermosyphon cooler pre-cools the condenser water leaving the steam surface condenser reducing the evaporative load on the cooling tower

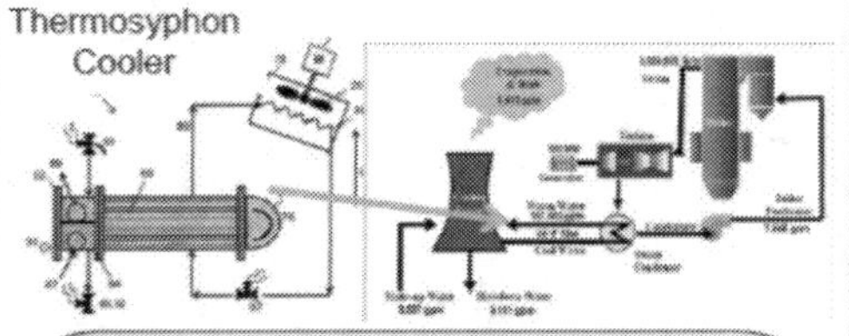

Project Scope

- Perform a thorough feasibility evaluation of a hybrid, wet/dry heat rejection system comprising recently developed, patent pending, thermosyphon coolers (TSC).
- Make comparisons in multiple climatic locations, to standard cooling tower systems, all dry systems using ACC's, hybrid systems using parallel ACC's, and air coolers replacing the thermosyphon coolers.
- Determine the most effective means to configure and apply the thermosyphon coolers .

Key Potential Benefits

- Potential for annual savings > 50% in cooling tower evaporation, make-up, chemical use, and blowdown
- Full plant output is available on the hottest days when the demand for electricity is the greatest
- Ease of retrofitting
  - Taps into existing condenser water cooling loop
  - Increased layout flexibility – just run water piping
  - Can be added incrementally in stages
- No increase in surface area exposed to primary steam
- Reduces operating concerns in sub freezing weather
- Broad application (hybrid, new, and existing cooling systems)

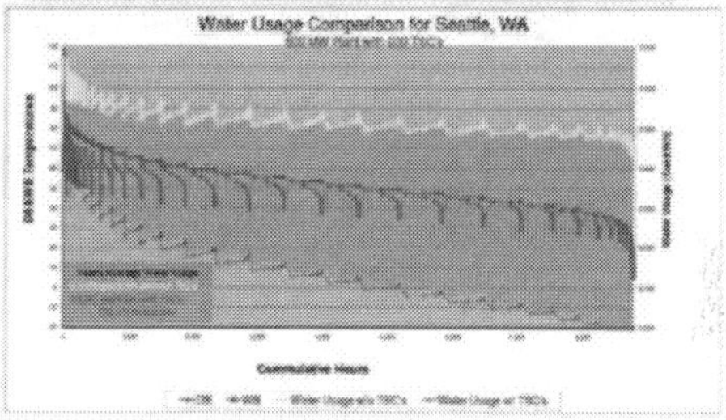

ENERGY and WATER

Conclusions

- Water availability concerns are growing internationally – climate change is exacerbating issues
- Water recycling within all industrial processes is becoming a major trend in the U.S. and worldwide
- Use of non-traditional water resources is becoming a bigger cooling water supply trend
- Cooling water availability, and consumptive use of cooling water is a growing issue associated with operational reliability challenges
- Most alternative approaches being considered have significant penalties
- New EPRI cooling tower research program is taking a look at a range of advanced approaches that could make some major improvements in cooling approaches that imrove system performance and reduce water consumption

ENERGY and WATER

UOP TA-30 Tatoray™ Catalyst for Xylene Production

Raymond Shih, Naiyl Rashid,
UOP LLC, A Honeywell Company, Des Plaines, Illinois 60017, USA

AIChE RPTC
Houston Texas
October 4, 2012

UOP A Honeywell Company

Agenda

UOP A Honeywell Company

- **Petrochemical Aromatics Complex**
  - Introduction to Aromatics complex
  - UOP Aromatic Complex Overview
- **UOP Tatoray unit and catalysts**
  - UOP Tatoray unit overview
  - UOP Tatoray Catalysts Background
  - Aromatic Process for TA-30
  - Commercial Data with UOP TA-30 catalyst
- **UOP Tatoray PTD™ process**
  - Overview of Tatoray PTD process
  - Benefits and Performance of Tatoray PTD process
- **Summary and Q&A**

UOP Aromatics BTX Technology

UOP A Honeywell Company

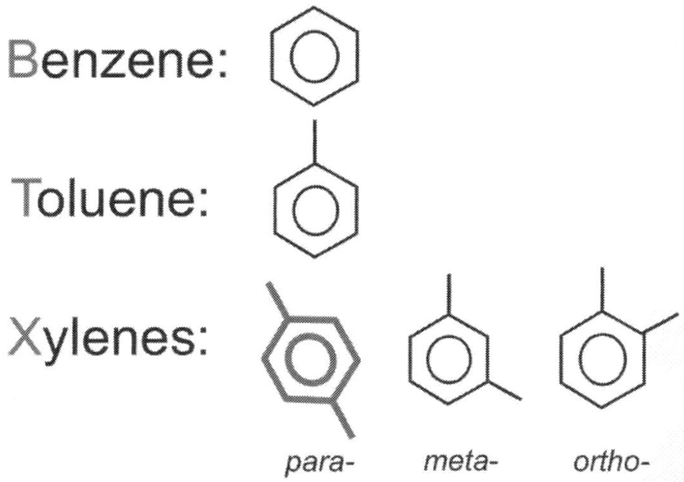

Global *p*X Demand will Double in a Decade

UOP A Honeywell Company

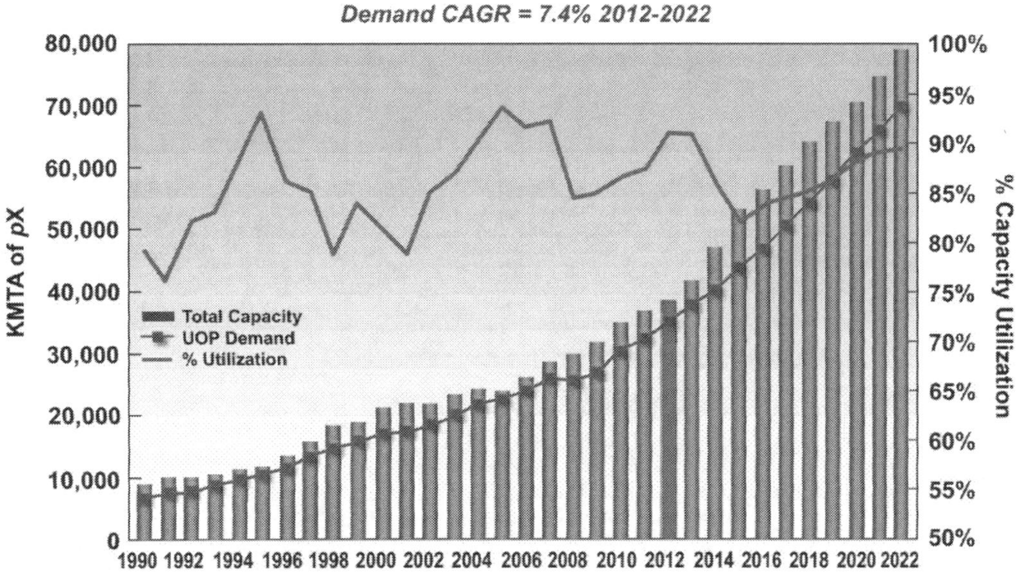

Source: UOP (SBA-CCI & PCI) 2012-06

UOP 5734B-4

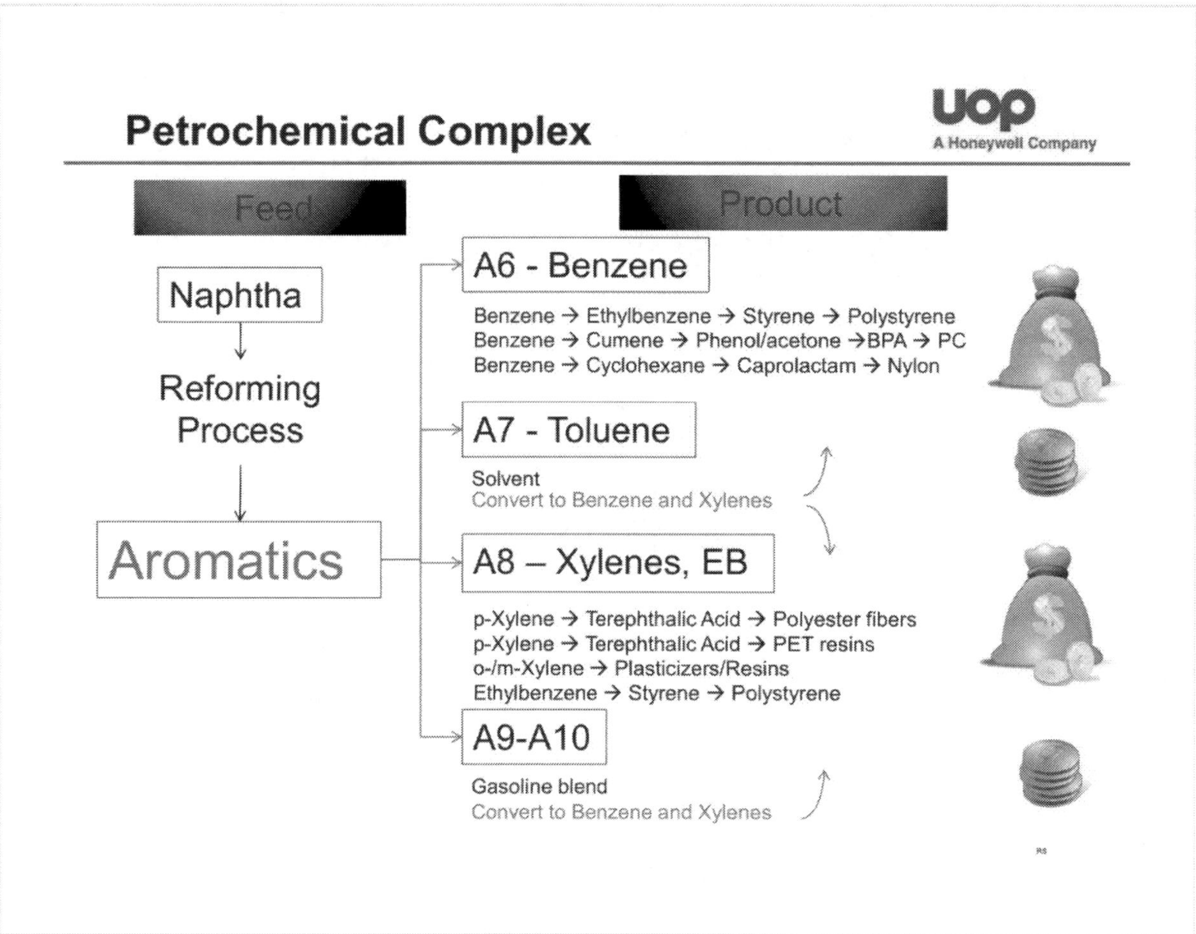
Petrochemical Complex
UOP
A Honeywell Company
Feed
Product
Naphtha
Reforming Process
Aromatics
A6 - Benzene
Benzene → Ethylbenzene → Styrene → Polystyrene
Benzene → Cumene → Phenol/acetone →BPA → PC
Benzene → Cyclohexane → Caprolactam → Nylon
A7 - Toluene
Solvent
Convert to Benzene and Xylenes
A8 – Xylenes, EB
p-Xylene → Terephthalic Acid → Polyester fibers
p-Xylene → Terephthalic Acid → PET resins
o-/m-Xylene → Plasticizers/Resins
Ethylbenzene → Styrene → Polystyrene
A9-A10
Gasoline blend
Convert to Benzene and Xylenes

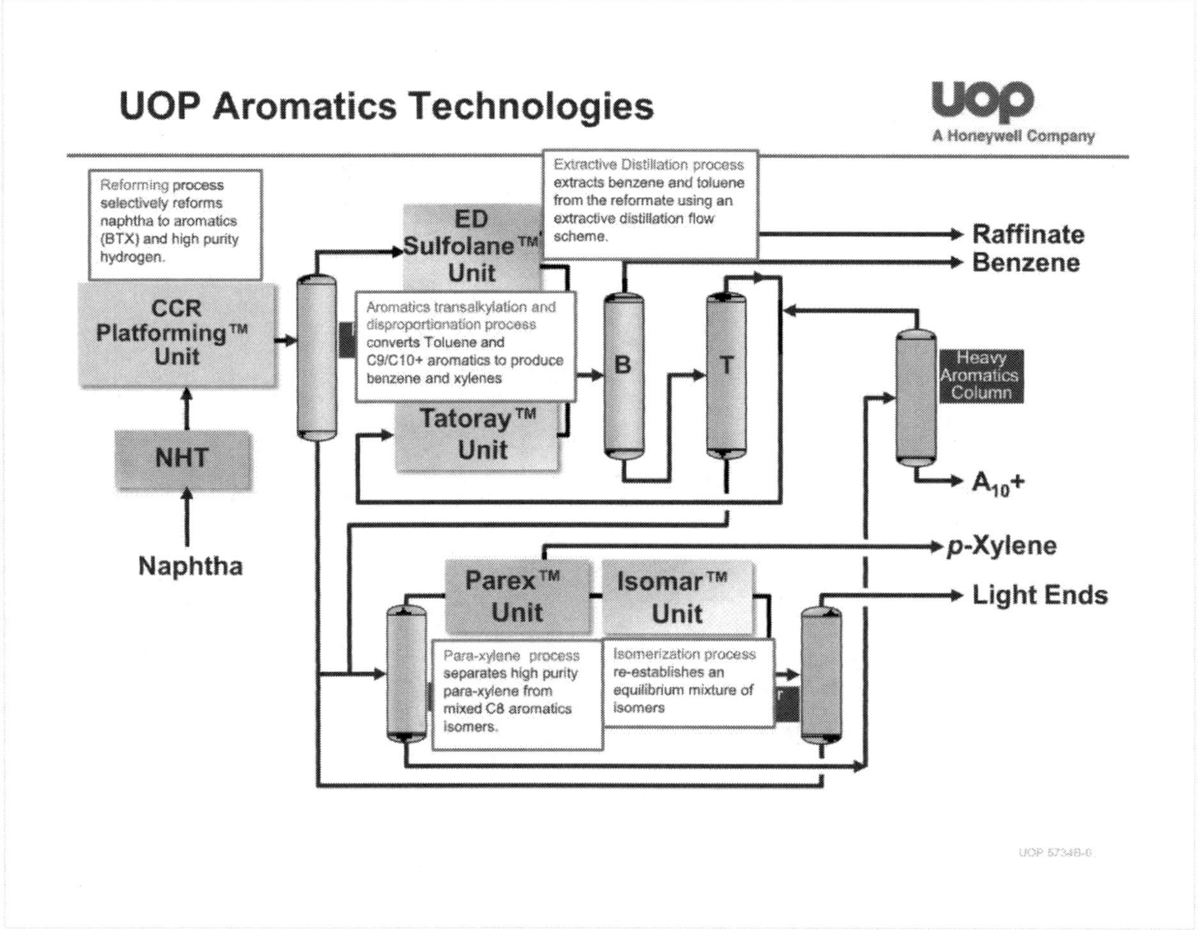
UOP Aromatics Technologies
UOP
A Honeywell Company
Reforming process selectively reforms naphtha to aromatics (BTX) and high purity hydrogen.
CCR Platforming™ Unit
NHT
Naphtha
Extractive Distillation process extracts benzene and toluene from the reformate using an extractive distillation flow scheme.
ED Sulfolane™ Unit
Aromatics transalkylation and disproportionation process converts Toluene and C9/C10+ aromatics to produce benzene and xylenes
Tatoray™ Unit
B
T
Heavy Aromatics Column
Raffinate
Benzene
A10+
p-Xylene
Light Ends
Parex™ Unit
Isomar™ Unit
Para-xylene process separates high purity para-xylene from mixed C8 aromatics isomers.
Isomerization process re-establishes an equilibrium mixture of isomers

UOP Tatoray Process Flow Diagram

UOP A Honeywell Company

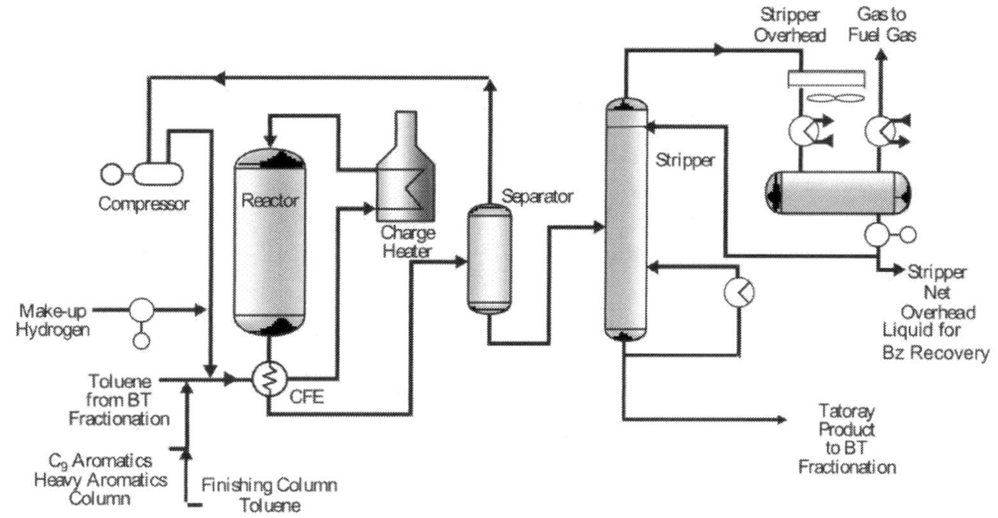

UOP Aromatics Catalysts and Adsorbents

UOP A Honeywell Company

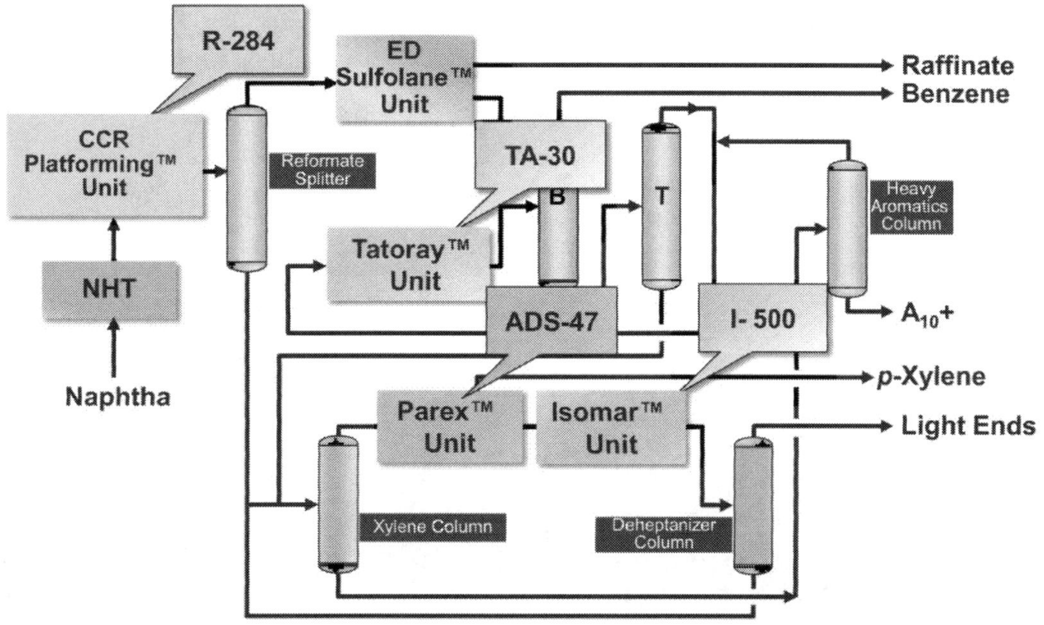

UOP 5734B-8

Xylene Production in the Aromatics Complex

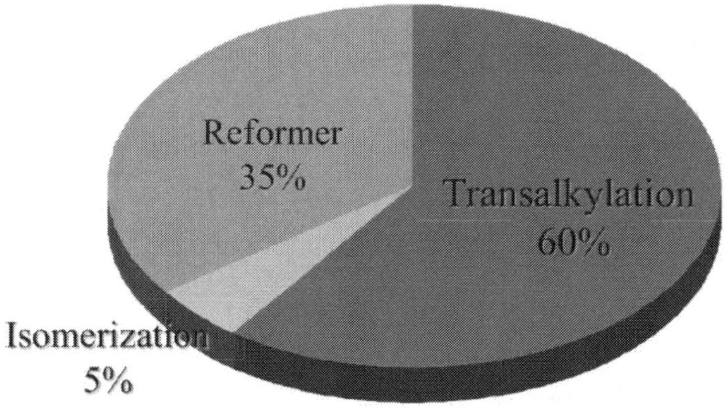

More than half of all xylenes formed by the Tatoray Unit

UOP 4097D-39

Tatoray Chemistry

- **Main Reactions**
  - **Transalkylation**
    Transalkylate Toluene and C9 aromatics to Xylenes

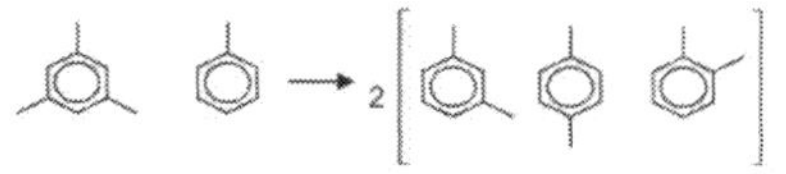

  - **Disproportionation**
    Disproportionate Toluene to Xylenes and Benzene

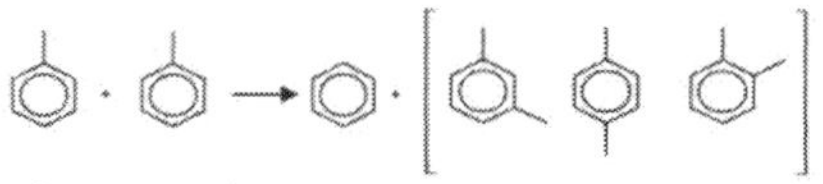

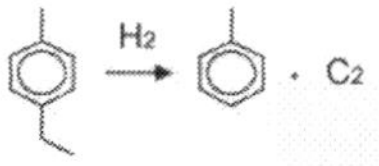

  - **Dealkylation**
    Dealkylate C9+ Aromatics to Toluene

- **Side Reactions**

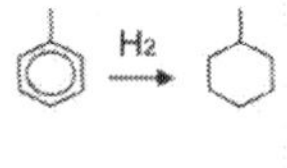

  - **Ring Saturation**
    Rings hydrogenated to form Naphthenes

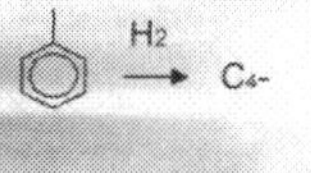

  - **Hydrocracking**
    Rings opened and cracked to form light ends

UOP Tatoray Catalyst History

UOP
A Honeywell Company

Future Development of new Tatoray Catalysts
UOP continues to deliver new and improved catalysts

2010 TA-30 Commercialized
Capable of processing heavy xylene column bottom
Currently operating in 2 units

2004 TA-20 Series Commercialized
Further increase benzene product purity
Currently operating in 11 units

2000 TA-5 Commercialized

TA-2, TA-3, TA-4 Commercialized

1969 First Tatoray Catalyst TA-1 Developed

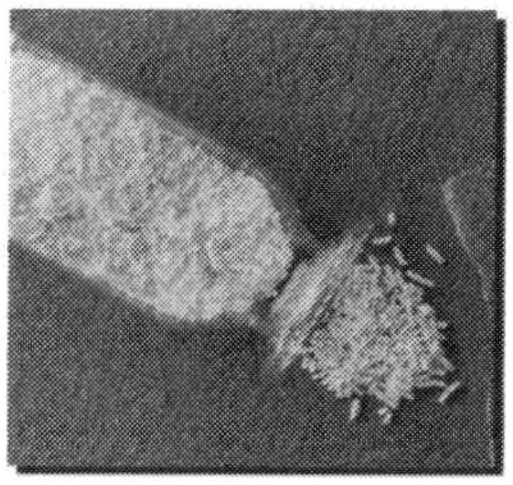

Heavy A9+ Aromatics

UOP
A Honeywell Company

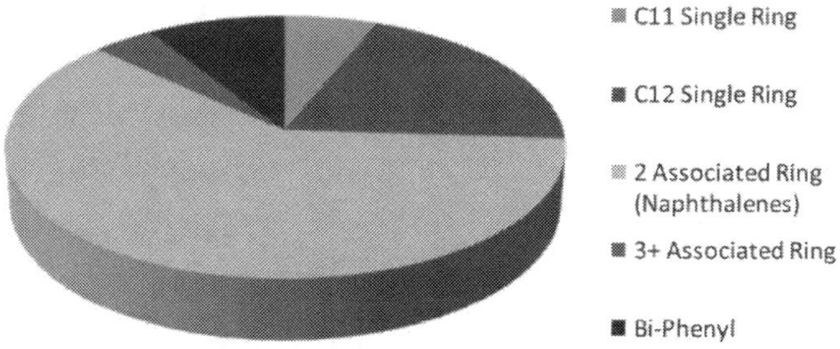

Typical heavy aromatics composition

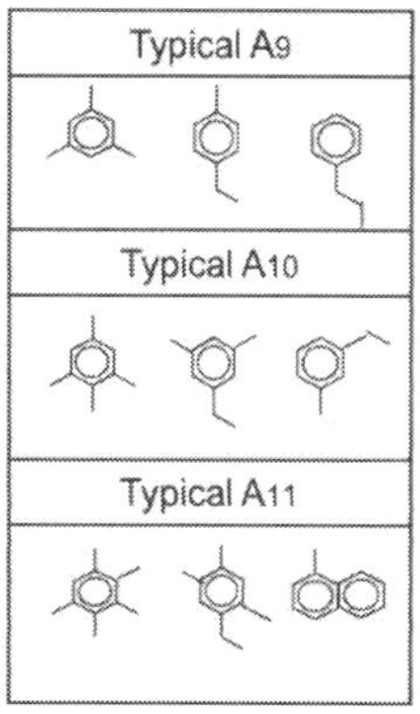

- Heavy aromatics are aromatic compounds with 9 or more carbon atoms.
- In an Aromatics complex, most A_9 and A_{10} heavy aromatics return to the Tatoray unit to be converted to more valuable product.
- Remaining A_{10+} becomes lower-value product. ~25% are single-ring aromatics, and ~60% are dual-ring aromatics (Naphthalenes).
- UOP TA-30 Tatoray Catalyst is designed to maximize the utilization of heavy aromatics

Tatoray Process in Aromatics Complex

UOP A Honeywell Company

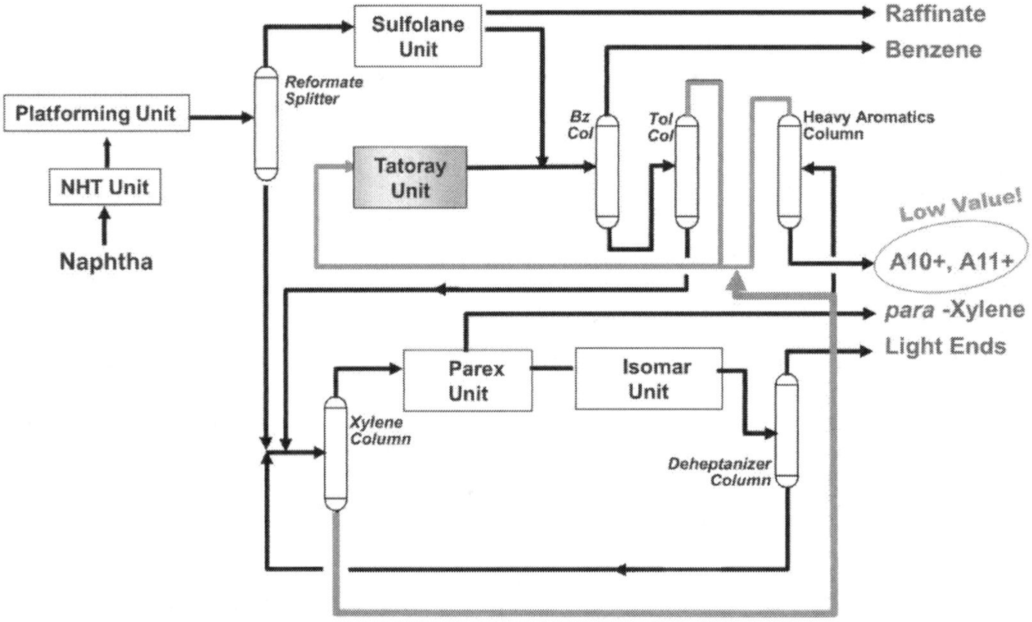

Benefits of TA-30 Catalyst

UOP A Honeywell Company

- **Compared to previous Tatoray Catalysts**
  - **Capable of processing heavier feeds**
  - **Higher activity and better stability**
  - **Improvement in xylene yield and ring retention**

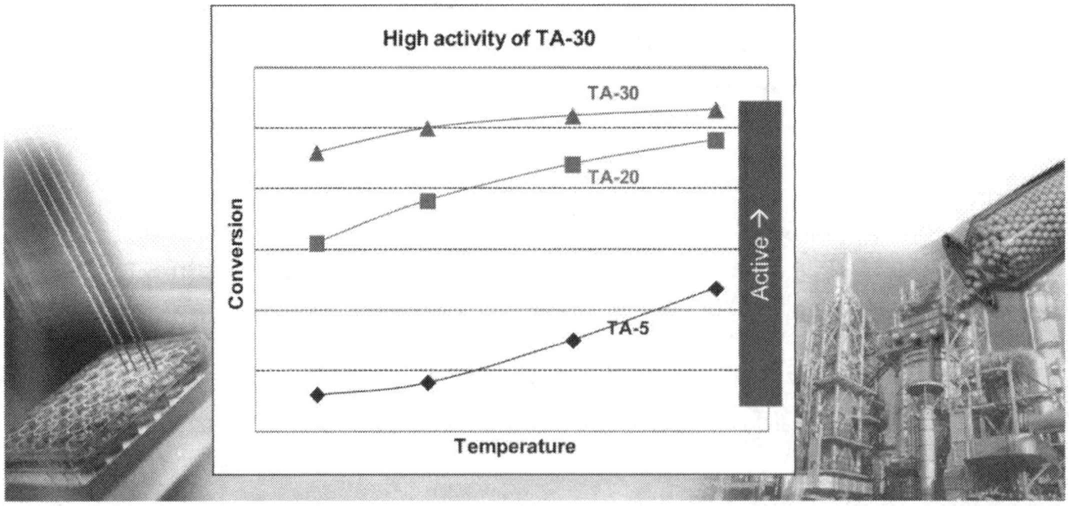

Comparing TA-30 with TA-20 Catalyst

UOP
A Honeywell Company

- **Comparing TA-30 and TA-20 Performance**
  - Following charts are based on identical feed
  - TA-30 catalyst is significantly more active and stable than TA-20 catalyst
  - TA-30 catalyst provides higher Xylene yield
  - TA-30 catalyst has less ring loss

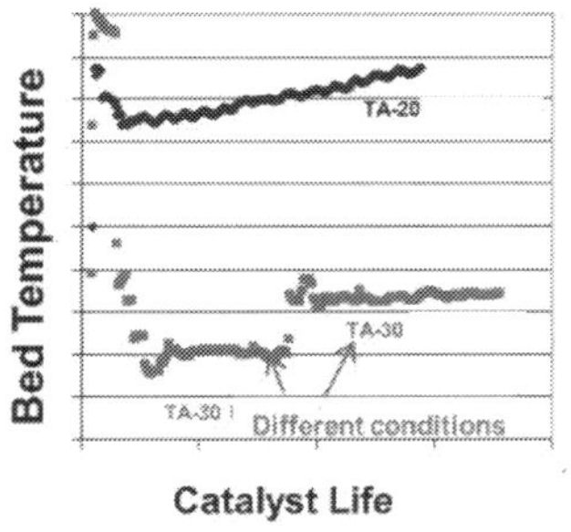

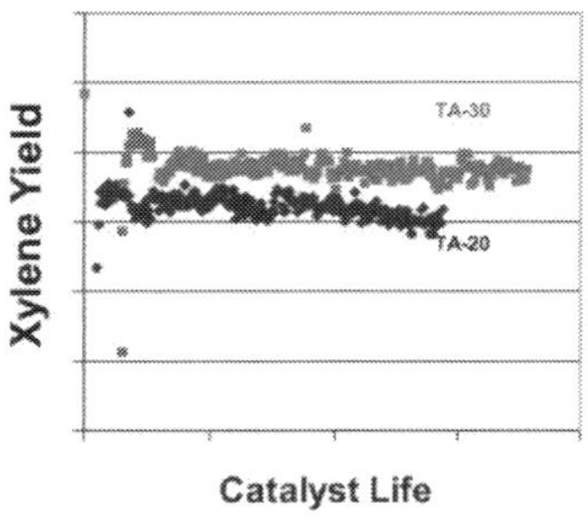

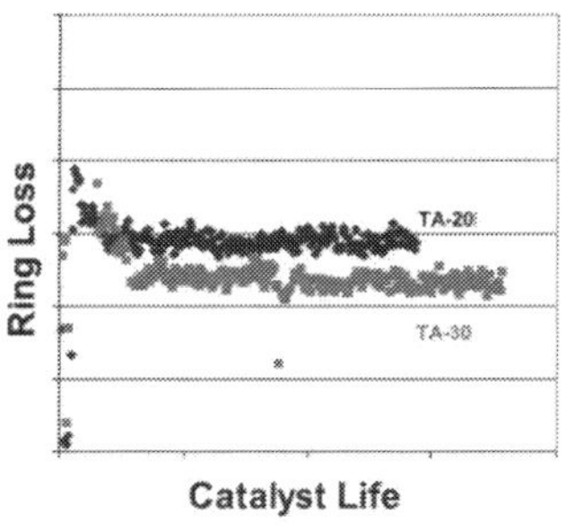

UOP Tatoray Catalyst

UOP
A Honeywell Company

- **UOP proprietary zeolites**
  - TA-30 catalyst uses state-of-the-art proprietary zeolite UOP developed specifically for this technology.
  - UOP proprietary zeolite's exceptional performance provides higher activity, lower ring loss, better stability, and higher xylene yield and selectivity
- **Metal**
  - Optimized metal function further improves catalyst, resulting in especially high stability with heavy feed.

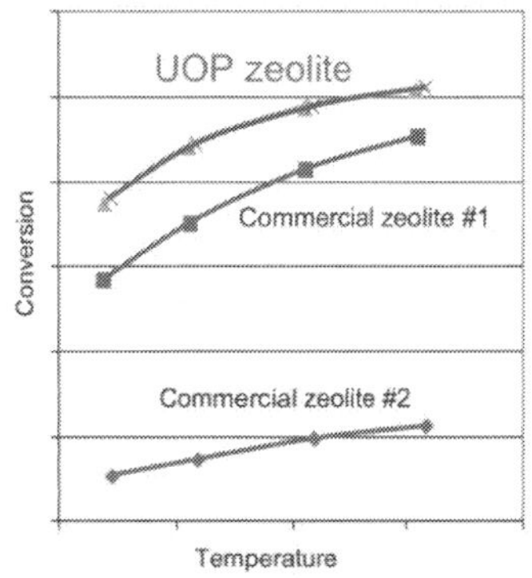

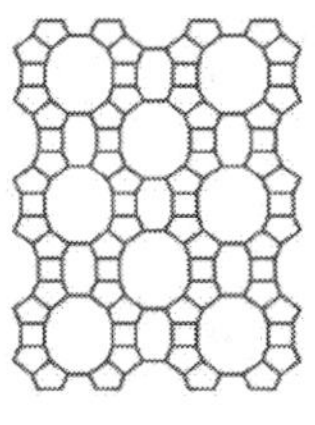

TA-30 Catalyst Success Story

UOP
A Honeywell Company

- **TA-30 catalyst successfully commercialized July 2010 in Asia**
- **Very heavy feed**
  - 25% Toluene
  - 65% A_9's
  - 10% A_{10+}
- **This heavy A_{9+} operation has demonstrated TA-30's Superior Yields, High Activity and Solid Stability**
- **Customer has loaded TA-30 catalyst into their second unit**

Customer A - Reactor Temperatures

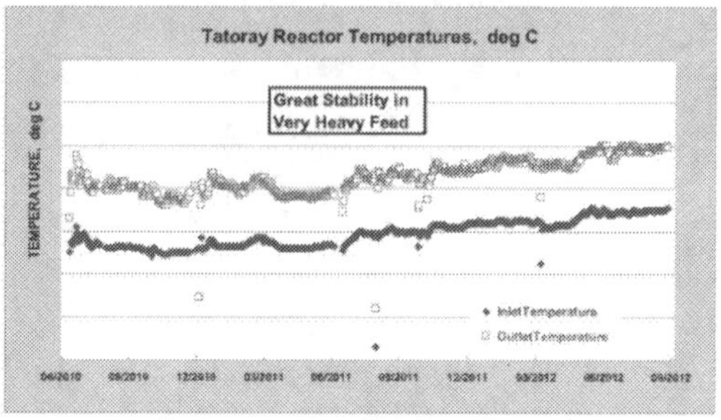

Customer A – C_8A and Bz Yields

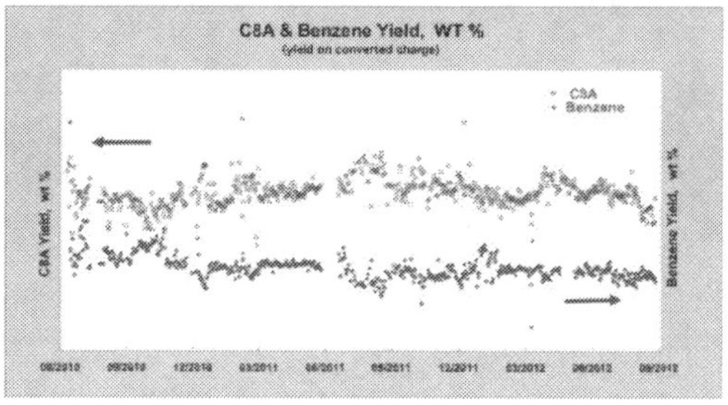

Tatoray PTD™ process

UOP
A Honeywell Company

- **The high stability and activity of the TA-30 catalyst allows for further optimization of the process conditions**
- **Benefits of Tatoray PTD™ process**
  - Reduction in capital and utility cost.
  - Tatoray PTD Process shows high catalyst stability **with** various feed composition.
  - Pilot plant testing has been performed to support the utilization of existing idle plant assets to lower the cost of implementing the Tatoray PTD Process.
- **Tatoray PTD™ process has been selected by several UOP customers for their new aromatic complexes**

Capital Cost Savings - Tatoray PTD Process

UOP A Honeywell Company

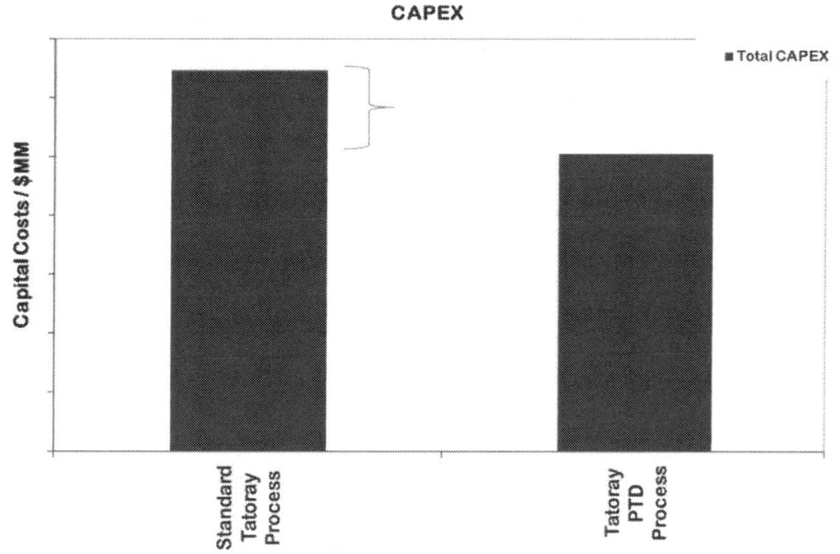

Significant reduction in total capital cost investment for Tatoray PTD Process over the typical Tatoray unit.

Tatoray PTD Process Stability

UOP A Honeywell Company

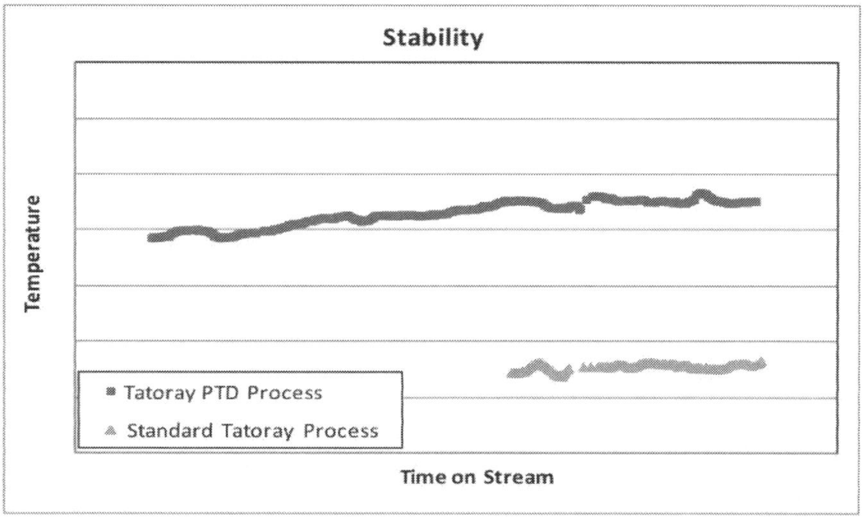

- Tatoray PTD Process demonstrates similar stability and catalyst life as he standard Tatoray process

Tatoray PTD vs. Standard Tatoray Process

UOP
A Honeywell Company

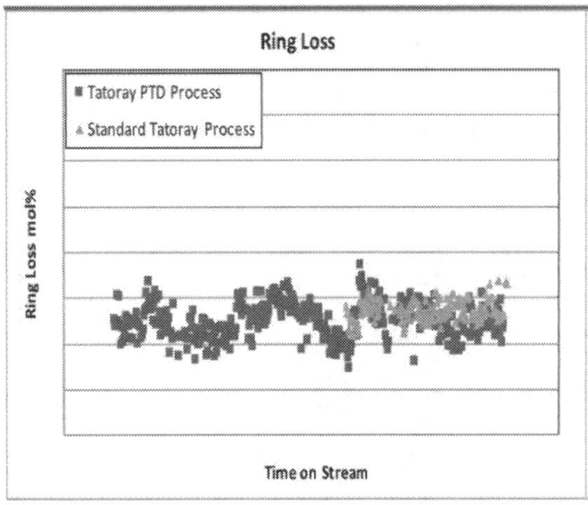

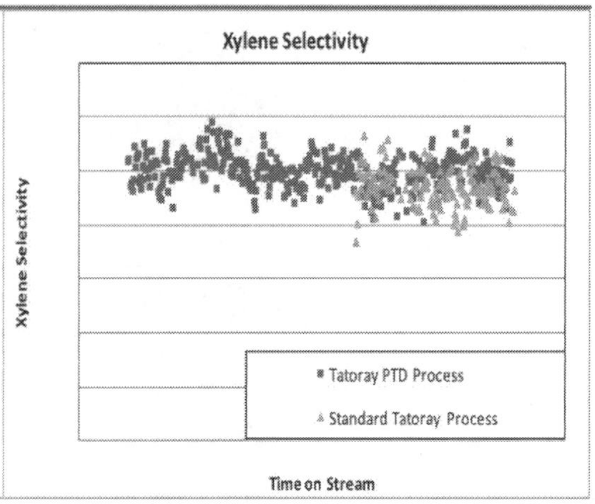

- Tatoray PTD process is able to achieve a performance similar to the standard Tatoray process at optimized operating conditions that lowers the capital cost.

Summary

UOP
A Honeywell Company

- **Tatoray TA-30 Catalyst**
  - **Superior catalyst life**
  - **Higher xylene yield**
  - **Lower feed requirement**
  - **Commercially Proven**
    - Two commercial units for two years of stable performance with minimal deactivation.

Novel & proprietary zeolite structure

- Tatoray PTD process
  - Optimized process condition allows processing of a wide range of feed composition to maximize value.
  - Reduced capital cost and utility cost
  - Demonstrated feasibility for new or revamped units

UOP 5734B-22

uop
A Honeywell Company

Q&A

Johnson Matthey

The Eight Steps to Specify a Catalyst Bed

Gary Gildert
New Business Technology Manager
Johnson Matthey Catalysts
gary.gildert@matthey.com
281-291-0709

October 4, 2012

chemical catalysts

Johnson Matthey

Johnson Matthey

- Leading supplier of high performance process catalysts and technology
- Continued development of technology for almost 200 years
- Leading position across the global catalyst market, with a history of delivering innovative, economic and sustainable solutions
- State of the art laboratories, manufacturing facilities, research and development centers
- Providing high quality products with high quality customer service
  - Working with customers to meet their timescales
  - Listening to customers needs
  - Developing and innovating for the future

2

chemical catalysts

The Eight Steps

Johnson Matthey

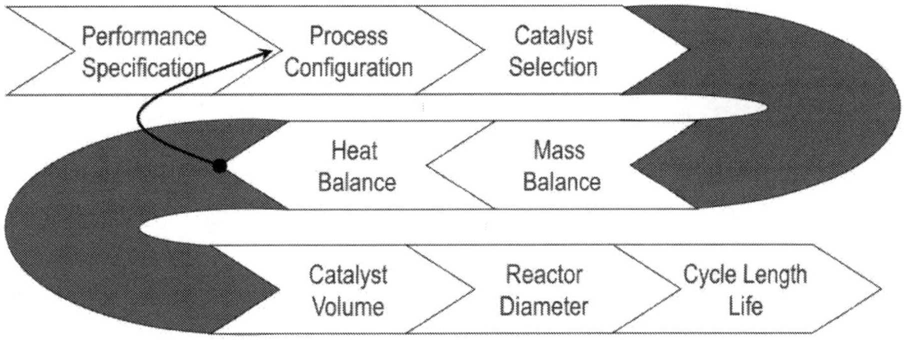

chemical catalysts

1. Performance Specification

Johnson Matthey

- Feed rate + margin
  - Maximum rate for sizing
  - Normal rate for life
- Stream composition / properties
  - Hydraulics
- Key concentrations for bed sizing – feed and product.
  - Limiting reactant
  - Basis for specification
- Hydrogen source
  - Purity
  - Pressure
- Poisons – assumptions vs. history

- Alternate cases – one governs sizing
- Units of measurement

chemical catalysts

Application Questionnaire

Johnson Matthey

Application Information							
Acetylene Converter, Tail-end							
Company:					Date:		
Location:					By:		
Process Information							
De-ethanizer Overheads		Normal	Maximum		**Feed Contaminants**		
Feed rate	kg/hr				H2S	ppm m	
Composition					COS	ppm m	
Methane	mol %				Arsine	ppb w	
Acetylene	mol %				water	ppm m	
Ethylene	mol %						
Ethane	mol %						
Propylene +	mol %						
Hydrogen Composition							
Hydrogen	mol %						
Methane	mol %						
Carbon Monoxide	mol %						
Ethane +	mol %						
Product Specification		Maximum	Typical				
Acetylene	ppm m						
Hydrogen	ppm m						

chemical catalysts

5

2. Process Configuration

Johnson Matthey

- Batch Reactor
- Continuous Stirred Tank Reactor
- Plug Flow Reactor
  - Adiabatic
  - Isothermal
- Fluidized Bed

chemical catalysts

6

2. Configuration - Batch

Johnson Matthey

- Birth of most reaction chemistry (Chemists)
- Fill time + reaction time + discharge time
- Catalyst active during product discharge
- Product heel

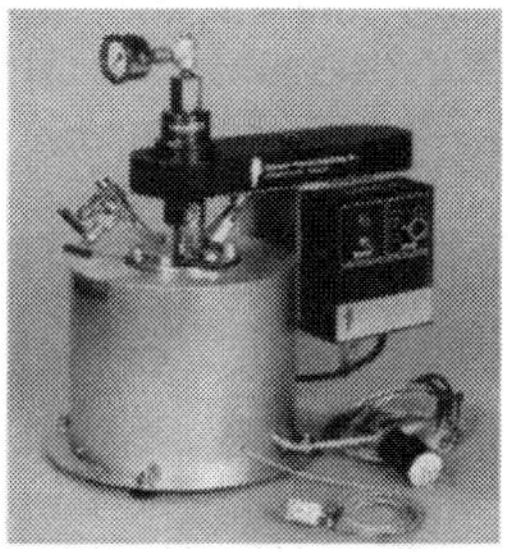

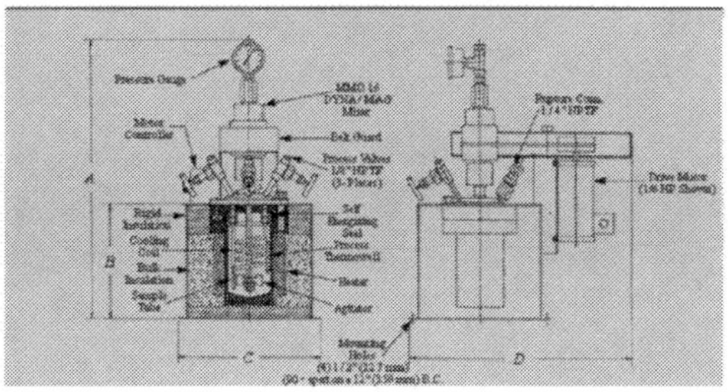

chemical catalysts

7

2. Configuration - CSTR

Johnson Matthey

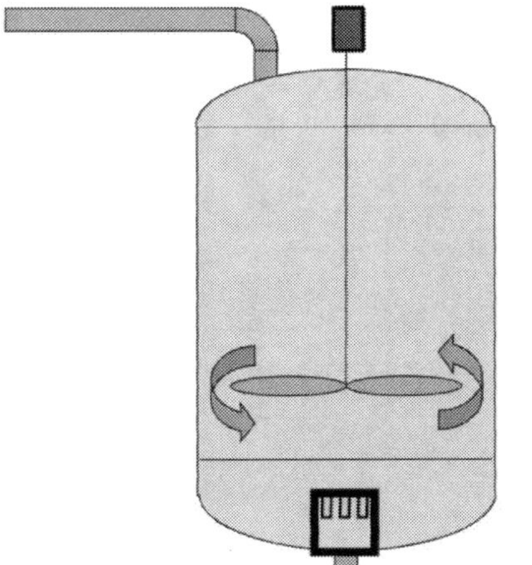

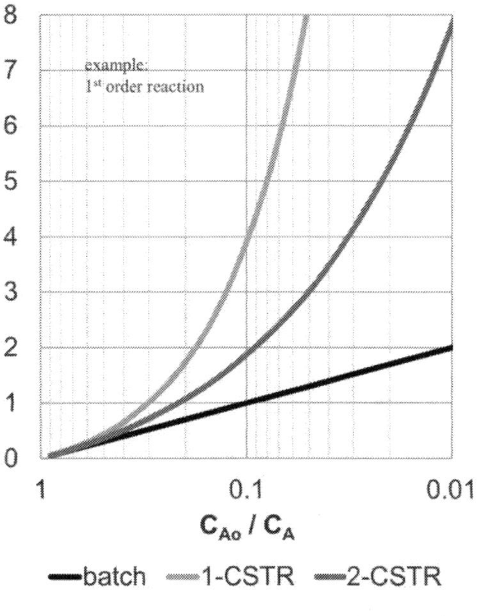

chemical catalysts

8

2. Configuration - PFR

Johnson Matthey

- Most Common Configuration
- Vapor Phase, Liquid Phase, Trickle Bed
- Many variations

chemical catalysts

9

Two Reactor Designs

Tracerco
Providing Insight Onsite

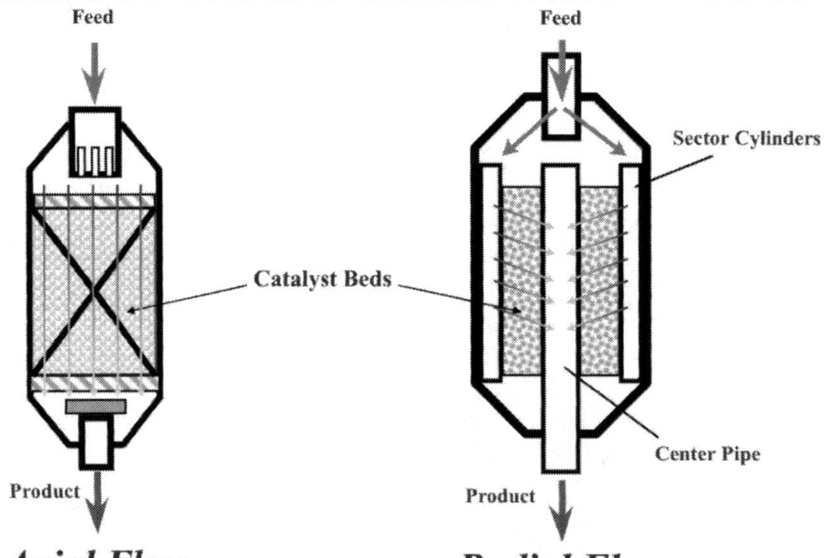

Axial Flow
(most common)

Radial Flow

July 2010 TruTec™ Scanning Services

Two More Reactor Designs

Tracerco
Providing Insight Onsite

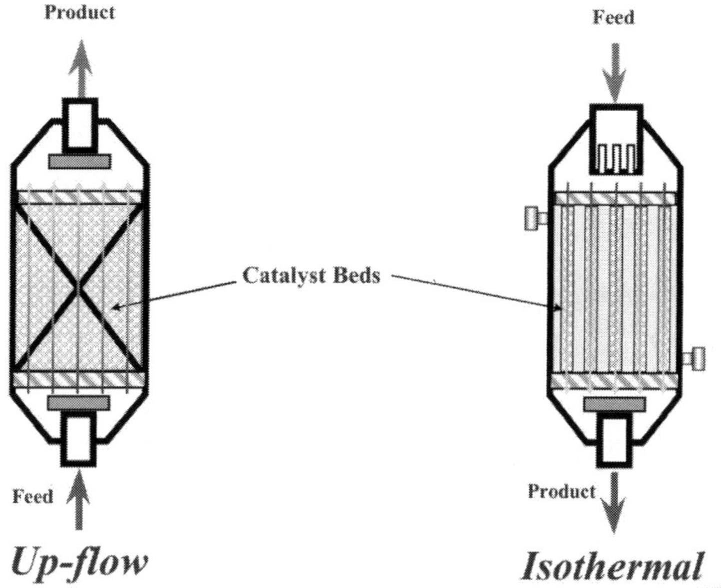

July 2010 TruTec™ Scanning Services

3. Catalyst Selection

Johnson Matthey

1. Active metal
   - i.e. hydrogenations: Pd, Pt, Ni, Cu, Co, Fe
2. Promoter
   - Depends on the required effect: Ag, Au, Mo, W, Pb, Sn
3. Carrier (Support)
   - Alumina, Silica, Zeolite, Carbon
4. Shape
   - sphere, extrusion, tablet
5. Size
   - 1 mm to 6 mm

- Standardized product by application

chemical catalysts

12

Catalyst Development

Johnson Matthey

- In-house state of the art testing facilities
- Gas and liquid phase testing capabilities specifically designed for different olefin streams
- Test work uses synthetic feed blends to mimic industrial compositions

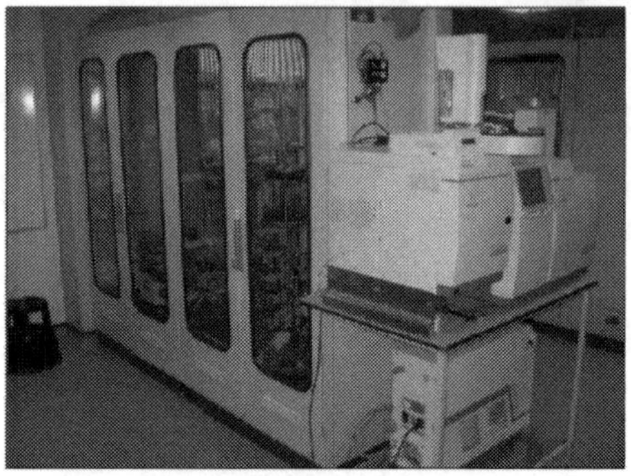

- Dedicated development team researching full and selective hydrogenation catalysts for various markets
- Proving on catalyst offerings under customer feed blends and process conditions

chemical catalysts

13

Catalysts for specific services

Johnson Matthey

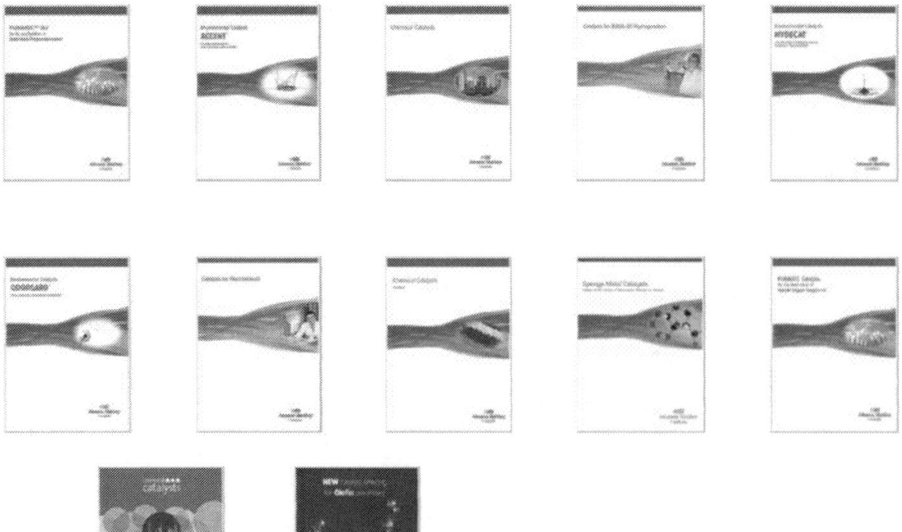

http://www.jmcatalysts.com/chemicalcatalysts/literature.asp

chemical catalysts

14

Acetylene Conversion

Johnson Matthey

Selective hydrogenation catalysts for acetylene conversion, suitable for a wide range of plant configurations and operating conditions

308 series of front-end catalysts

- acetylene removal (to <1ppm)
- minimise ethylene hydrogenation

408 series of tail-end catalysts

- high selectivity
- acetylene removal (to <1ppm)
- minimise green oil

Acetylene conversion catalysts

		PRICAT PD 308/1	PRICAT PD 308/4	PRICAT PD 308/6	PRICAT PD 408/5	PRICAT PD 408/7
Active metal		Pd	Pd	Pd	Pd	Pd
Promoted		yes	yes	yes	yes	yes
Size (mm)		3 x 3	3 x 3	3 x 3	2 - 4	2 - 4
Shape		tablet	tablet	tablet	sphere	sphere
Support		alumina	alumina	alumina	alumina	alumina
Service						
Front-end	de-ethaniser	yes		yes		
	depropaniser	yes	yes	yes		
	raw gas		yes			
Tail-end	lead bed				yes	yes
	middle bed				yes	yes
	clean-up bed					yes

PRICAT PD 308/6 is suitable for low CO concentrations

PRICAT PD 308/1 is suitable for high CO concentrations

chemical catalysts

15

The Eight Steps

Johnson Matthey

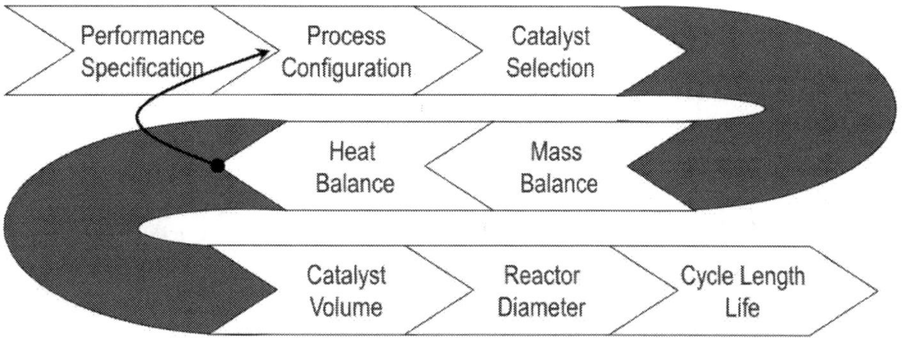

chemical catalysts

16

4. Material Balance

Johnson Matthey

- Ch.E. 101
- Moles!
- Conversion for spec component
- Account for every reaction
- Equilibrium limits
- Amount of "reactant" (hydrogen, oxygen, fuel)
  - i.e. H2:Ac, scfh per bbl
  - Excess reactant
  - % conversion
- Recycle composition
- Vent
- Spreadsheet
- Process simulation

17

chemical catalysts

5. Heat Balance

Johnson Matthey

- Required temperature
  - Minimum inlet
  - WABT
  - EIT
- Heat of Reaction
  - Heat of formation
  - Heat of combustion
- Pressure Effects
  - Dew point
  - Bubble point
  - V / L split
- VLE Data
- Spreadsheet?
- Process simulation

18

chemical catalysts

Heat and Material Balance Issues

Johnson Matthey

- Poor performance if temperature rise is greater than 75˚F (42˚C) per bed
  - Activity & selectivity issues
  - Increase recycle
  - Add another bed in series
- Must be at least 15°C (25°F) above the dew point to prevent condensation on catalyst
  - Feed superheat
  - Intercooler operation
- Hydrogen solubility issues
  - Choose thermo package carefully
  - 2-phase feed more complicated than single phase
- Vaporization due to heat of reaction
  - Channeling
  - Hot spots

chemical catalysts

19

The Eight Steps

Johnson Matthey

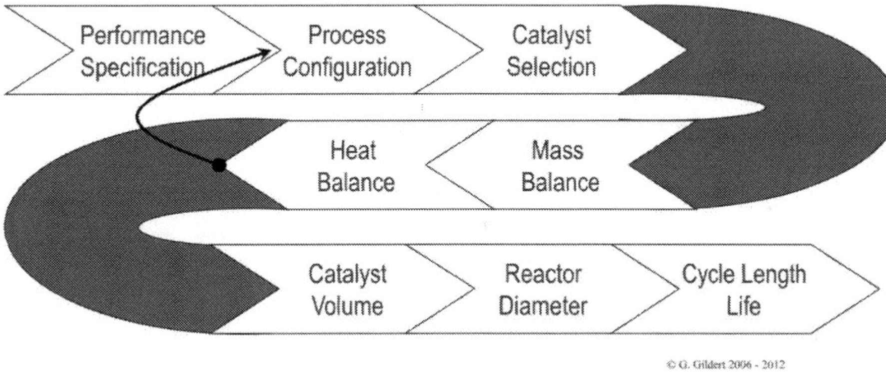

chemical catalysts

20

6. Catalyst Volume - Practice

Johnson Matthey

- Required contact time (τ) determined by laboratory performance testing and experience
- Kinetic theory can provide interpolation for alternate conditions.
- Rates are normally transformed for nominal conditions
  - SI = 1 atm, 0°C = Nm^3/hr per m^3 of catalyst
  - USCU = 14.7 psia, 60°F = scfh per ft^3 of catalyst

$$GHSV = gas\ hourly\ space\ velocity \propto \frac{1}{\tau}$$

$$LHSV = liquid\ hourly\ space\ velocity \propto \frac{1}{\tau}$$

hr^{-1}

chemical catalysts

21

7. Reactor Diameter

Johnson Matthey

- L/D = ½ to 5
  - Radial distribution of short beds
  - Wall effects on tall beds
- Bed Height Limits
  - Maximum based on catalyst crush strength, loading, channeling
  - Minimum based on history, conversion
- Superficial Velocity
  - Maximize for mass transfer
  - Limited by channeling for 2-phase systems
  - Turbulence via Re
  - Mass Transfer via Sh
- Pressure Drop / Flow Regime
  - Process design to minimize
  - High cost & system limits
  - 2-phase flow regime for improved mass transfer = reaction rate

chemical catalysts

22

8. Catalyst Cycle and Life

Johnson Matthey

- Cycle Length
  - **Experience**
  - Temperature
  - Heavies in feed

- Life
  - Number of regenerations
  - **Accumulation of poisons**
    - Average feed rate * average concentration vs. capacity

chemical catalysts

23

The Eight Steps

Johnson Matthey

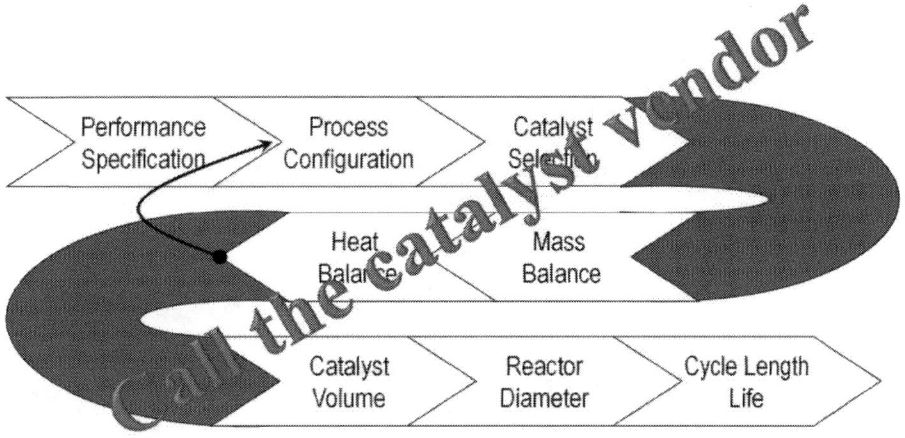

chemical catalysts

24

Technical Support from JM

Johnson Matthey

The technical service function offers:

- Catalyst selection
- Operating conditions and life predictions
- Start up, shut down and regeneration procedures
- Routine plant monitoring.
- On site technical assistance
- Trouble shooting
- Basic process flow sheets for new applications
- Reactor design (process vessel sketches)
- Chemical and physical analysis of spent catalyst

Highly experienced Johnson Matthey personnel have extensive knowledge of catalyst technology, applications and operations available to the customer.

chemical catalysts

25

Johnson Matthey

Thank You.
Questions?

Gary Gildert
New Business Technology Manager
Johnson Matthey Catalysts
gary.gildert@matthey.com
281-291-0709

chemical catalysts

"Hydroprocessing Catalysts and Processes"

2012 AIChE Process Technology Conference
Catalysis and Reaction Engineering Section

Presented by:

Paul R Robinson, PhD
Senior Engineer, Hydrocracking
Criterion Catalysts & Technologies/Zeolyst International

October 4, 2012

CRITERION
CATALYSTS & TECHNOLOGIES

Outline

- Purpose of this presentation
  - To give an overview of the role of hydroprocessing in petroleum refining
- Crude oil to pure products
  - Properties of crude oil
  - Petroleum refining overview
  - Chemicals in refinery streams
  - Reactions & mechanism
- Catalysis fundamentals
- Hydroprocessing catalysts

2

Refining Crude Oil into Pure Products

Crude Oils	Desired Products
▪ Conventional and synthetic ▪ Wide range of crudes • API gravity: 8 to >45 • Sulfur: >5 wt% to <0.03 wt% • Nitrogen: 7400 wppm to <0.1 wppm ▪ Chemists have indentified >7000 unique molecules in crude oil. There are probably more than a million.	▪ Coke (sponge, needle) ▪ Waxes (three major types) ▪ Asphalts (four major classes) ▪ Lubricants (>800 different products) ▪ Fuel oils (six major types) ▪ ULSD ▪ Kerosene, Jet A /B, JP-4, -5, -6, -8) ▪ Low-sulfur gasoline ▪ LPG ▪ Petrochemical feedstocks ▪ Pure sulfur
	▪ Some petrochemical monomers must be >99.9% pure ▪ Even coke and asphalt have tight specifications

3

Role of Hydroprocessing

- Hydrotreating removes contaminants. Reactions include:
  - Saturation of Olefins (OSAT)
  - Saturation of Aromatics (ASAT)
  - Hydrodesulfurization (HDS)
  - Hydrodenitrogenation (HDN)
  - Hydrodeoxygenation (HDO)
- Hydrocracking breaks C-C bonds.
- Most reactions are exothermic. Heat control is crucial, especially in hydrocrackers!
  - Exception: coke formation/condensation reactions are endothermic. They deactivate catalysts

4

Hydroprocessing in a Typical Refinery

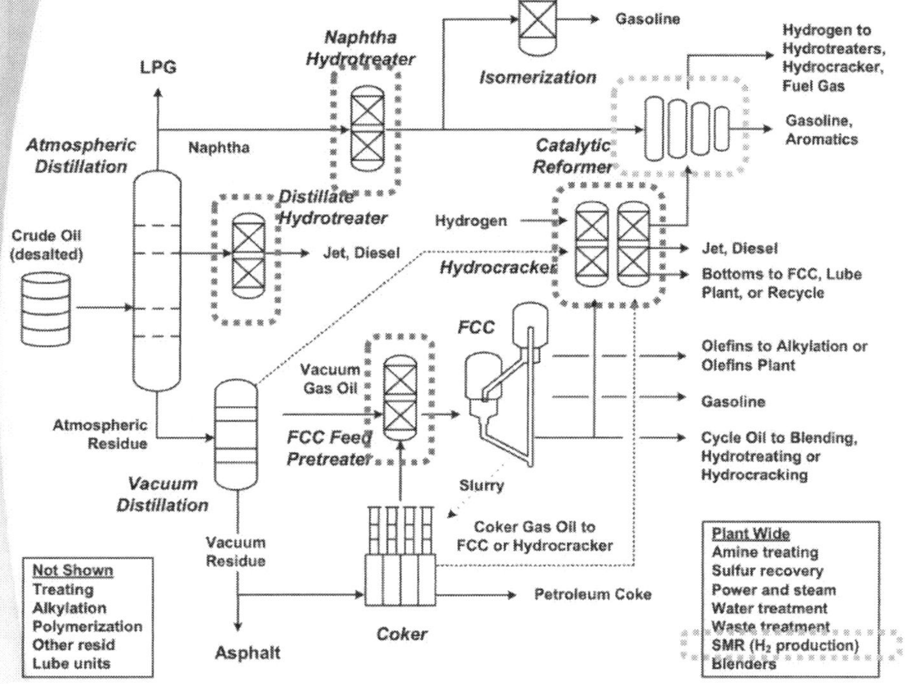

P.R. Robinson, Chapter 18 in Kent and Riegel Handbook of Industrial Chemistry, Springer (2012)

5

Straight-run Yields for Selected Crude Oils

Source field Country	Brent Norway	Bonny Light Nigeria	Green Canyon USA	Ratawi Mid East
API gravity	38.3	35.4	30.1	24.6
Specific gravity	0.8333	0.8478	0.8752	0.9065
Sulfur, wt%	0.37	0.14	2.00	3.90
		Yields, wt% feed		
Light ends	2.3	1.5	1.5	1.1
Light naphtha	6.3	3.9	2.8	2.8
Medium naphtha	14.4	14.4	8.5	8.0
Heavy naphtha	9.4	9.4	5.6	5.0
Kerosene	9.9	12.5	8.5	7.4
Atmospheric gas oil	15.1	21.6	14.1	10.6
Light VGO	17.6	20.7	18.3	17.2
Heavy VGO	12.7	10.5	14.6	15.0
Vacuum residue	12.3	5.5	26.1	32.9
Total naphtha	30.1	27.7	16.9	15.8
Total middle distillate	25.0	34.1	22.6	18.0

P.R. Robinson, Chapter 18 in Kent and Riegel Handbook of Industrial Chemistry, Springer (2012)

6

Hydrotreating Process Flow: Fixed Bed

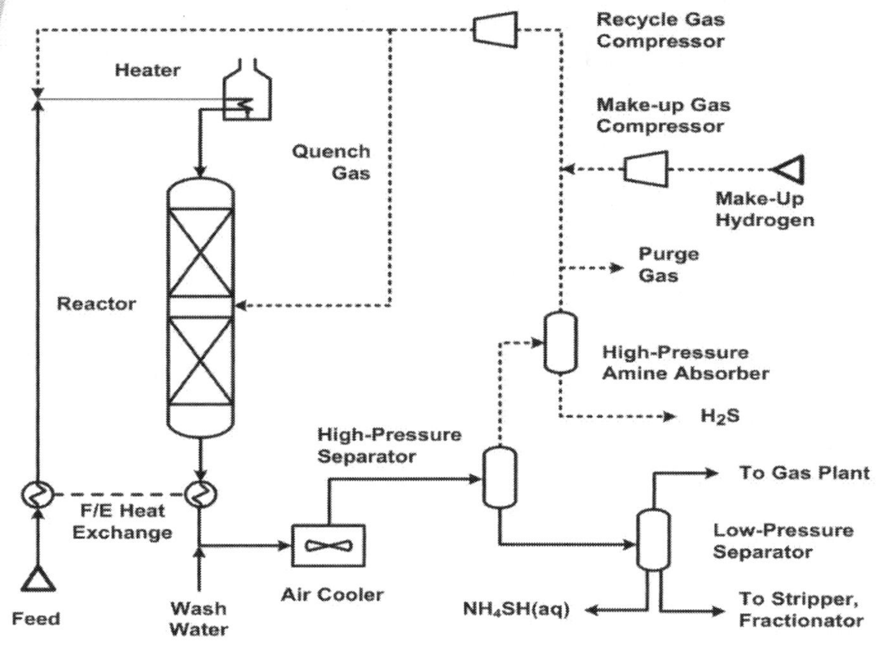

P.R. Robinson, Chapter 18 in Kent and Riegel Handbook of Industrial Chemistry, Springer (2012)

7

Hydrocracking Process Flow: Fixed Bed

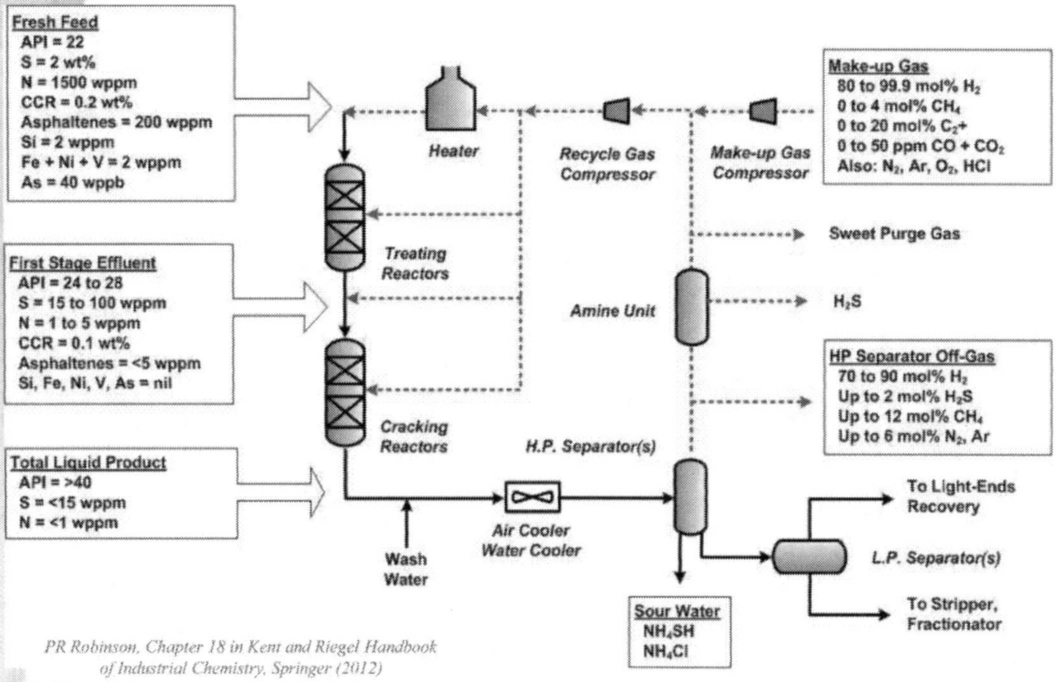

PR Robinson, Chapter 18 in Kent and Riegel Handbook of Industrial Chemistry, Springer (2012)

8

Fixed-Bed Reactor Internals

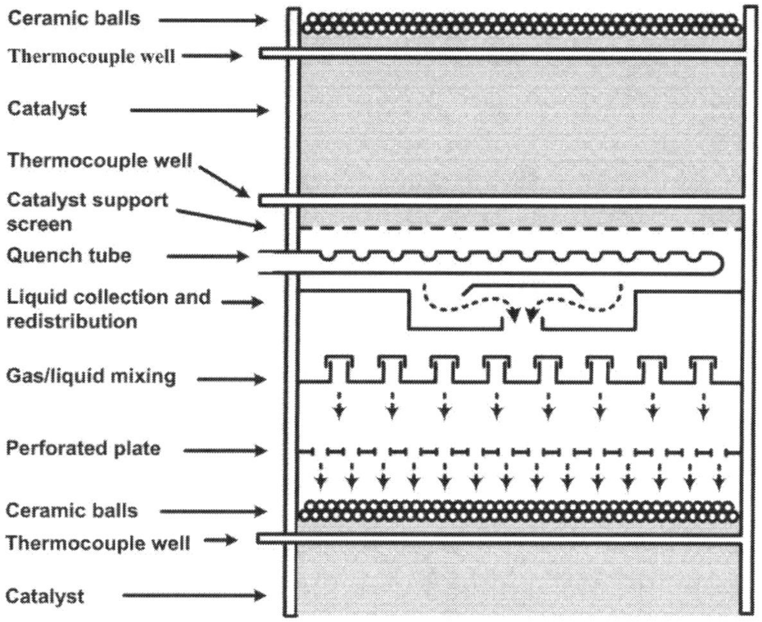

P.R. Robinson and G.E. Dolbear, Chapter 7 in Practical Advances in Petroleum Processing, Springer (2006)

9

Ebullated Bed Hydrocracking Reactor

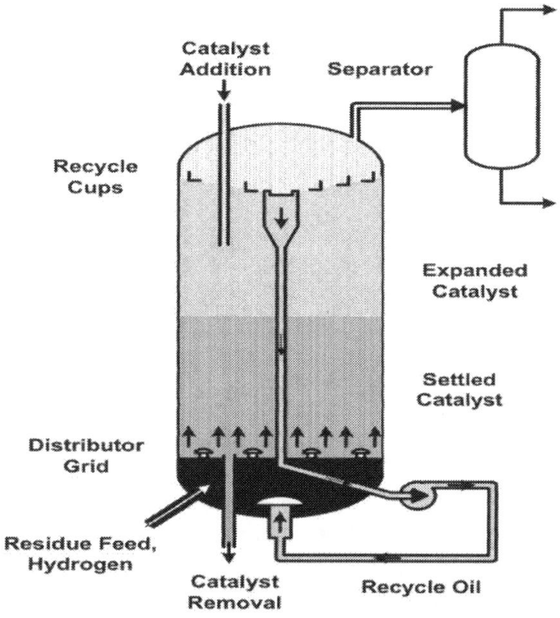

P.R. Robinson, Chapter 18 in Kent and Riegel Handbook of Industrial Chemistry, Springer (2012)

10

Slurry-Phase Hydrocracking: VCC Process

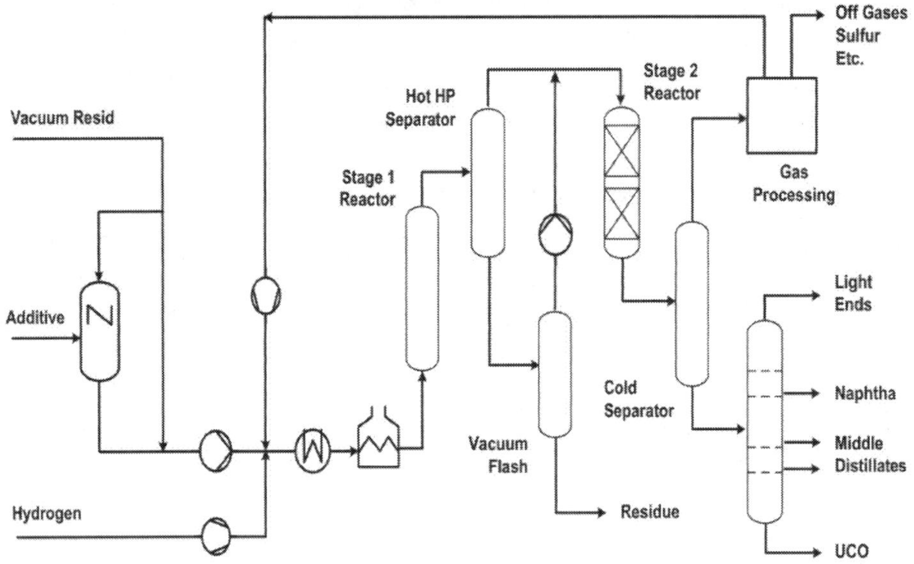

Source: KBR Technology. Used with kind permission from KBR Inc.

11

Hydroprocessing Reactions: Types of Molecules in Petroleum

- Hydrocarbons: Carbon and Hydrogen Only
  - Paraffins (methane, ethane, iso-octane, etc.)
  - Olefins (ethylene, propylene, etc.)
  - Aromatics (toluene, naphthalene, etc.)
  - Naphthenes (cyclohexane, etc.)
- Compounds with Heteroatoms
  - Sulfur, nitrogen, oxygen
  - Trace contaminants
    - (Fe, Ni, V, Si, As)

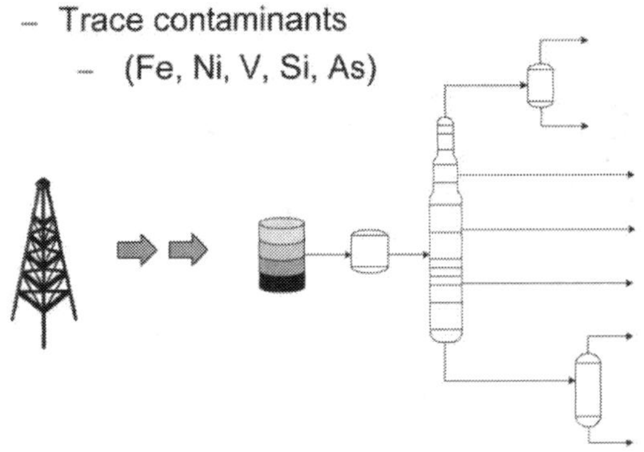

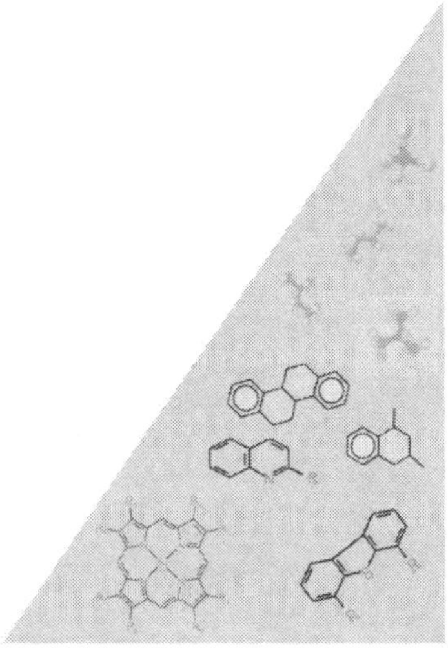

12

Hydrocarbon Ring Compounds: Examples

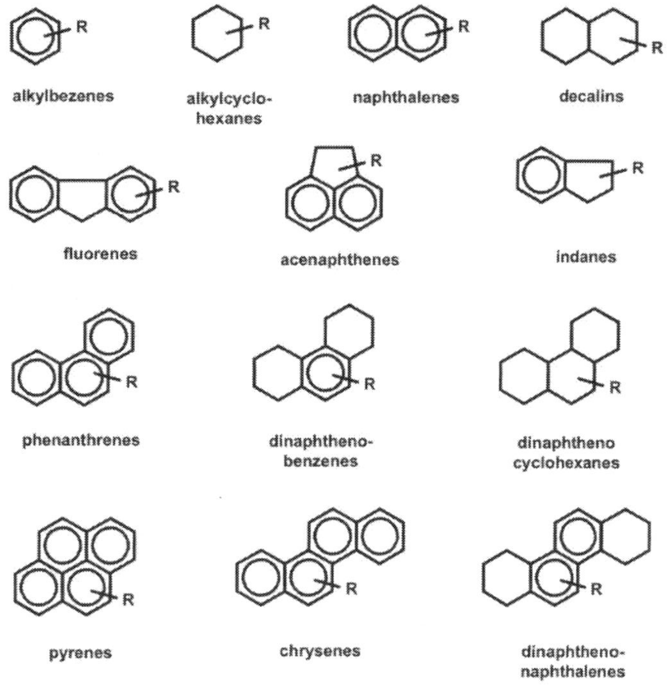

13

Chemical and Physical Properties within Functional Groups Vary by Carbon Number

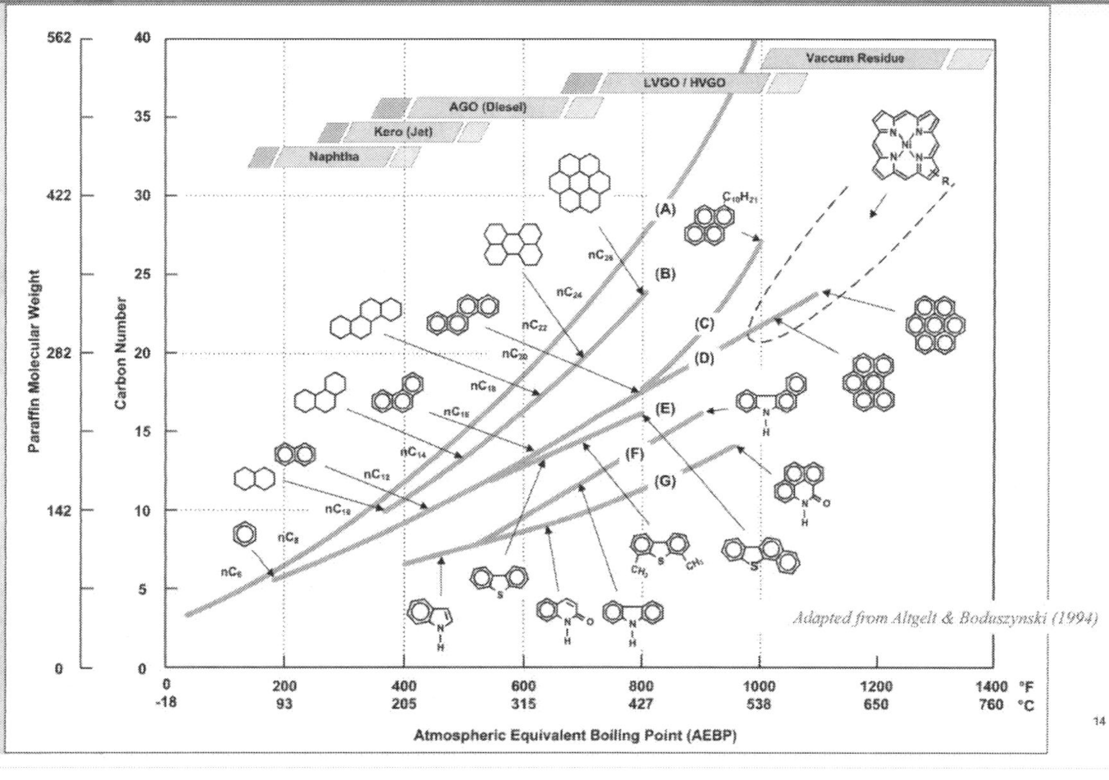

14

Density vs Molecular Weight for Different Types of Hydrocarbons

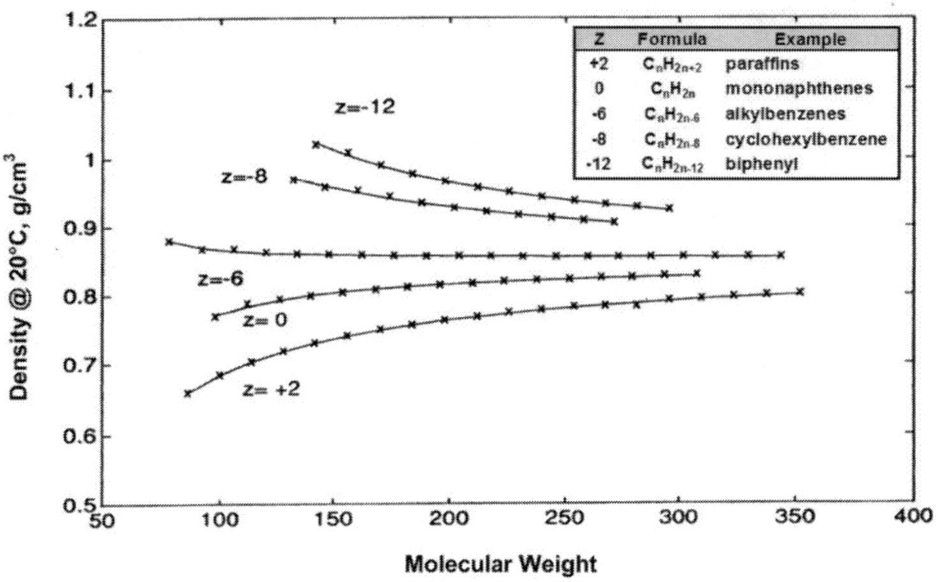

Used with kind permission from James W Bunger

15

Hetero-Atom Compounds

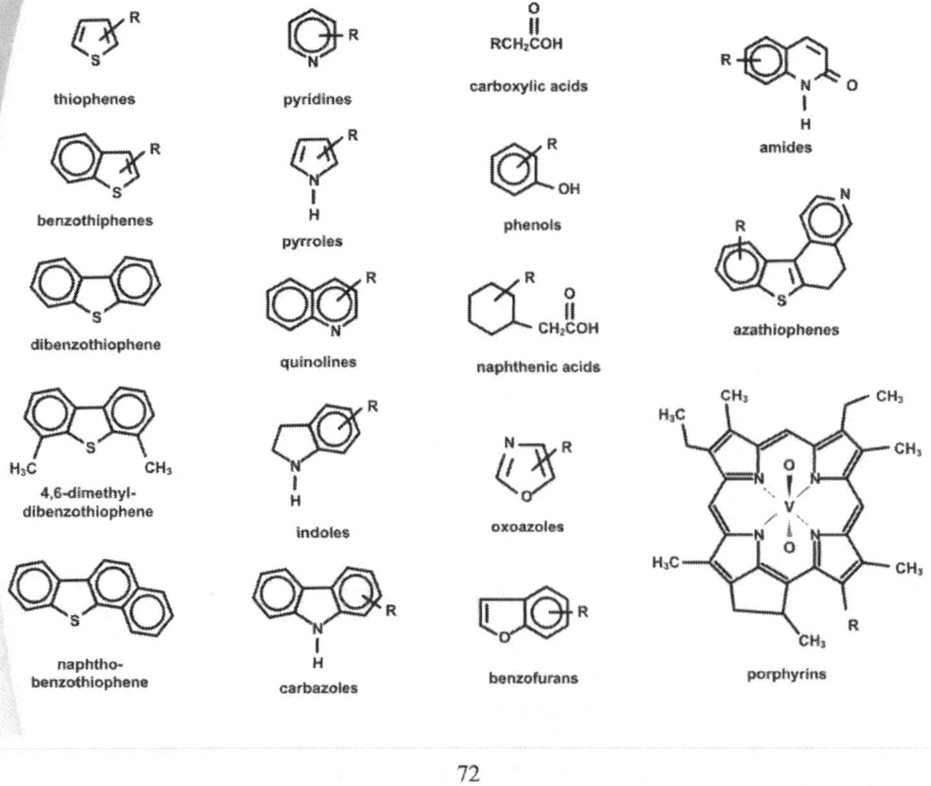

16

Biomarkers Show Similarities between Molecules of Life and Molecules in Petroleum

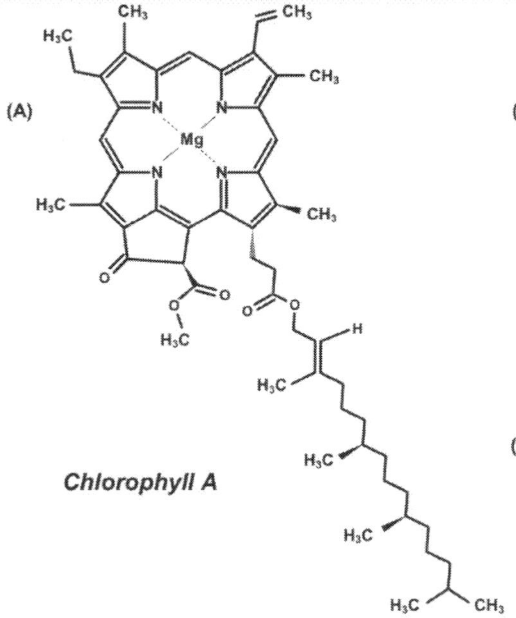

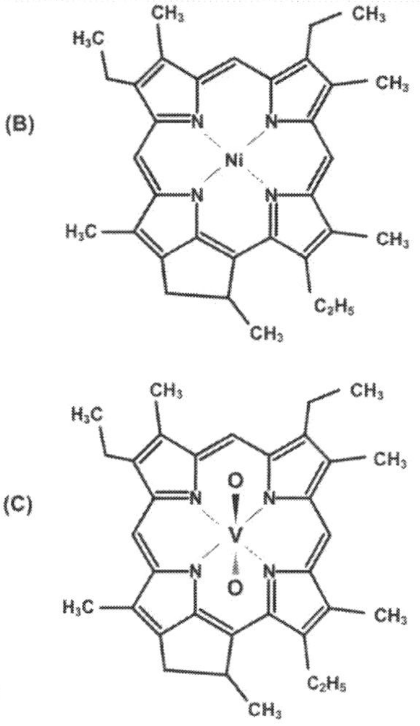

PR Robinson, Chapter 18 in Kent and Riegel Handbook of Industrial Chemistry, Springer (2012)

Porphyrins in Petroleum

17

Hydroprocessing Reactions

Reaction Type	Heat of Reaction * Range	Avg. or Typical
Olefin Saturation	-25.5 to -29.9	-27.8
Aromatics Saturation	-14.3 to -16.9	-15.9
Hydrodeoxygenation of ethers (HDO)	-12.5	-12.5
Hydrodesulfurization (HDS)	-6.4 to -13.4	-12.6
Hydrodenitrogenation (HDN)	-14.4 to -17.4	-15.3
Hydrocracking	-10.0 to -15.4	-12.2
Isomerization**	-2.1 to -2.3	-2.3

* ΔH_R per mole of H_2 consumed @ 25°C and 1 bar (kcal/mole)

** No hydrogen consumed or produced

Raw data came from the NIST Chemistry WebBook http://webbook.nist.gov
The table comes from GW Hampton and PR Robinson, "Controlling Hydrocracker Temperature Excursions," NPRA Paper PD-11-01. Used courtesy of AFPM.

18

Olefin Saturation (OSAT) Examples

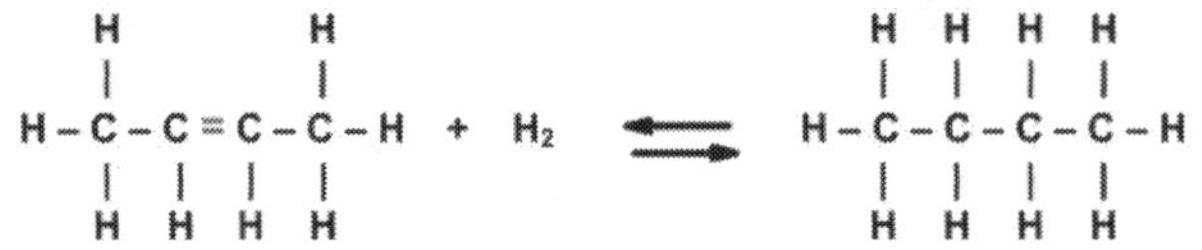

2-butene (cis- or trans-) butane

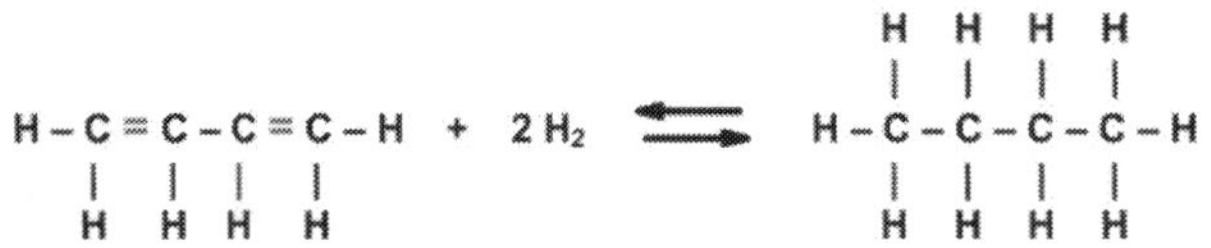

1,4-butadiene (cis- or trans-) butane

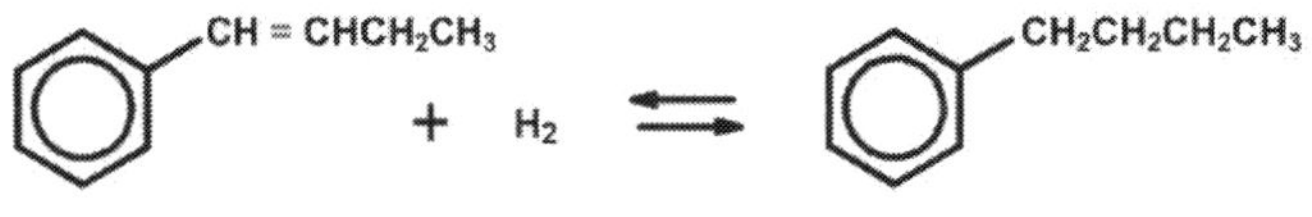

1-Butenyl benzene (cis- or trans-) butyl benzene

19

Aromatic Saturation (ASAT) Examples

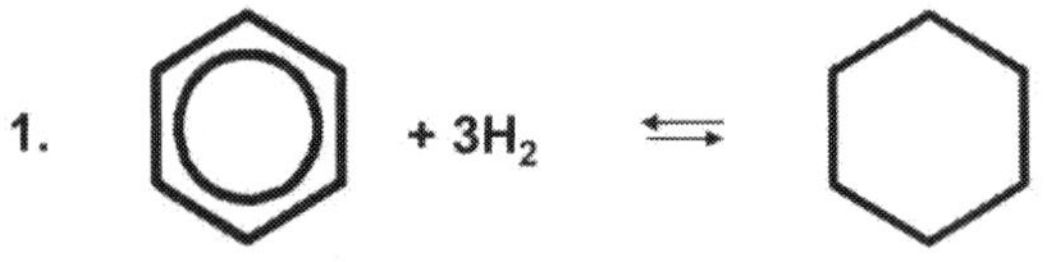

Benzene Cyclohexane

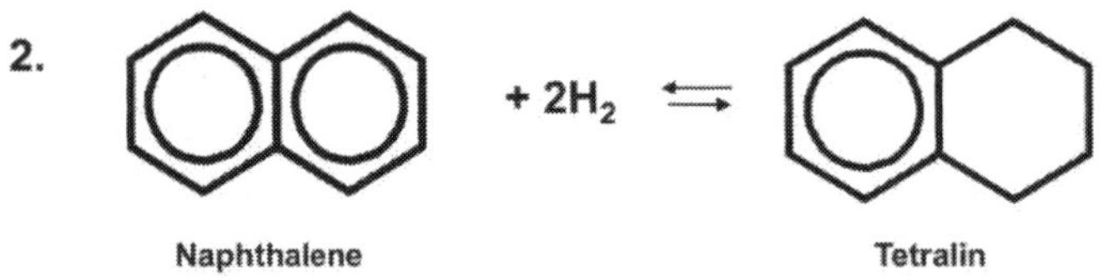

Naphthalene Tetralin

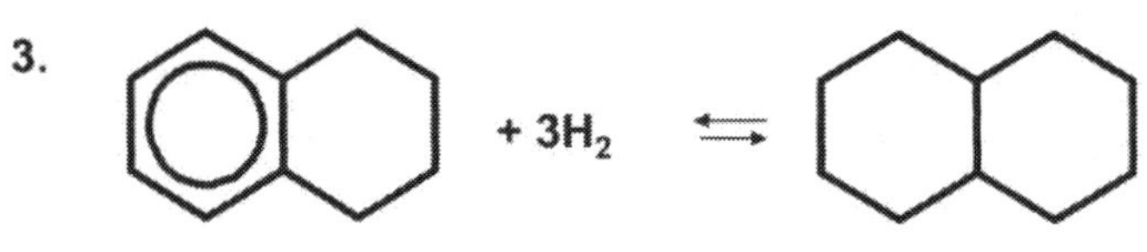

Tetralin Decalin

multi-ring aromatic = polycyclic aromatic = PCA = PNA = polynuclear aromatic

20

Saturation of Poly Aromatics is Sequential

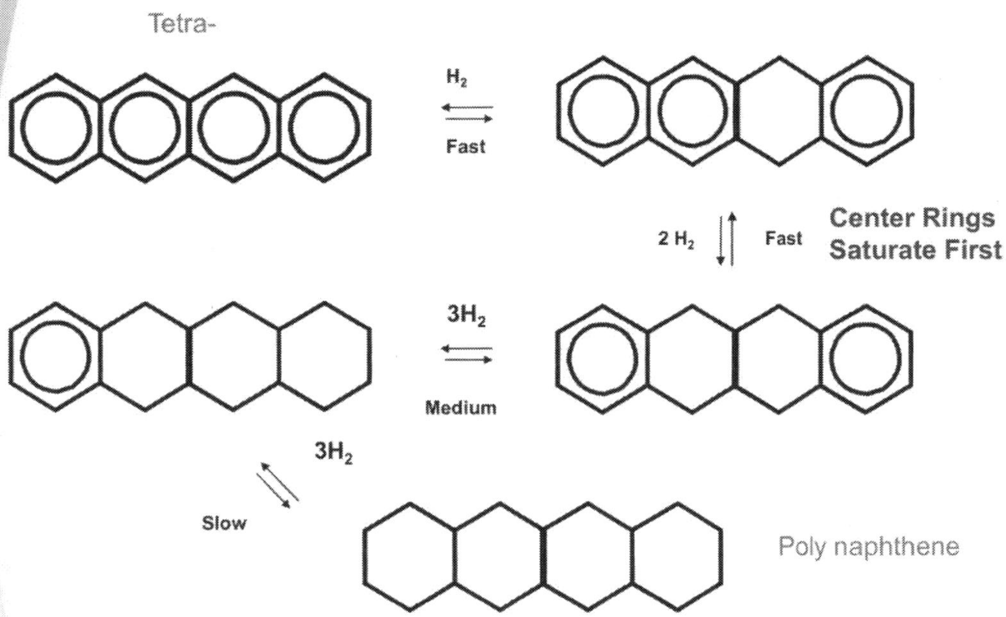

21

ASAT Dependence on T and P: Crossover Effect

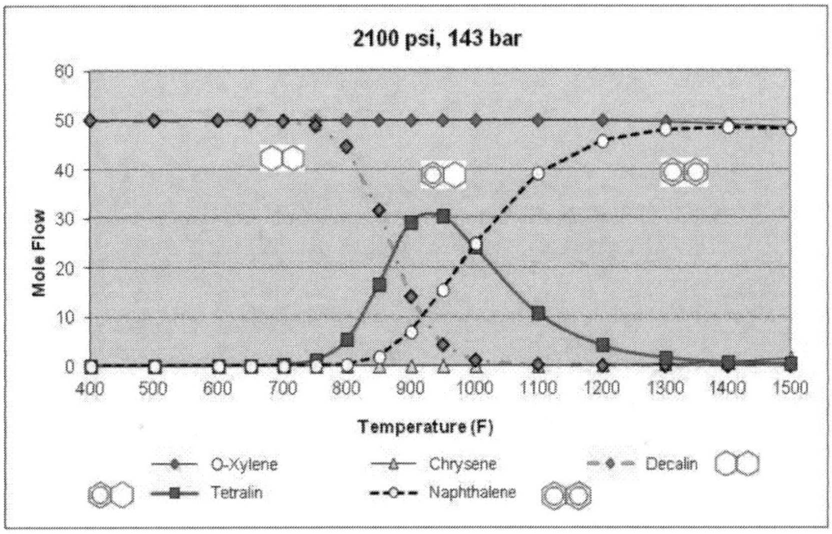

°F	°C
100	38
200	93
300	149
400	204
500	260
600	316
700	371
800	427
900	482
1000	538
1200	649
1400	760

Adapted from PR Robinson and LS Kraus, Thermochemistry of Coking in Hydroprocessing Units, AIChE Spring Meeting, Houston, 2004

ASAT Dependence on T and P: Crossover Effect

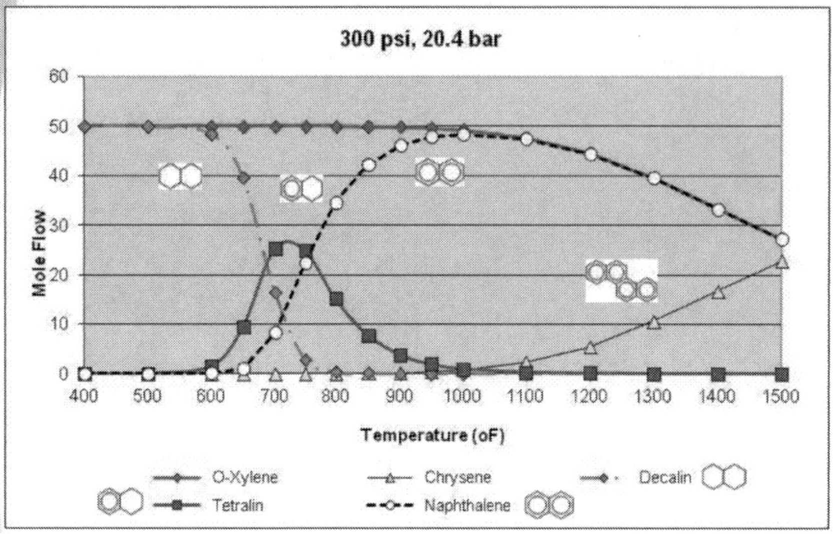

°F	°C
100	38
200	93
300	149
400	204
500	260
600	316
700	371
800	427
900	482
1000	538
1200	649
1400	760

Adapted from PR Robinson and LS Kraus, Thermochemistry of Coking in Hydroprocessing Units, AIChE Spring Meeting, Houston, 2004

23

Crossover Is Predicted by Thermo Kinetic Regime at Low T Confirmed by Impact of LHSV

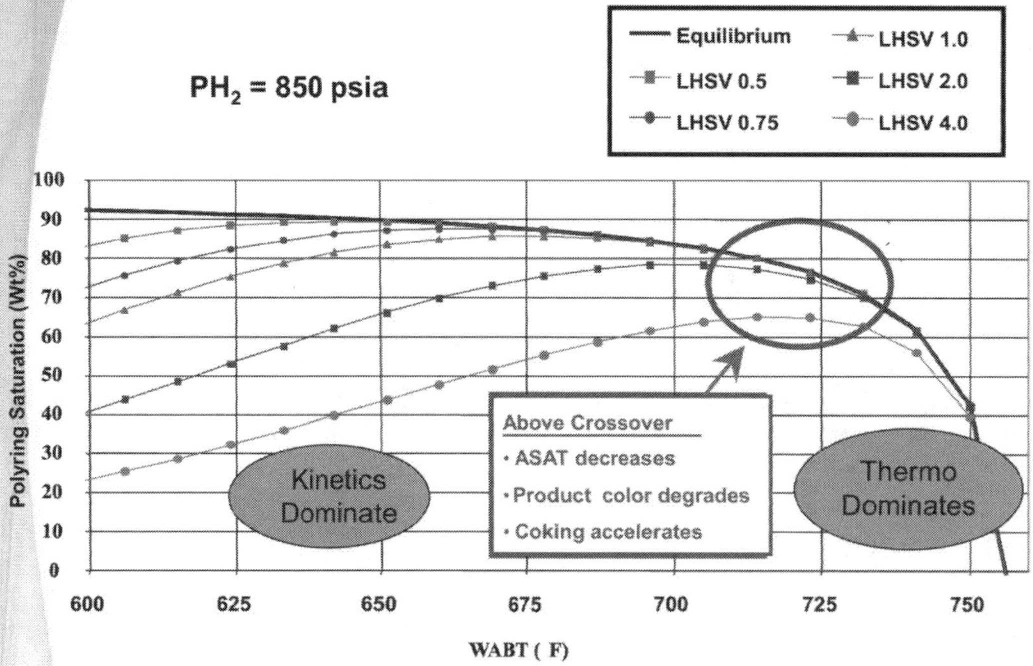

24

Naphthenoaromatics

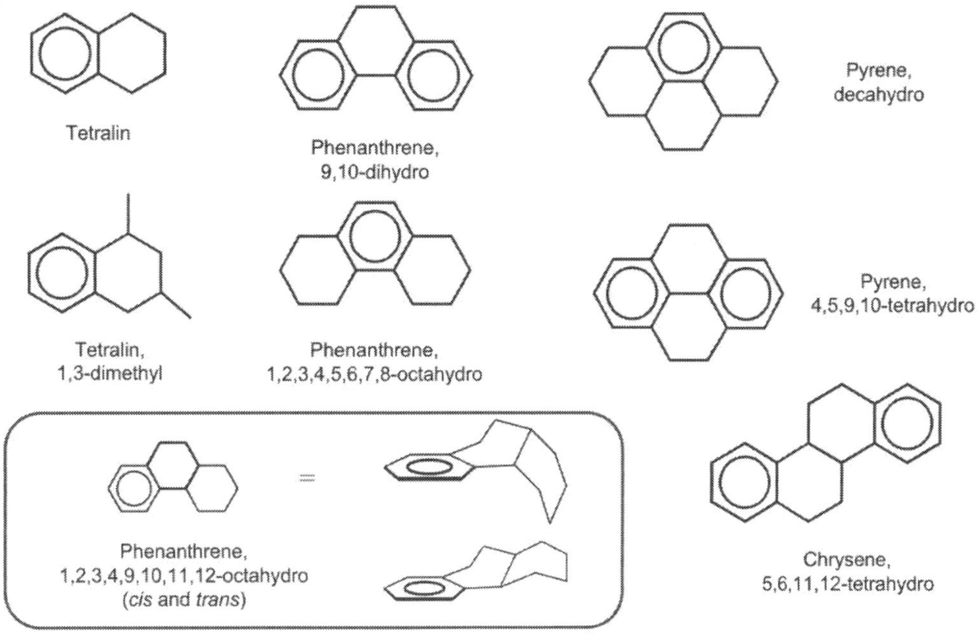

25

HDS of Distillate Sulfur Species: Relative Reactivity

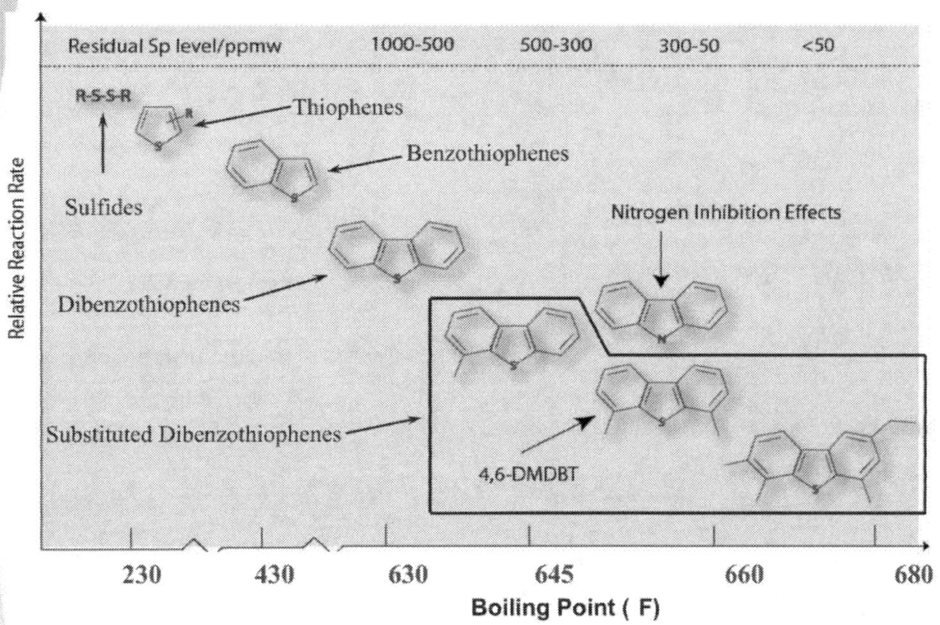

26

HDS Reactions: "Direct" Mechanism

$CH_3\text{-}S\text{-}C_4H_9$ (t-butylmethylsulfide) $\xrightarrow[\text{"Direct"}]{2\,H_2}$ CH_4 (methane) + C_4H_{10} (isobutane) + H_2S

$CH_3\text{-}S\text{-}S\text{-}CH_3$ (dimethyldisulfide) $\xrightarrow[\text{"Direct"}]{3\,H_2}$ $2\,CH_4$ (methane) + H_2S

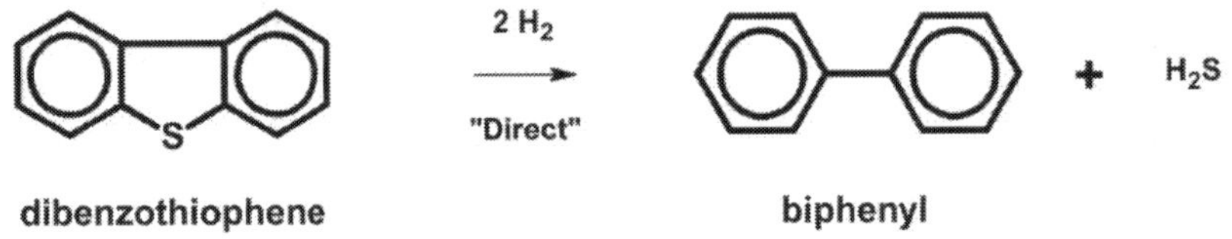

27

HDS Reactions: "Indirect" Mechanism

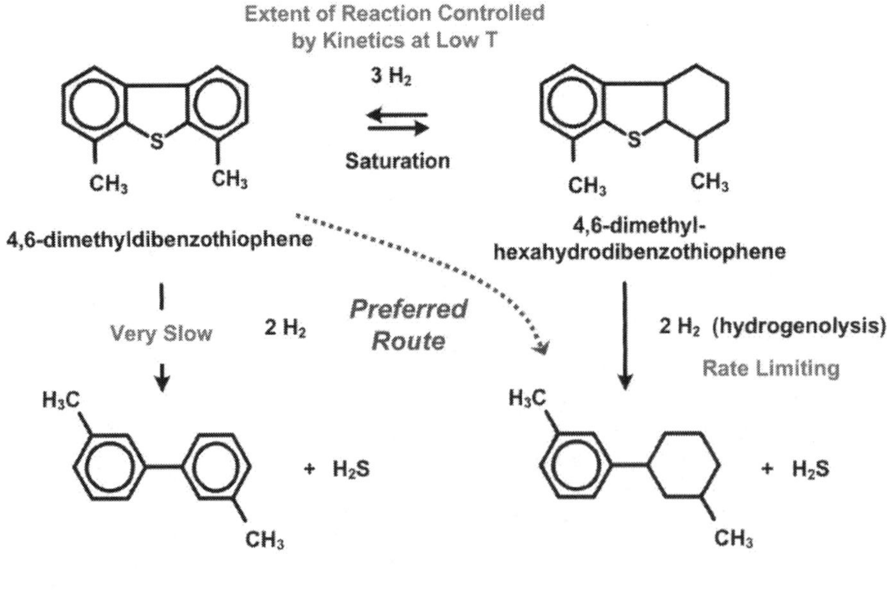

28

Hydrodenitrogenation (HDN) Reactions

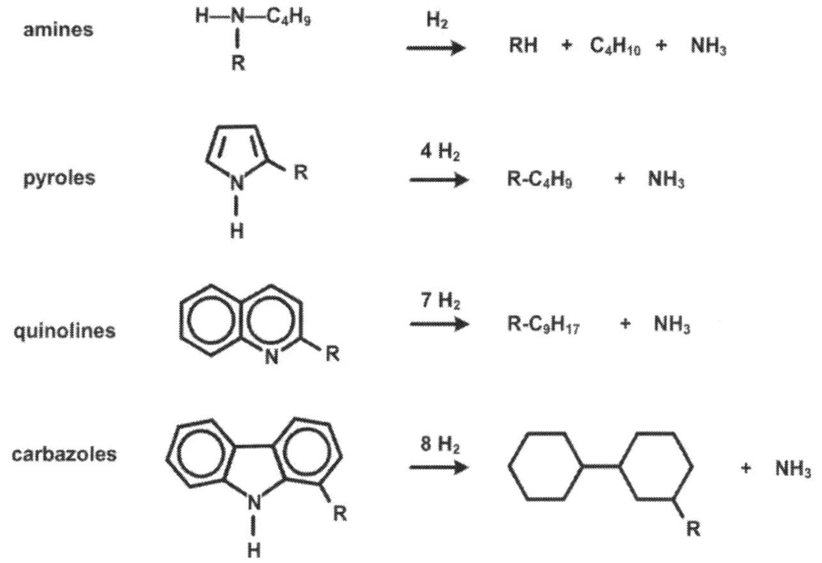

29

Typical Nitrogen Species: Relative Reactivities

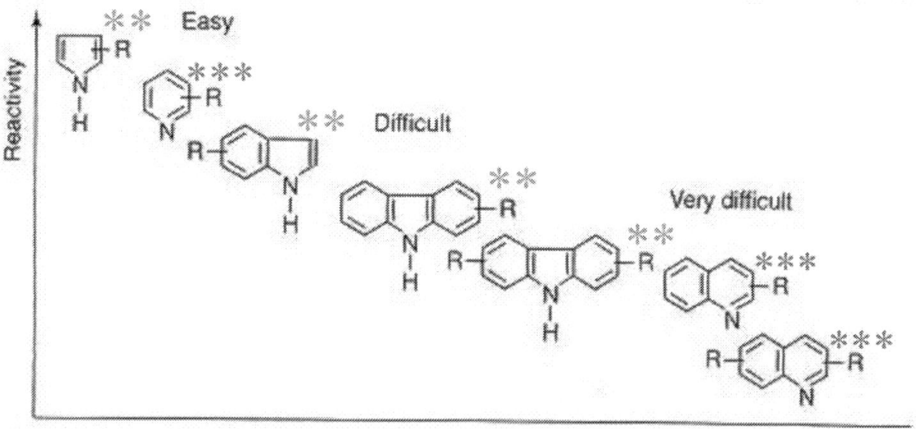

** = Neutral (*Non-Basic*) N species *** = Basic N species

30

Mechanism for Quinoline HDN

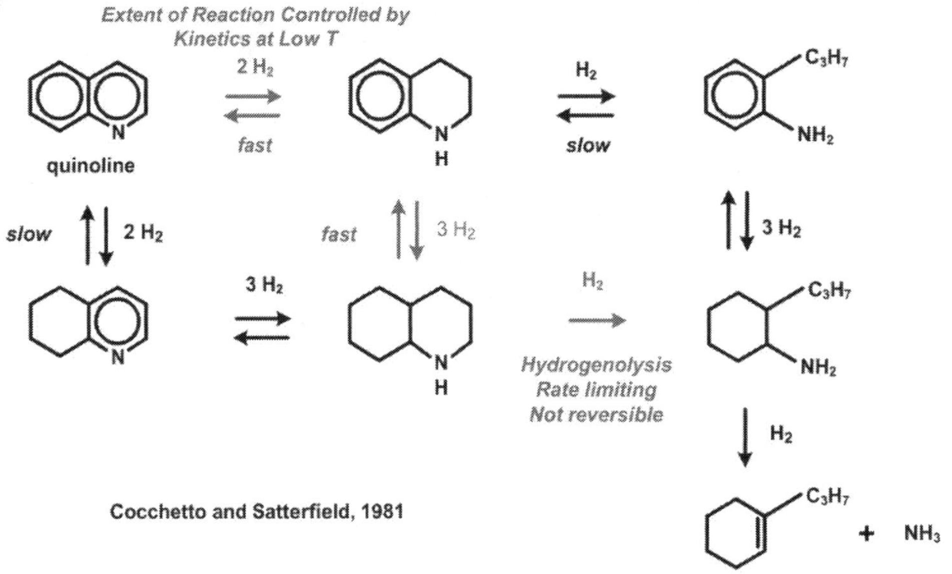

Cocchetto and Satterfield, 1981

31

Thermodynamic Limitations

- Deep HDS and HDN require prior saturation of aromatics
- Aromatics saturation crossover creates a “floor” where raising temperature has less impact on reaction rates
- A collateral observation: Product color degrades due to polyaromatics

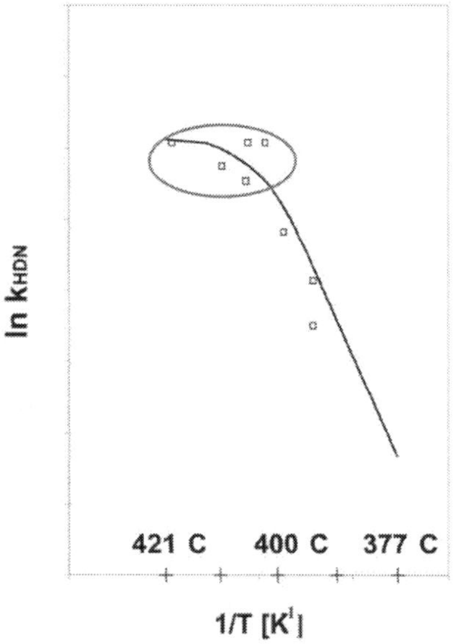

32

Hydrocracking Reactions

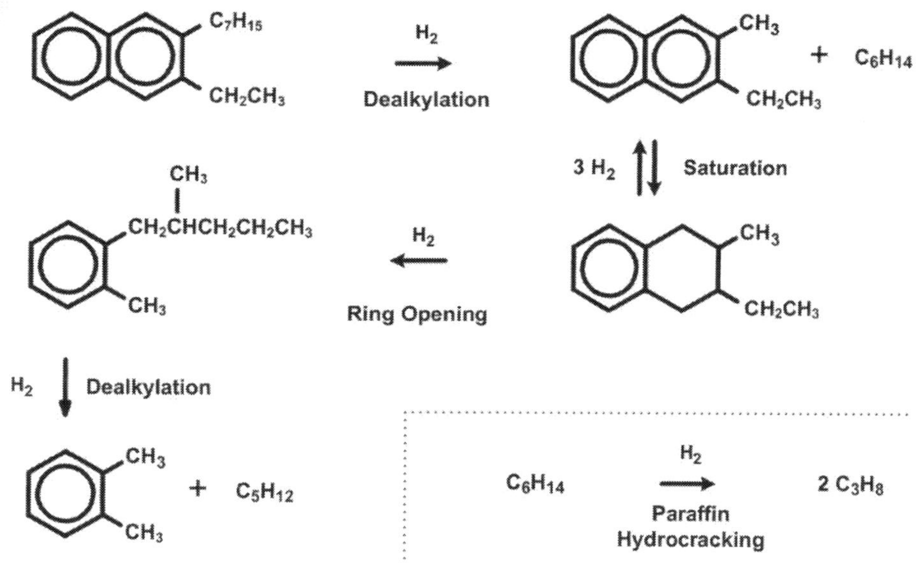

33

Catalytic Hydrocracking Involves Carbenium Ions

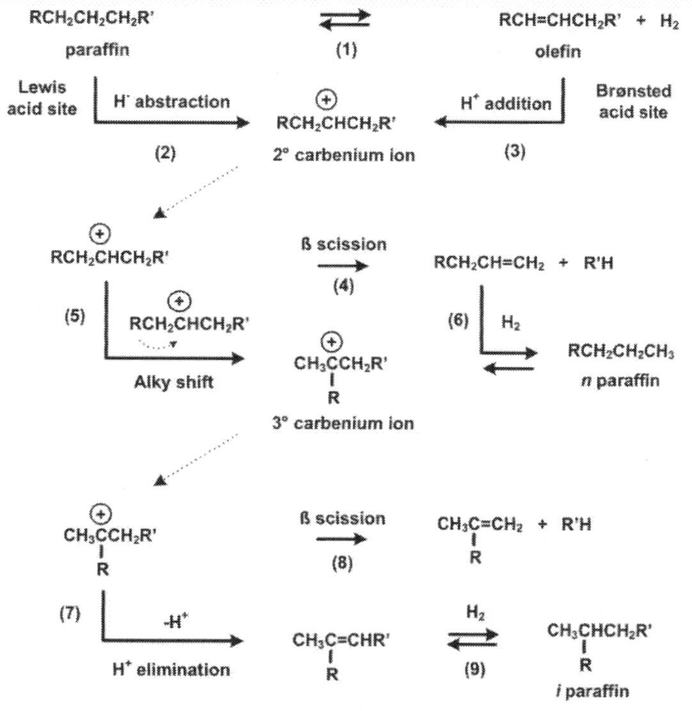

Chapter 18 in Kent and Riegel Handbook of Industrial Chemistry, Springer (2012)

34

Acids Catalyze Carbenium Ion Formation

Alkane (Hydride Extraction):

$R-CH_2-CH_2-CH_2-CH_3 + LA^+ \longrightarrow R-CH_2-CH_2-\overset{+}{C}H-CH_3 + LAH$

LA^+ = Lewis Acid Site

Alkene (Proton Addition):

$R-CH_2-CH=CH-CH_3 + BA^+ \longrightarrow R-CH_2-CH_2-\overset{+}{C}H-CH_3$

BA^+ = BrØnsted Acid Site

35

Markovnikov's Rule: Reactions Proceed via the Most Stable Intermediate

Carbenium Ion Stability:

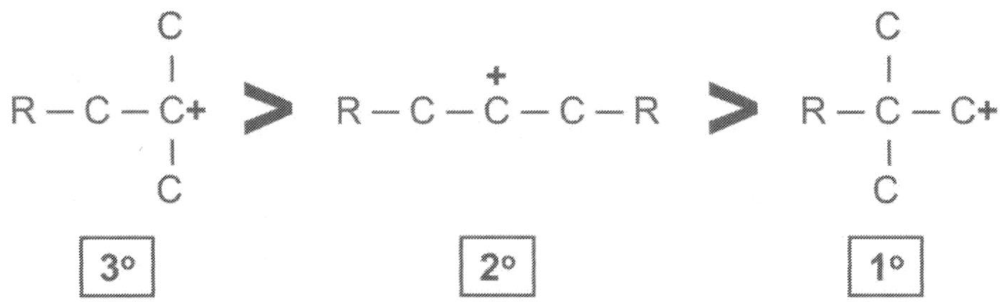

- Branched intermediates are more stable. This drives rearrangement of less stable carbenium ions to more stable forms: **1° → 2°, 2° → 3°, 1° → 3°**
- This results in a predominance of branched products.

36

Other Carbenium Ion Reactions: Making Big Ones Out of Little Ones

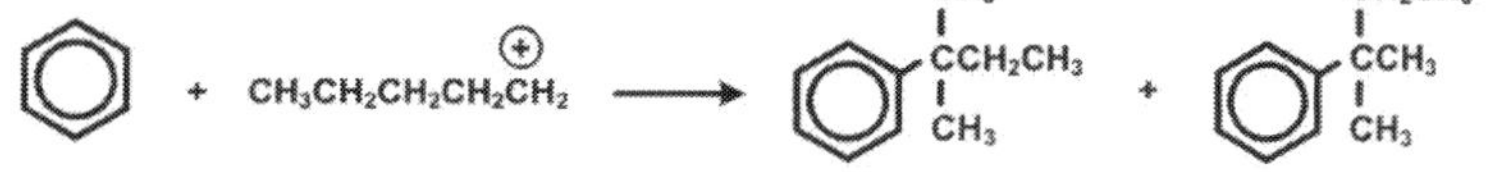

Ring alkylation

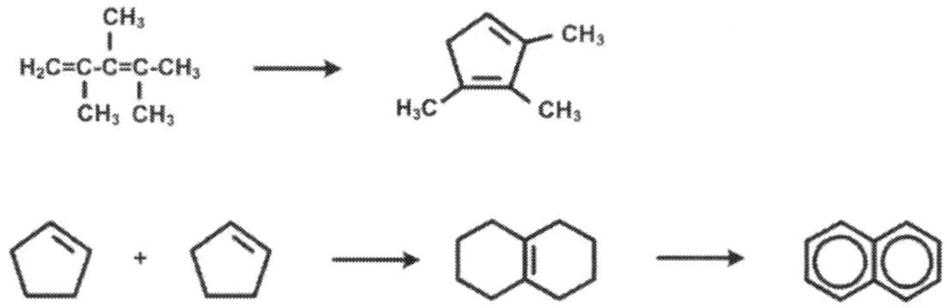

Cyclization and Condensation

H. Pines, The Chemistry of Catalytic Hydrocarbon Conversions" (1981) Academic Press, New York

37

Catalytic Cracking Is Driven by Acid Sites of Amorphous Silica Alumina (ASA) and/or Zeolites

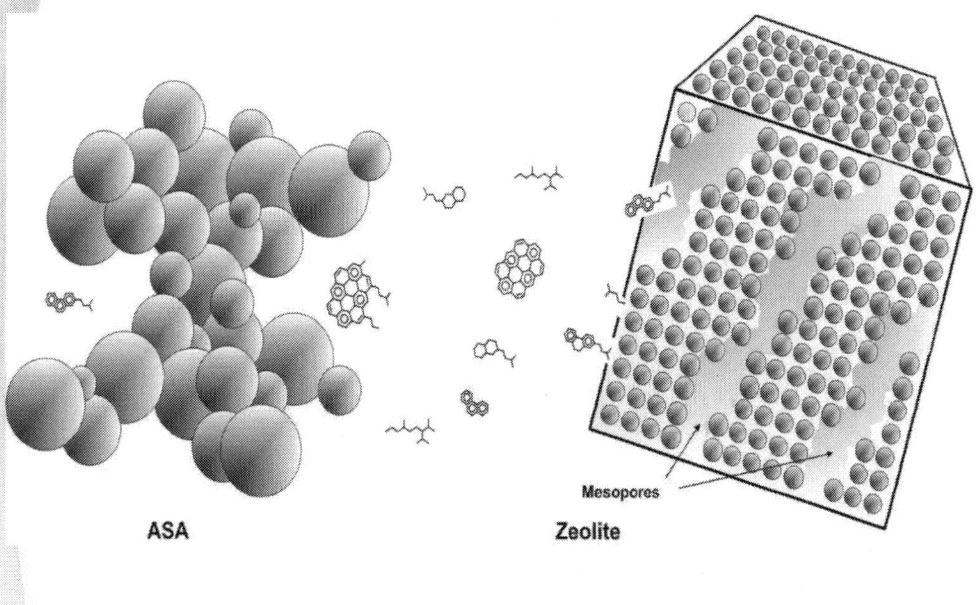

PR Robinson, Chapter 18 in Kent and Riegel Handbook of Industrial Chemistry, Springer (2012)

38

Zeolite Structures

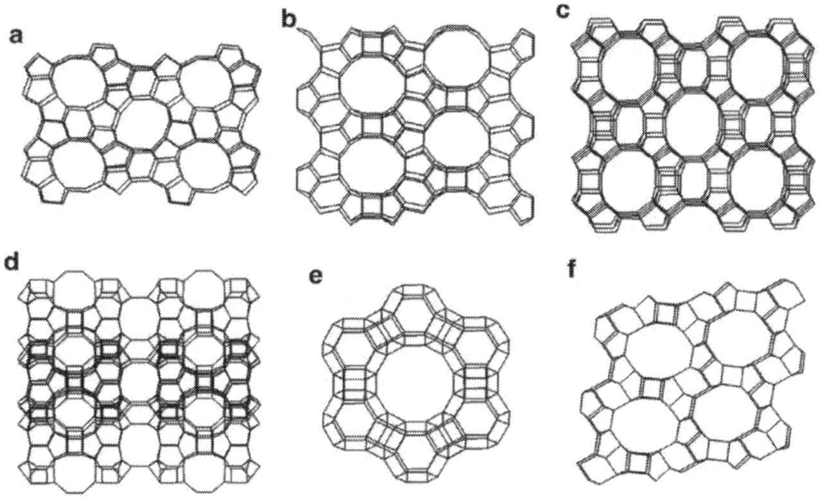

Structures of zeolites :ZSM-5 (a), mordenite (b), beta (c), MCM-22 (d), zeolite Y (e), and zeolite L (f)
Used with kind permission from Springer Images

39

Thermal Cracking Reactions Involve Free Radicals

Initiation:

(1) $R-CH_2-CH_2-CH_2-CH_3 \longrightarrow R-CH_2-CH_2\bullet + CH_3-CH_2\bullet$

(2) $R-CH_2-CH_2-CH_2-CH_3 \longrightarrow R-CH_2-CH_2-CH_2\bullet + CH_3\bullet$

Propagation - Hydrogen Abstraction:

$R'-CH_2-CH_2\bullet + R''-CH_2-CH_3$

$\longrightarrow R'-CH_2-CH_3 + R''-CH_2-CH_2\bullet$

Propagation - Radical Decomposition:

$R'''-CH_2-CH_2-CH_2-CH_2\bullet \longrightarrow R'''-CH_2-CH_2\bullet + CH_2=CH_2$

40

Thermal Cracking Reaction Mechanism (2)

Propagation Radical Addition:

(1) $R'''' - CH_2 - CH_2 - CH_2 - CH_2\bullet \quad + \quad CH_2 = CH_2$

$$\longrightarrow R'''' - CH_2 - CH_2 - CH_2 - CH_2 - CH_2 - CH_2\bullet$$

(2) $R^* - CH_2 - CH_2\bullet \quad + \quad H - H \longrightarrow R^* - CH_2 - CH_3 \quad + \quad H\bullet$

Termination:

(1) $R^* - CH_2 - CH_2\bullet \quad + \quad R^{**} - CH_2 - CH_2\bullet$

$$\longrightarrow R^* - CH_2 - CH_2 - CH_2 - CH_2 - R^{**}$$

(2) $R^* - CH_2 - CH_2\bullet \quad + \quad H\bullet$

$$\longrightarrow R^* - CH_2 - CH_3$$

41

Comparison: Thermal and Catalytic Cracking Processes

Process	Type	Characteristics
Hydrocracking	Catalytic Cracking in the Presence of Hydrogen	Minimal C_1 & C_2 Major Products: C_3+ Significant Product Branching Minimal Olefin Formation Minimal Coke Formation
FCC	Catalytic Cracking	Minimal C_1 & C_2 Major Products: C_3+ Significant Product Branching Significant Aromatics Formation Moderate Olefin Formation Significant Coke formation
Coking	Thermal Cracking	Significant C_1 & C_2 Major Gaseous Product: C_2 Significant Olefin Formation

42

Catalyst Deactivation

- Feed contamination (trace metals, resid)
- Coke Formation
- Process Upsets

43

Coke Formation

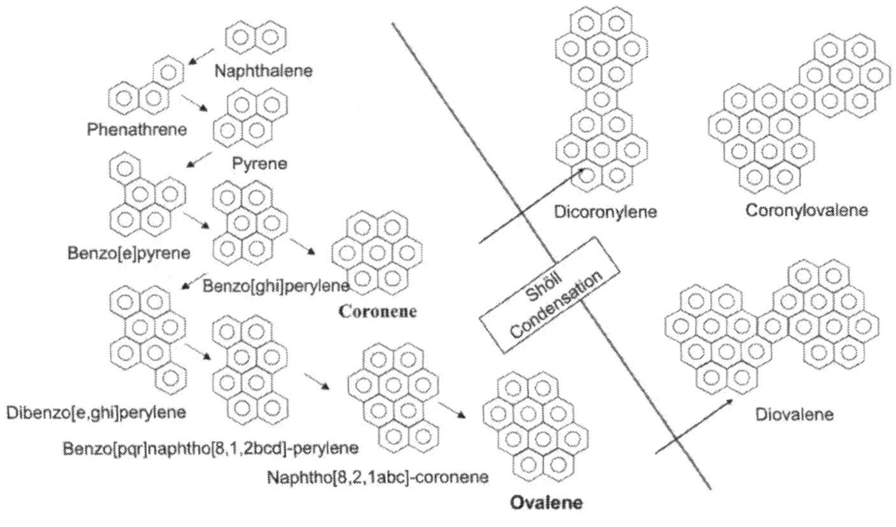

Sullivan et al. 1989

Fetzer and Lammel, 1993

44

Principles of Catalysis

"A catalyst increases the rate of chemical reactions but remains unchanged in the process"

- Commercial Implications:
  - Faster Reactions Rates
  - Smaller equipment and lower capital costs
  - Higher production rates and lower production costs
  - Lower Operating Temperatures
    - Lower Energy Costs
    - Better Selectivity
    - Reduced Metallurgy Requirements

45

Catalyst Impacts the Activation Energy

Ammonia Synthesis: $N_2 + 3H_2 \rightleftharpoons 2NH_3$

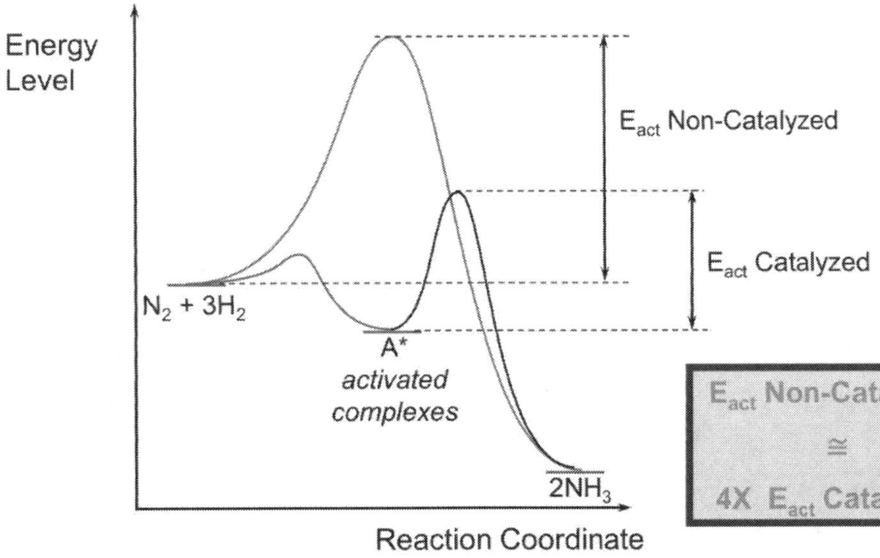

E_{act} Non-Catalyzed $\cong$ 4X E_{act} Catalyzed

46

Catalytic Ammonia Synthesis (CAS)

- First commercial example of modern chemical engineering
- Catalytic Ammonia Synthesis: $N_2 + 3H_2 \rightleftharpoons 2NH_3$

CAS Development:

Chemistry:

Fritz Haber

(Man of Science)

Process:

Carl Bosch

(Man of Action)

47

Catalytic Ammonia Synthesis (CAS)

Catalytic Ammonia Synthesis: $N_2 + 3H_2 \rightleftharpoons 2NH_3$

Interesting CAS Facts:

- Fritz Haber Inventor / Carl Bosch Developer
- Fritz Haber - Nobel Prize Chemistry – 1918
- "Most Significant Technical Discovery of 20th Century"
- Requires ~ 1% of Total World Energy Consumption
- Annual Production: 250 - 285 MMM LBS/YR

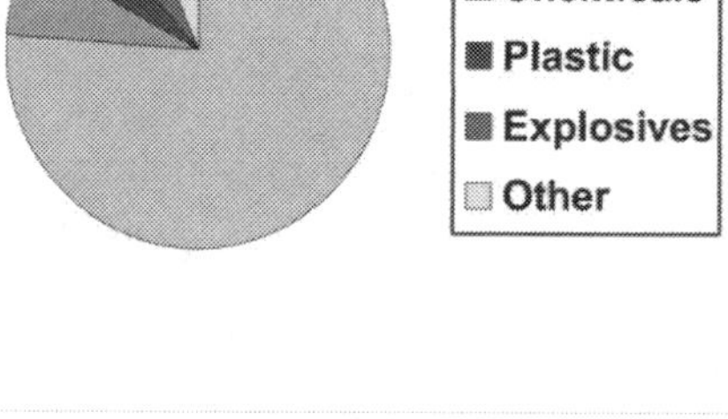

48

Reactant Absorption on Catalyst Surface

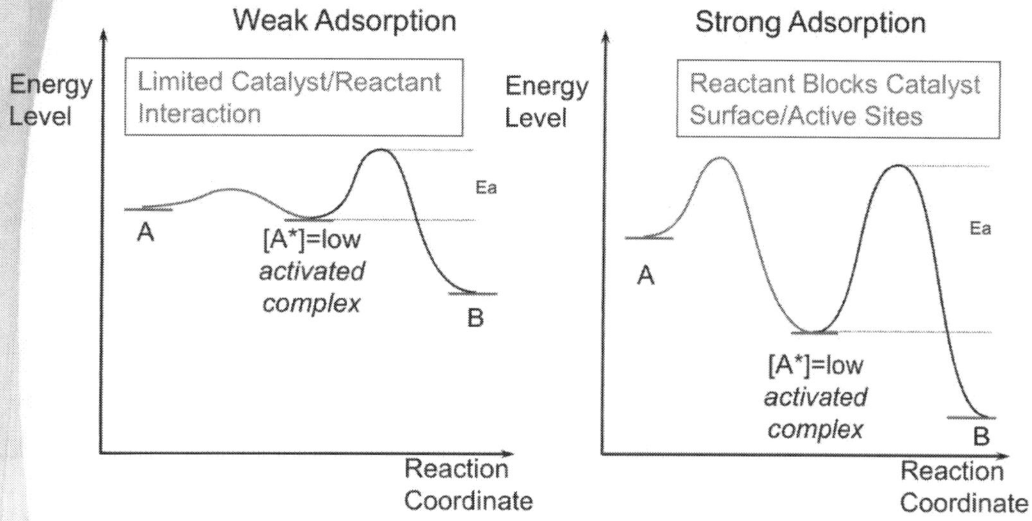

Sabatier Principle: Moderate adsorption makes for the best catalysis

49

HDS Catalyst Requirements

- Requirements for HDS Active Sites:
  - Adsorb Sulfur Compounds (Reactants)
  - Activate H_2
  - Hydrolyze C-S Bonds
  - Saturate Carbon Bonds
  - Convert S to H_2S
  - Desorb H_2S & Hydrocarbon
- Commercial Aspects:
  - Suitable for Commercial Operation (Activity)
  - Economical/Practical

50

Metal Sulfides Catalyze HDS Reactions

Very Active & Expensive Exotic Sulfides

Metal Sulfides Commercial HDS catalysts

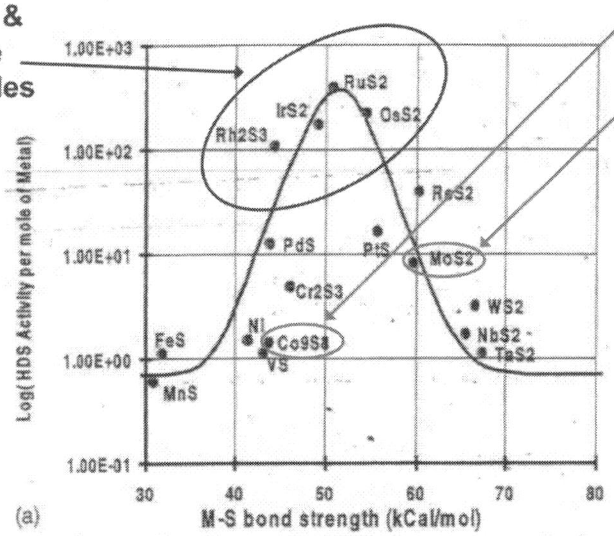

51

Promoters Provide Optimal Performance

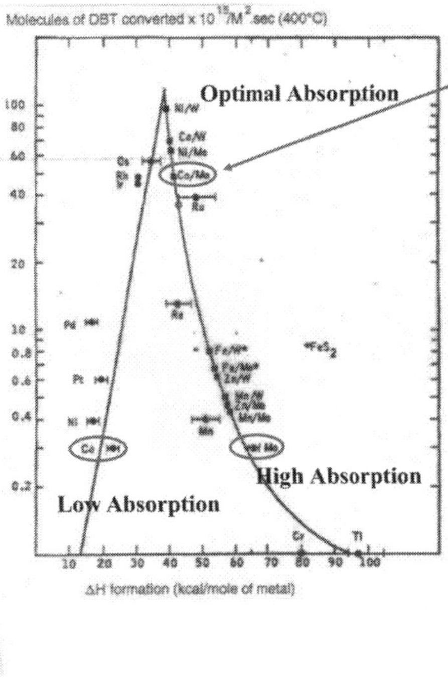

-Combining metals to achieve useful/practical HDS catalyst, CoMoS. Metals with weak and strong adsorption properties are combined

-Co acts as promoter by electron donation to Mo in CoMoS.

- Effect is that the Co-S bond strength in CoMoS becomes intermediate to the pure Co-S and the Mo-S bond strengths

- S vacancy in CoMoS molecule results

- Same mechanism in NiMoS catalysts

52

Composition of Commercial HDS Catalysts

Two Main Components

- 1 – Active Phase (The Actual Catalyst)
  - High activity in HDS reactions
  - Minimizes undesirable side reactions and yield loss
  - Resistance to coking and catalyst deactivation

- 2 - The Carrier
  - High surface area enabling monolayer distribution of active phase
  - High void fraction to enable reactants to get to active sites
  - Controlled activity or inert to prevent undesirable reactions
  - Good strength to enable operation in a reactor without breakage

53

Catalyst Utilization / Particle Size / Particle Shape

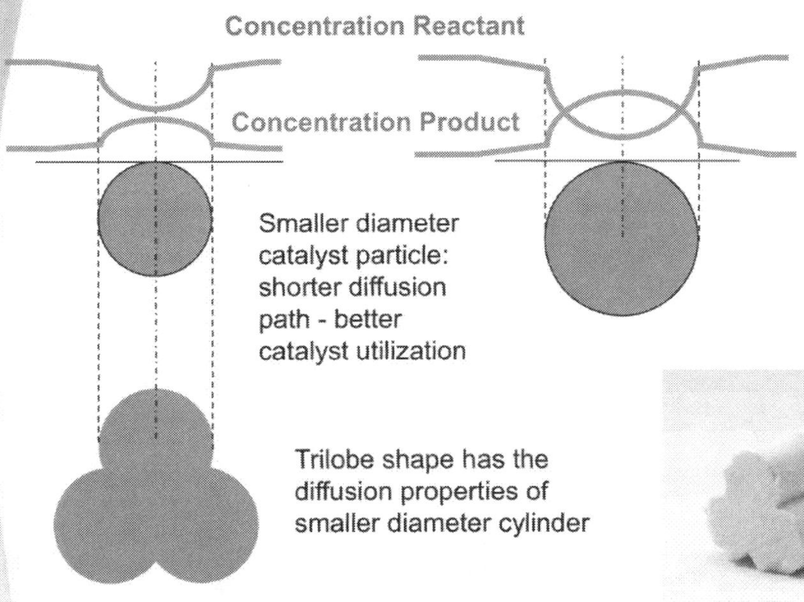

54

Catalyst Manufacturing Steps

- Crystallization
- Precipitation
- Separation: filtration or centrifugation
- Washing and drying
- Forming
- Calcining
- Impregnation and/or co-mulling
- Activation

- Other steps, such as kneading, mulling, grinding, and/or sieving also may be used.

55

Zeolite Manufacturing Steps

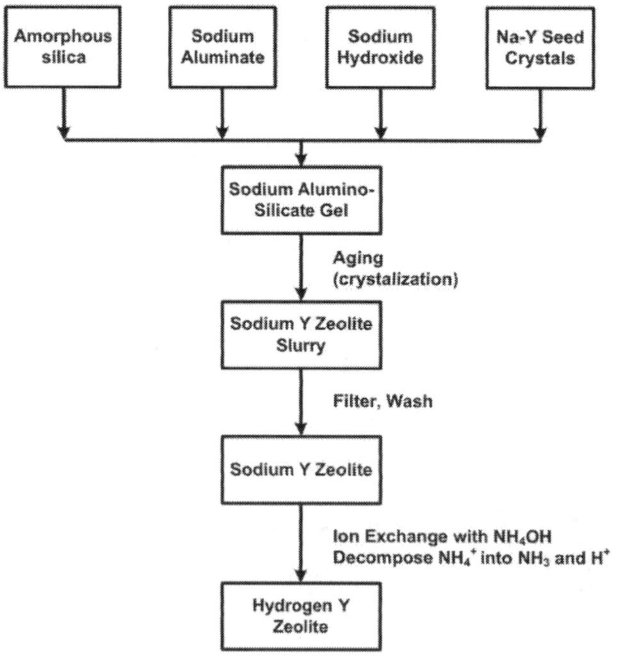

56

Catalyst Shapes and Grading: Example

Size grading provides "deep bed filtration"

- 16 mm wagon wheels: inert material for hold-down and removal of large particulates
- 8 mm hollow cylinders: Removal of smaller particulates
- 4.8 mm hollow cylinders: Additional size grading, iron removal, low activity to remove highly reactive molecules
- 2.5 mm trilobes: metal-trap catalysts, moderate activity
- 1.3 mm trilobes: high-activity main-bed catalysts
- Support balls keep small-size catalyst particles from leaving the bed

Source: Criterion Catalysts & Technologies, LLC. Used with kind permission.

57

Summary

- Purpose of this presentation
  - To give an overview of the role of hydroprocessing in petroleum refining
- Crude oil to pure products
  - Properties of crude oil
  - Petroleum refining overview
  - Chemicals in refinery streams
  - Reactions & mechanism
- Catalysis fundamentals
- Hydroprocessing catalysts

58

CRITERION/ZEOLYST DISCLAIMERS

59

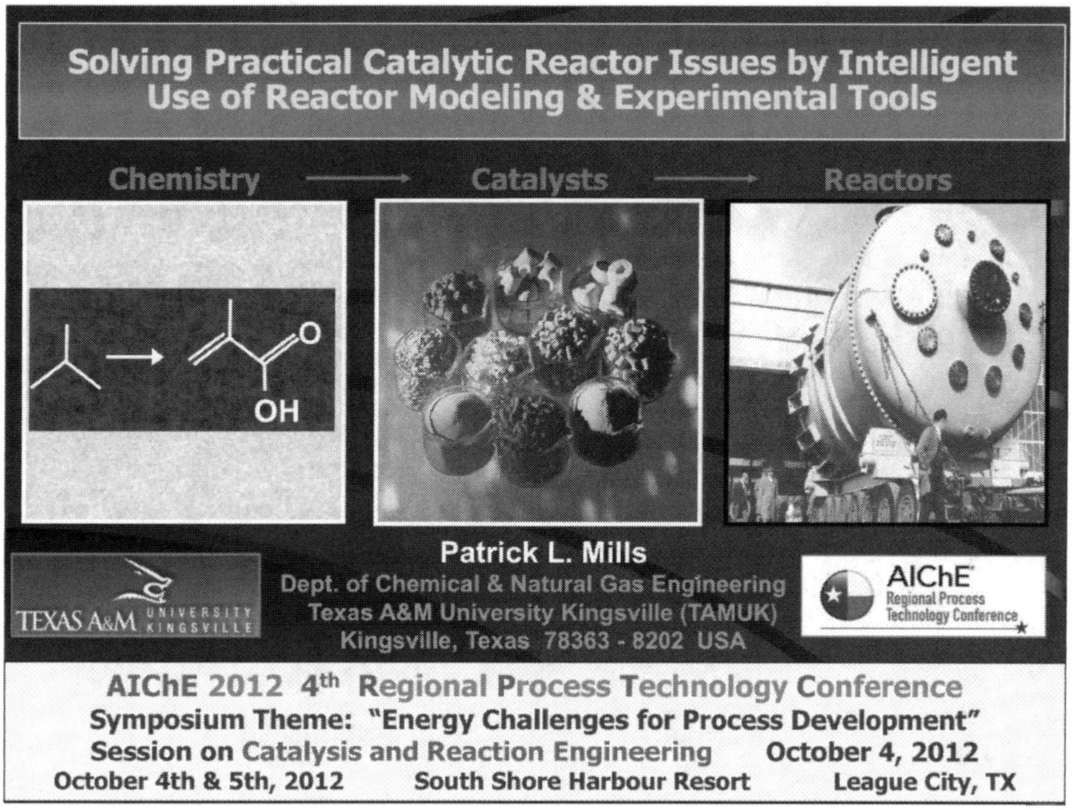
Solving Practical Catalytic Reactor Issues by Intelligent Use of Reactor Modeling & Experimental Tools
Chemistry
Catalysts
Reactors
OH
Patrick L. Mills
Dept. of Chemical & Natural Gas Engineering
Texas A&M University Kingsville (TAMUK)
Kingsville, Texas 78363 - 8202 USA
TEXAS A&M UNIVERSITY KINGSVILLE
AIChE
Regional Process Technology Conference
AIChE 2012 4th Regional Process Technology Conference
Symposium Theme: "Energy Challenges for Process Development"
Session on Catalysis and Reaction Engineering October 4, 2012
October 4th & 5th, 2012 South Shore Harbour Resort League City, TX

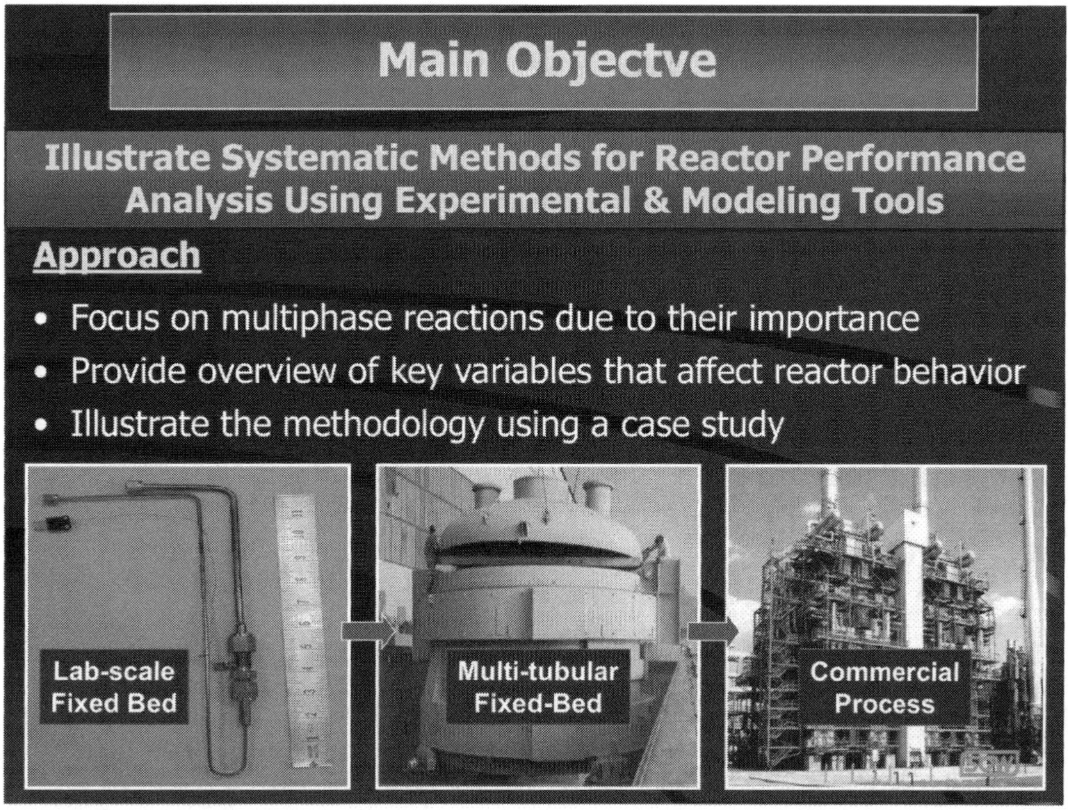
Main Objectve
Illustrate Systematic Methods for Reactor Performance Analysis Using Experimental & Modeling Tools
Approach
• Focus on multiphase reactions due to their importance
• Provide overview of key variables that affect reactor behavior
• Illustrate the methodology using a case study
Lab-scale Fixed Bed
Multi-tubular Fixed-Bed
Commercial Process

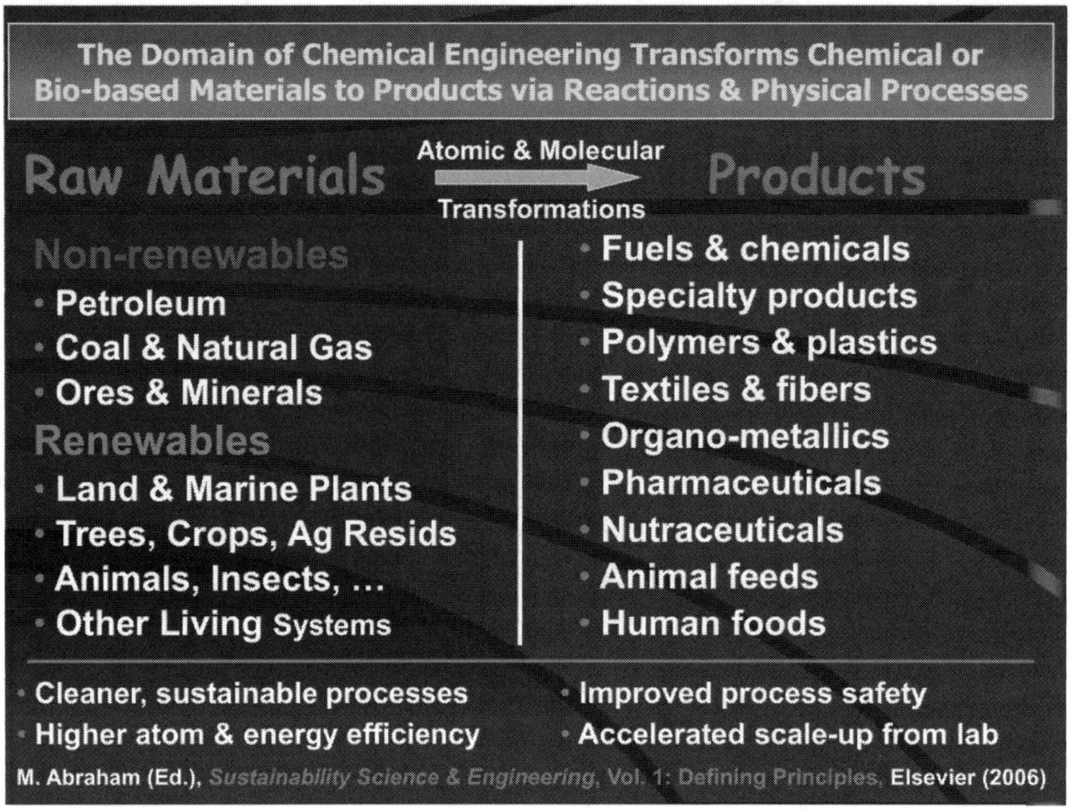
The Domain of Chemical Engineering Transforms Chemical or
Bio-based Materials to Products via Reactions & Physical Processes
Raw Materials
Atomic & Molecular
Transformations
Products
Non-renewables
• Petroleum
• Coal & Natural Gas
• Ores & Minerals
Renewables
• Land & Marine Plants
• Trees, Crops, Ag Resids
• Animals, Insects, …
• Other Living Systems
• Fuels & chemicals
• Specialty products
• Polymers & plastics
• Textiles & fibers
• Organo-metallics
• Pharmaceuticals
• Nutraceuticals
• Animal feeds
• Human foods
• Cleaner, sustainable processes
• Higher atom & energy efficiency
• Improved process safety
• Accelerated scale-up from lab
M. Abraham (Ed.), Sustainability Science & Engineering, Vol. 1: Defining Principles, Elsevier (2006)

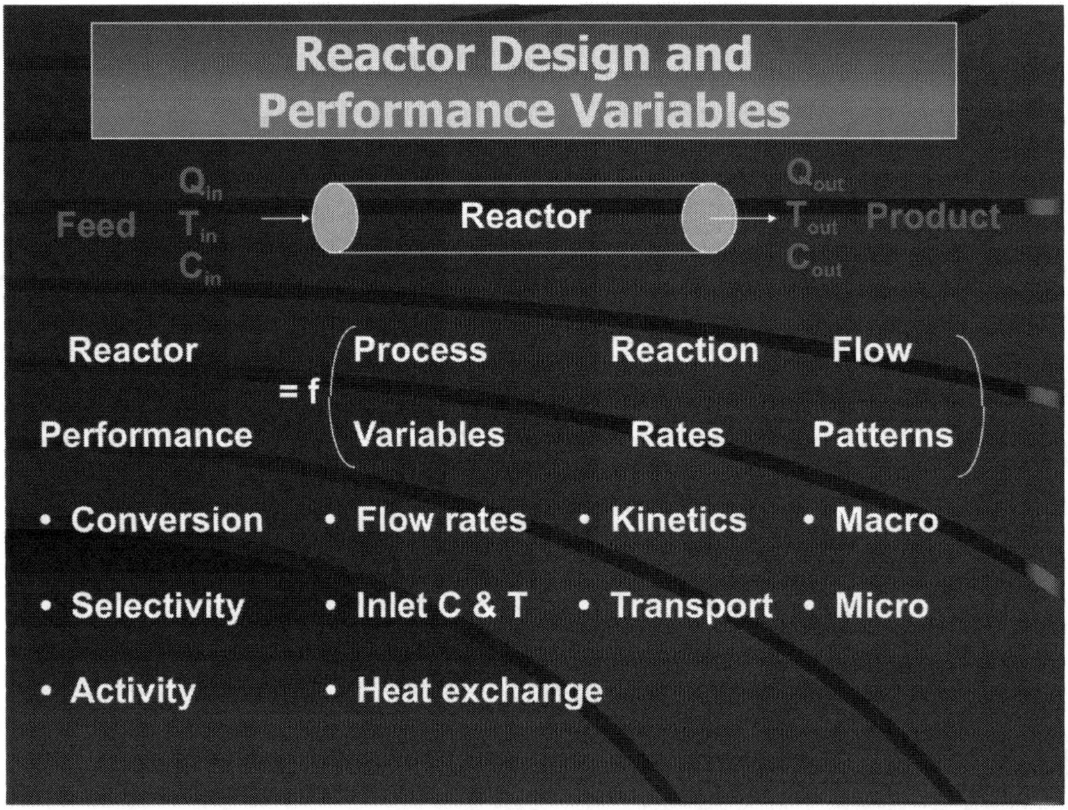
Reactor Design and
Performance Variables
Feed
Q_{in}
T_{in}
C_{in}
Reactor
Q_{out}
T_{out}
C_{out}
Product
Reactor Performance = f (Process Variables, Reaction Rates, Flow Patterns)
• Conversion
• Selectivity
• Activity
• Flow rates
• Inlet C & T
• Heat exchange
• Kinetics
• Transport
• Macro
• Micro

Commonly Used Multiphase Reactors

- Mechanically agitated tanks
- Multistage agitated columns
- Bubble columns
- Draft-tube reactors
- Loop reactors

Soluble catalysts & Powdered catalysts

- Packed columns
- Trickle-beds
- Packed bubble columns
- Ebullated-bed reactors

Soluble catalysts & Pelleted catalysts

Multiphase Reactor Supermarket

- Reaction type will eliminate some reactors.
- Systematic approach needed to evaluate reactor performance and process options.

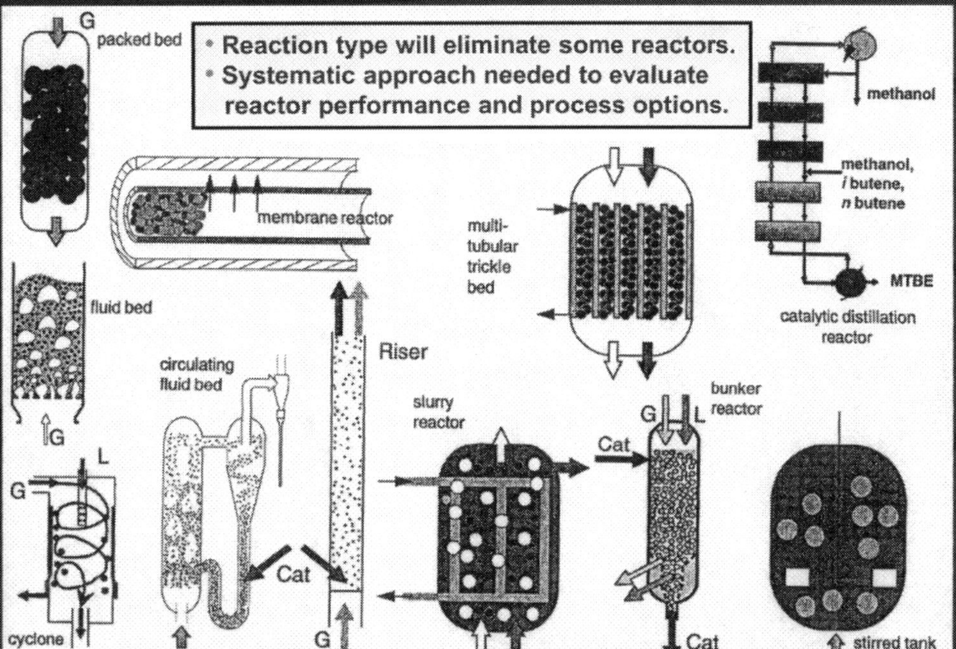

Fixed-Bed
Gas-Solid Reactors
Moving-Bed
Gas-Solid Reactors
Fixed-Bed
Gas-Liq-Solid Reactors
Multi-tubular
Fixed-Bed
Gas-Solid
Circulating Fluid Bed
Hydrotreater
Adiabatic
Fixed-Bed
Gas-Solid
Fluidized-Bed
Internals
http://www.topsoe.com

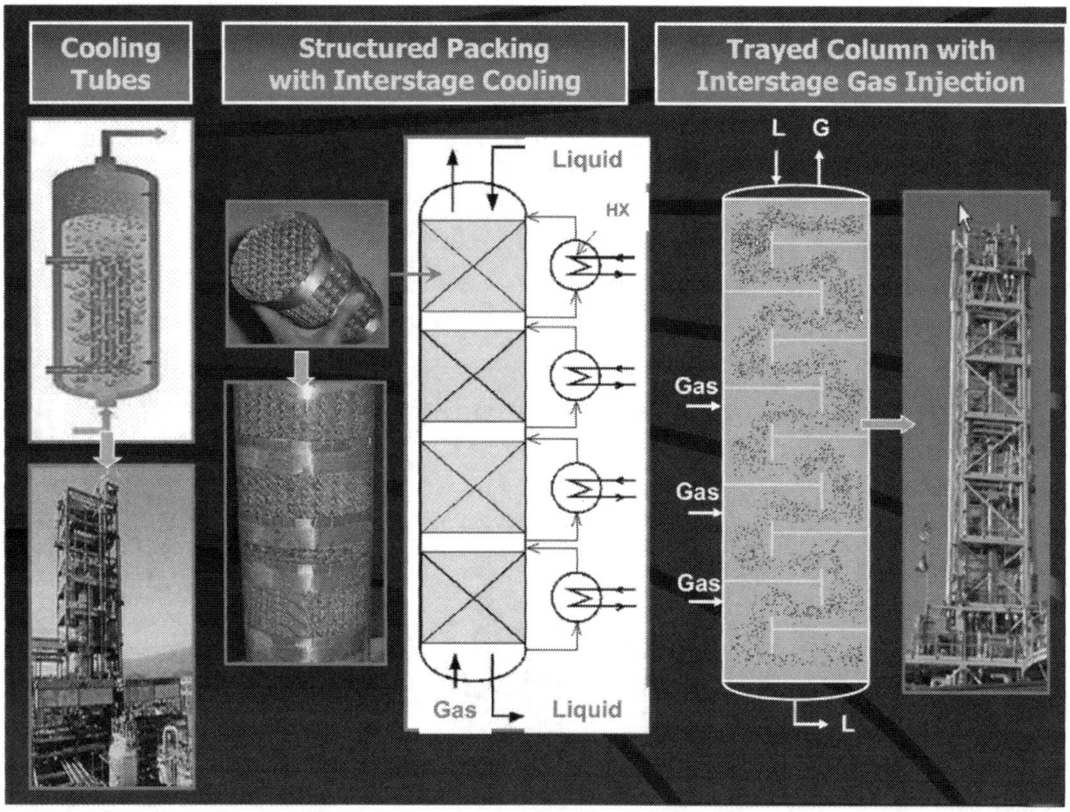
Cooling
Tubes
Structured Packing
with Interstage Cooling
Trayed Column with
Interstage Gas Injection
Liquid
HX
Gas
Liquid
L
G
Gas
Gas
Gas
L

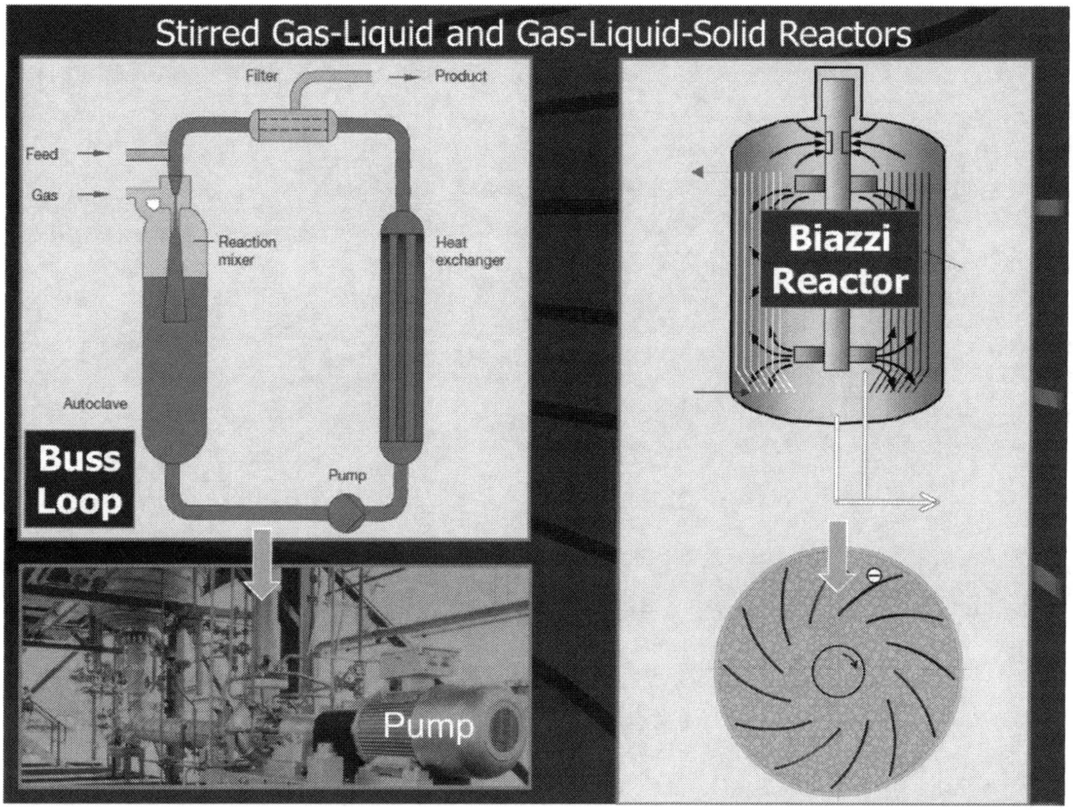
Stirred Gas-Liquid and Gas-Liquid-Solid Reactors
Filter
Product
Feed
Gas
Reaction mixer
Heat exchanger
Autoclave
Pump
Buss Loop
Pump
Biazzi Reactor

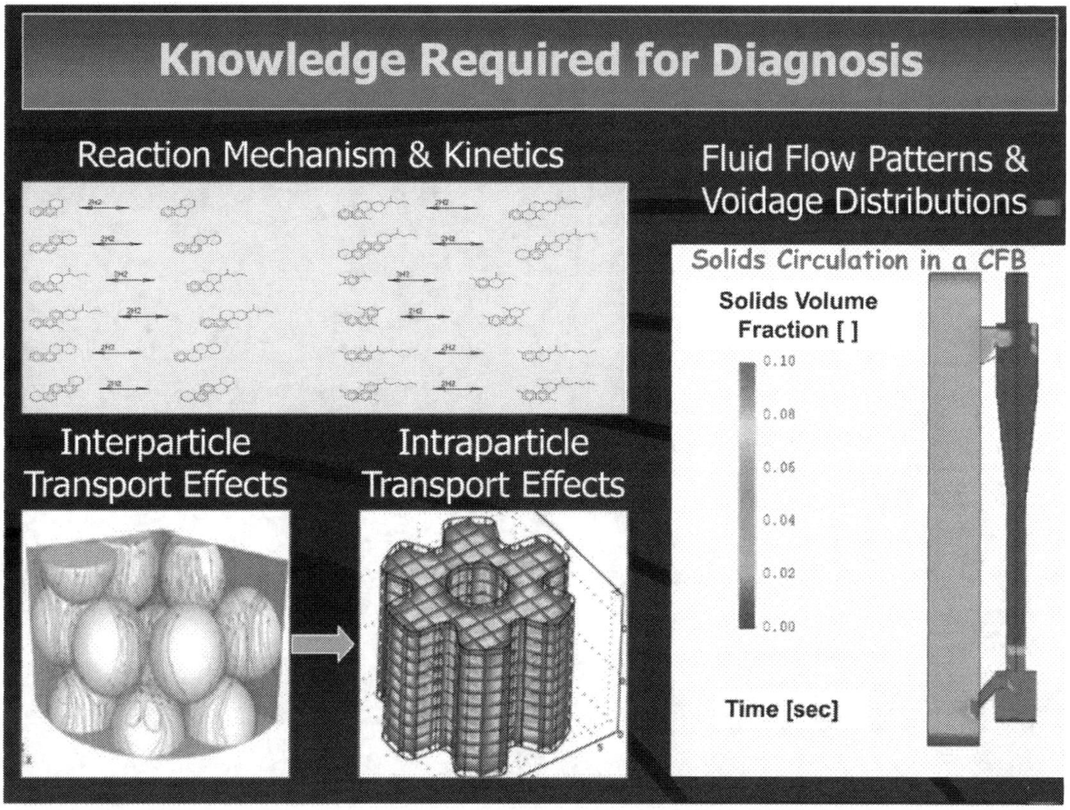
Knowledge Required for Diagnosis
Reaction Mechanism & Kinetics
Fluid Flow Patterns & Voidage Distributions
Solids Circulation in a CFB
Solids Volume Fraction [ ]
0.10
0.08
0.06
0.04
0.02
0.00
Time [sec]
Interparticle Transport Effects
Intraparticle Transport Effects

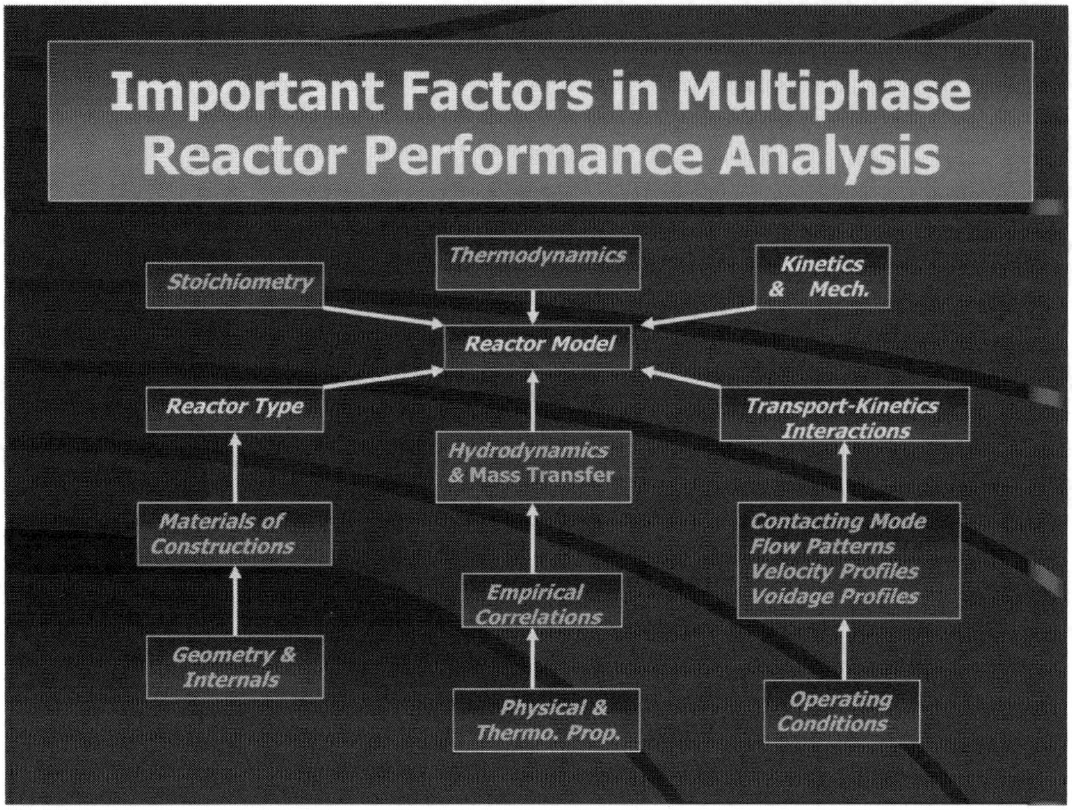
Important Factors in Multiphase Reactor Performance Analysis
Thermodynamics
Stoichiometry
Kinetics & Mech.
Reactor Model
Reactor Type
Transport-Kinetics Interactions
Hydrodynamics & Mass Transfer
Materials of Constructions
Contacting Mode
Flow Patterns
Velocity Profiles
Voidage Profiles
Empirical Correlations
Geometry & Internals
Physical & Thermo. Prop.
Operating Conditions

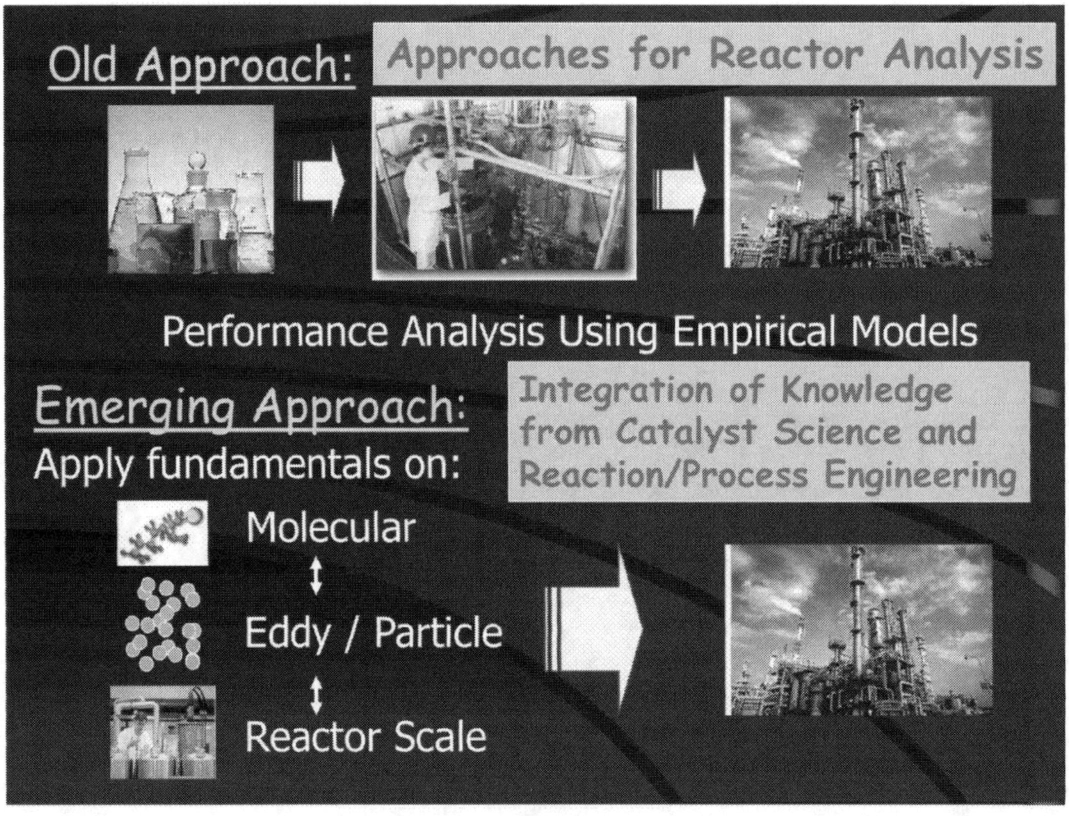
Old Approach:
Approaches for Reactor Analysis
Performance Analysis Using Empirical Models
Emerging Approach:
Apply fundamentals on:
Integration of Knowledge from Catalyst Science and Reaction/Process Engineering
Molecular
Eddy / Particle
Reactor Scale

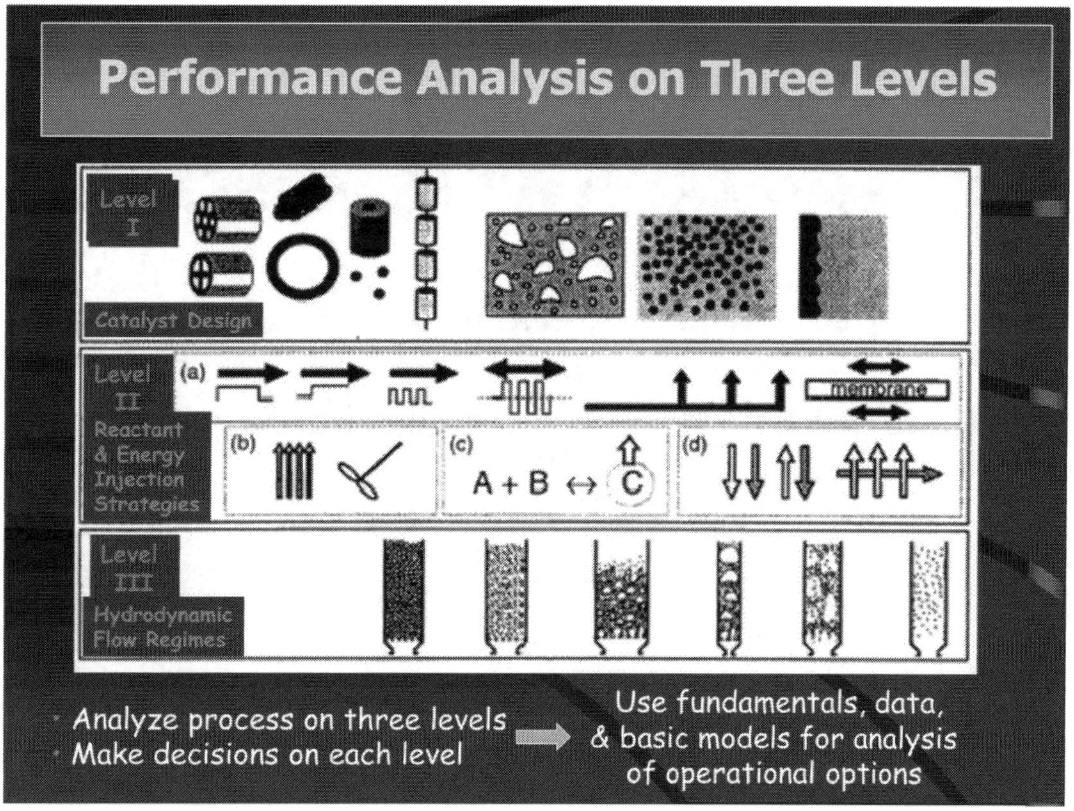
Performance Analysis on Three Levels
Level I
Catalyst Design
Level II
Reactant & Energy Injection Strategies
(a)
membrane
(b)
(c)
A + B ↔ C
(d)
Level III
Hydrodynamic Flow Regimes
Analyze process on three levels
Make decisions on each level
Use fundamentals, data, & basic models for analysis of operational options

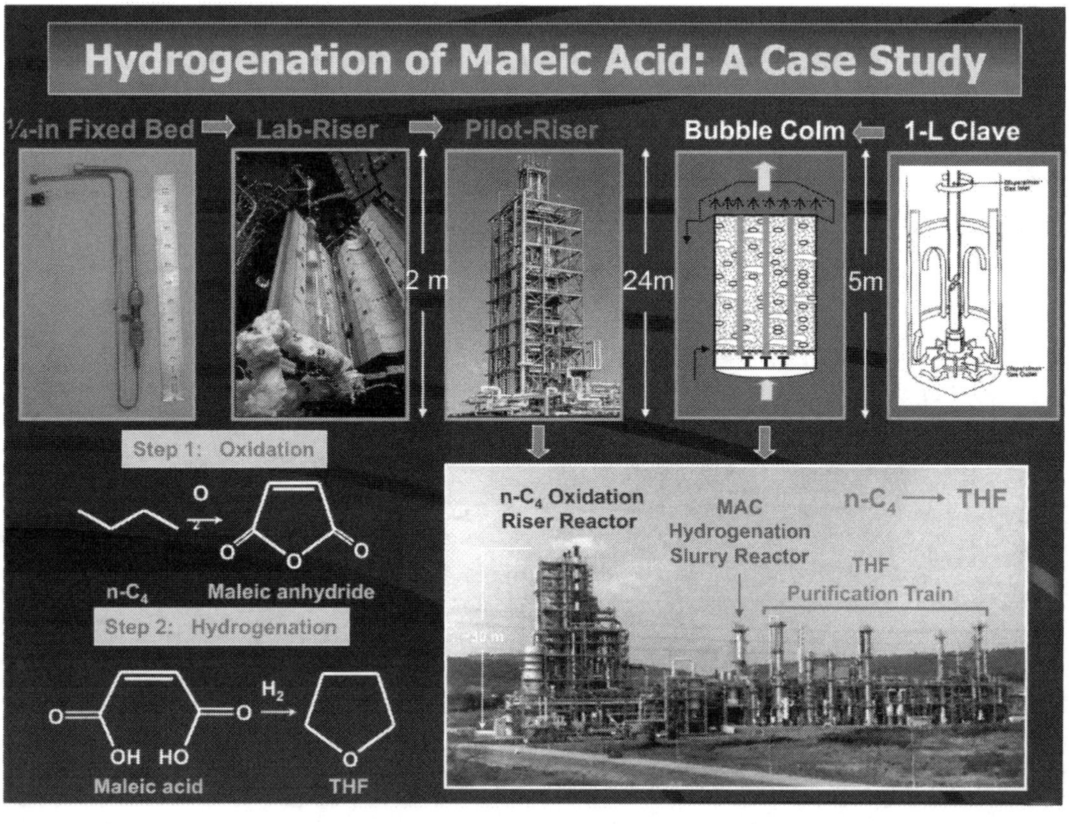
Hydrogenation of Maleic Acid: A Case Study
¼-in Fixed Bed
Lab-Riser
Pilot-Riser
Bubble Colm
1-L Clave
2 m
24m
5m
Step 1: Oxidation
O
n-C4
Maleic anhydride
Step 2: Hydrogenation
H2
OH HO
Maleic acid
THF
n-C4 Oxidation Riser Reactor
MAC Hydrogenation Slurry Reactor
n-C4 → THF
THF Purification Train

Kinetics of Hydrogenation of Maleic Acid

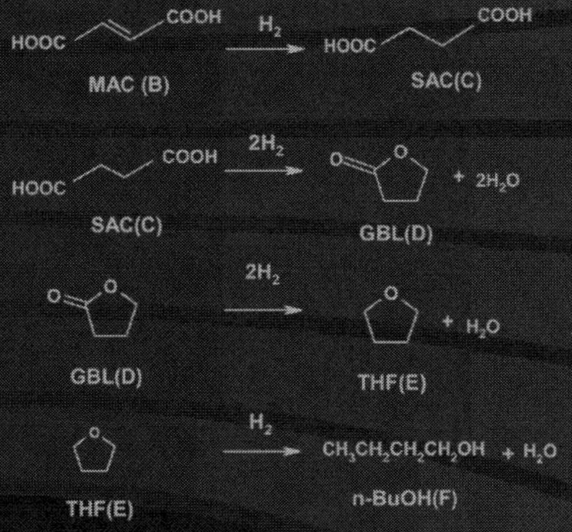

$$r_1 = \frac{w k_1 A_{sj} B_{sj}}{\left(1 + K_A A_{sj} + K_B B_{sj} + K_C C_{sj} + K_D D_{sj}\right)^2}$$

$$r_2 = \frac{w k_2 A_{sj} C_{sj}}{\left(1 + K_A A_{sj} + K_B B_{sj} + K_C C_{sj} + K_D D_{sj}\right)^2}$$

$$r_3 = \frac{w k_3 A_{sj} D_{sj}}{\left(1 + K_A A_{sj} + K_B B_{sj} + K_C C_{sj} + K_D D_{sj}\right)^2}$$

$$r_4 = \frac{w k_4 A_{sj} E_{sj}}{\left(1 + K_A A_{sj} + K_B B_{sj} + K_C C_{sj} + K_D D_{sj}\right)^2}$$

T	503-543K
Catalyst:	Ru-Re/C
	(20-100 kg/m³)
P_{H2}	14 - 20 MPa
Reactor	Batch Slurry Reactor

Overall rate of hydrogenation

$$R_A = \frac{\eta_c\, w\left(k_1 B_s + k_2 C_s + k_3 D_s + k_4 E_s\right) A_s}{\left(1 + K_A A_s + K_B B_s + K_C C_s + K_D D_s\right)^2}$$

Analysis of Initial Rate Data - Effect of Agitation -

This ensures the kinetic data represent the intrinsic rate

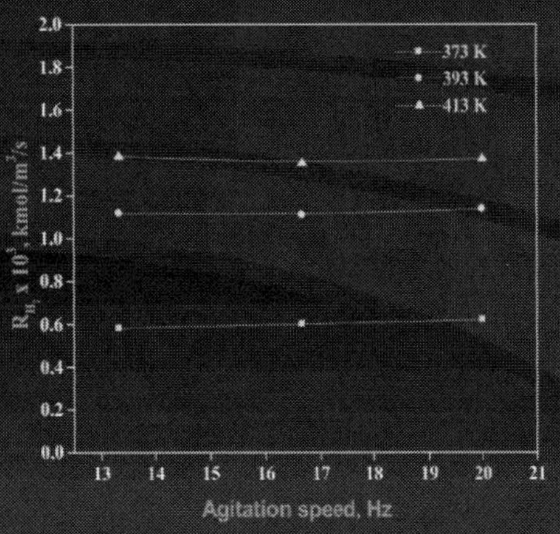

Gas-liquid mass transfer resistance is negligible if

$$\alpha_1 = \frac{R_{H_2}}{k_l a_b\, C_{A*}} \prec 0.1$$

Liquid-solid mass transfer resistance is negligible if

$$\alpha_2 = \frac{R_{H_2}}{k_s a_p\, C_{A*}} \prec 0.1$$

Pore diffusion resistance is negligible if

$$\phi_{exp} = \frac{d_p}{6}\left[\frac{(m+1)}{2D_e}\frac{\rho_p\, R_{H_2}}{w\, C_{A*}}\right]^{1/2} \prec 0.2$$

Experimental *vs* Model Predictions

Slurry Reactor Kinetics

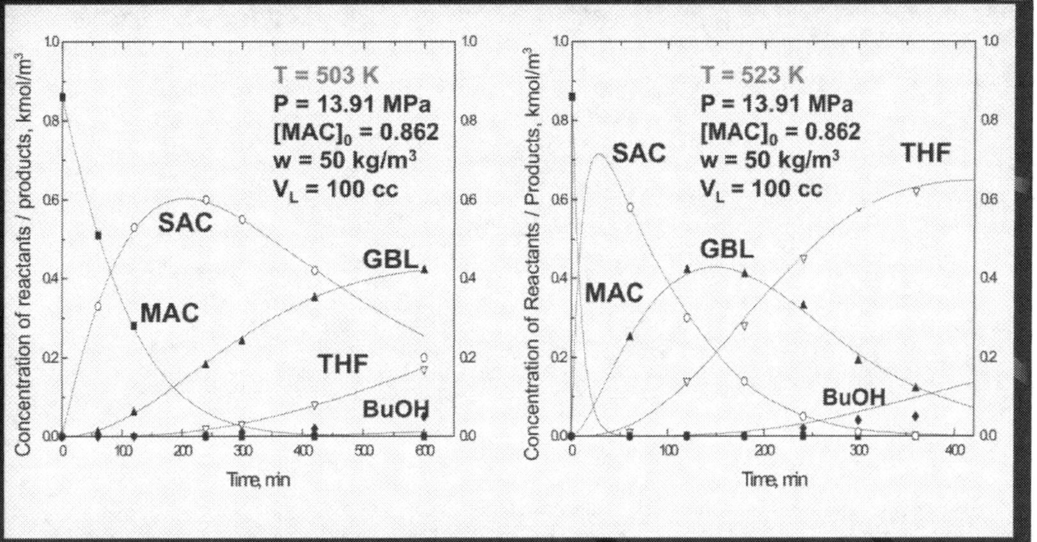

Hydrogenation of Maleic Acid: Bubble Column Slurry Reactor Model

Assumptions:

- The BCSR is consisting of N cells of stirred tanks in series liquid is completely back-mixed and gas is in plug flow.
- G-L, L-S, & intraparticle mass transfer resistances are significant for hydrogen only.
- For the non-volatile liquid components, the intraparticle diffusion has been assumed to be negligible
- The catalyst particles are uniformly dispersed throughout the reactor
- Interphase and intraparticle heat transfer is negligible
- N approximated as Pe/2

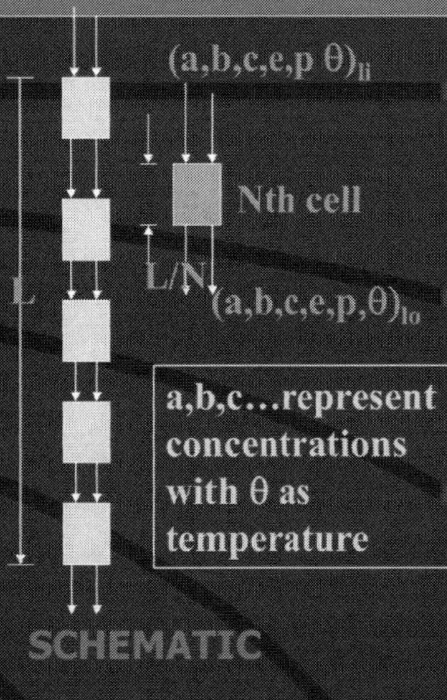

Hydrogenation of Maleic Acid: Bubble Column Slurry Reactor Model - Parameters

Parameters & Range of Operating Conditions

Catalyst	1%Re-6%Ru/C
Catalyst loading, W	10-100-200 kg/m^3 (1%-20% w/w)
Initial conc. of MAC, B_{l}	0.862-3.448 $kmol/m^3$, (10% 40%w/w)
H_2 pressure, P	15 Mpa
Temperature	503-543 K
Solvent	Water
Liquid velocity, U_l	2-10 x 10^{-4} m/s
Gas velocity, U_g	2-10 x 10^{-2} m/s
Reactor diameter, d_t	0.25m
Total reactor length, L	4m
Particle diameter, d_p	30 microns
Density of catalyst, ρ_p	2000 kg/m^3

Hydrogenation of Maleic Acid: Bubble Column Slurry Reactor Model - Model

For species A (Hydrogen):

Gas phase:

$$-\frac{da_g}{dz} = \alpha_A\left[a_g - a_{l\,j}\right]$$

Liquid phase:

$$\int_0^1 \alpha_A \beta_A \left[a_{g\,j} - a_{l\,j}\right] dz = \left[a_{l\,j} - a_{l\,j-1}\right] + \frac{\alpha_{s,A}}{N}\left[a_{l\,j} - a_{s\,j}\right]$$

and

$$\alpha_{s,A}\left[a_{l\,j} - a_{s\,j}\right] = \alpha_R \eta_c \left[\frac{b_{s\,j} + 2k_{21}c_{s\,j} + 2k_{31}d_{s\,j} + k_{41}e_{sj}}{\left(1 + k_a a_{sj} + k_b b_{s\,j} + k_c c_{s\,j} + k_d d_{s\,j}\right)^2}\right] a_{s\,j}$$

For species B (MAC):

Liquid phase:

$$\left[b_{l\,j-1} - b_{l\,j}\right] = \frac{\alpha_s}{N}\left(b_{l\,j} - b_{s\,j}\right)$$

$$\alpha_s\left(b_{l\,j} - b_{s\,j}\right) = \frac{\alpha_R}{q_B}\eta_c\left[\frac{a_{s\,j} b_{s\,j}}{\left(1 + k_a a_{sj} + k_b b_{s\,j} + k_c c_{s\,j} + k_d d_{s\,j}\right)^2}\right]$$

For species C (SAC):

Liquid phase:

$$\left[c_{l\,j} - c_{l\,j-1}\right] = \frac{\alpha_s}{N}\left[c_{l\,j} - c_{s\,j}\right]$$

$$\alpha_s\left[c_{l\,j} - c_{s\,j}\right] = \frac{\alpha_R}{2q_B}\eta_c\left(\frac{2b_{s\,j} - k_{21}c_{s\,j}}{\left(1 + k_a a_{sj} + k_b b_{s\,j} + k_c c_{s\,j} + k_d d_{s\,j}\right)^2}\right) a_{s\,j}$$

Hydrogenation of Maleic Acid: Bubble Column Slurry Reactor Model - (cont'd)

For species D (GBL):

In the liquid phase:
$$\left[d_{l\,j} - d_{l\,j-1}\right] = \frac{\alpha_s}{N}\left[d_{l\,j} - d_{s\,j}\right]$$

and
$$\alpha_s\left[d_{l\,j} - d_{s\,j}\right] = \frac{\alpha_R}{2q_B}\eta_c\left(\frac{k_{21}c_{s\,j} - k_{31}d_{s\,j}}{(1 + k_a a_{sj} + k_b b_{s\,j} + k_c c_{s\,j} + k_d d_{s\,j})^2}\right)a_{s\,j}$$

For species E (THF):

In the gas phase:
$$\frac{de_g}{dz} = \alpha_E\left[q_{bE}\, e_{l\,j} - e_{g\,j}\right]$$

in the liquid phase:
$$\int_0^1 \frac{\alpha_E\,\beta_E}{q_{BE}}\left[q_{BE}e_{l\,j} - e_{g\,j}\right]dz = \left[e_{l\,j-1} - e_{l\,j}\right] + \alpha_{s,A}\left[e_{l\,j} - e_{s\,j}\right]$$

and
$$\alpha_s\left[e_{lj} - e_{s\,j}\right] = \frac{\alpha_R}{2q_B}\eta_c\left(\frac{k_{31}d_{s\,j} - 2k_{41}e_{sj}}{(1 + k_a a_{sj} + k_b b_{s\,j} + k_c c_{s\,j} + k_d d_{s\,j})^2}\right)a_{sj}$$

For species F (n-BuOH)

In the liquid phase:
$$\left(f_{lj} - f_{lj-1}\right) = \frac{\alpha_s}{N}\left(f_{sj} - f_{l\,j}\right)$$

and
$$\alpha\left(f_{sj} - f_{lj}\right) = \eta_c\,\frac{\alpha_R}{q_B}\,\frac{k_{4l}\,e_{sj}\,a_{sj}}{\left(1 + \sum_{i=a}^{d} k_i\, i_{sj}\right)^2}$$

Hydrogenation of Maleic Acid: Bubble Column Slurry Reactor Model - Results

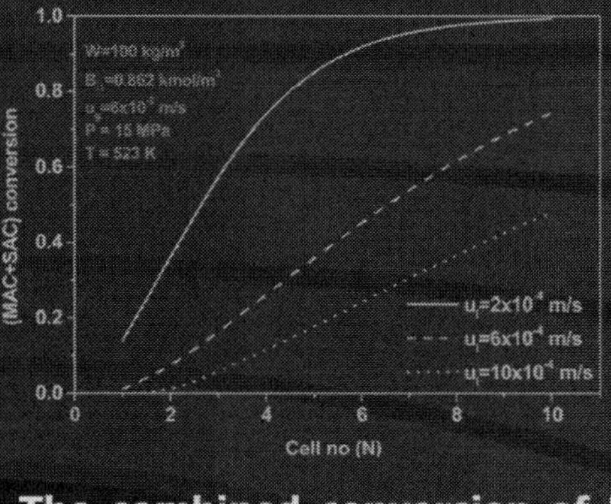

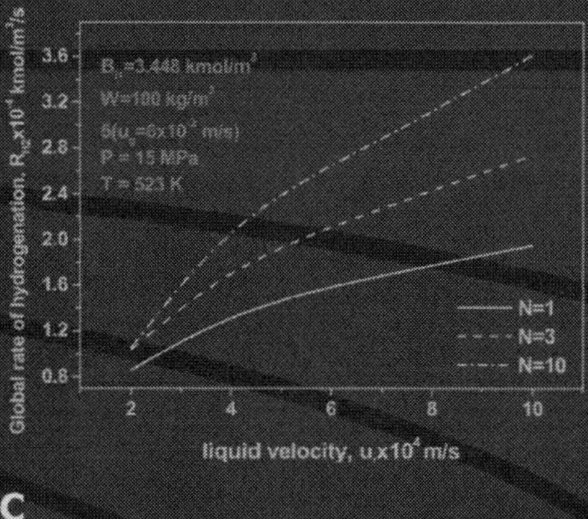

The combined conversion of MAC and SAC increased with the number of mixing cells indicating that higher conversion is achieved under plug flow conditions. This effect is significant only at higher catalyst loadings

The global rate of hydrogenation increases with liquid velocity

Hydrogenation of Maleic Acid: Bubble Column Slurry Reactor Model - Key Learnings

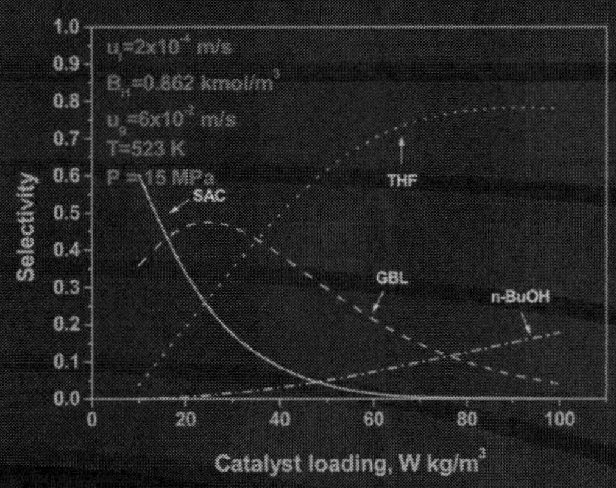

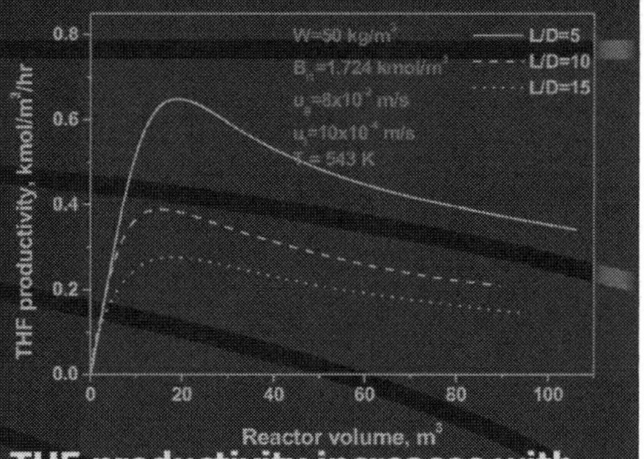

The THF selectivity generally increases with catalyst loading

THF productivity increases with reactor volume at lower catalyst loading. However, at higher catalyst loading, the productivity passed through a maximum due to increase in n-BuOH formation

Advanced Bubble Column Modeling

- CARPT/CT and other measurements are used to develop a phenomenological reactor flow and mixing model.
- CFD generated data are used to assess model parameters at pilot plant or plant conditions.
- Reactor flow and mixing model are coupled with the kinetic information to provide reactor performance
  Degaleesan et al., Chem. Eng. Sci., 51, 1967(1996); I&EC Research, 36,4670 (1997);
  Gupta et al., Chem. Eng. Sci., 56, 1117 (2001)

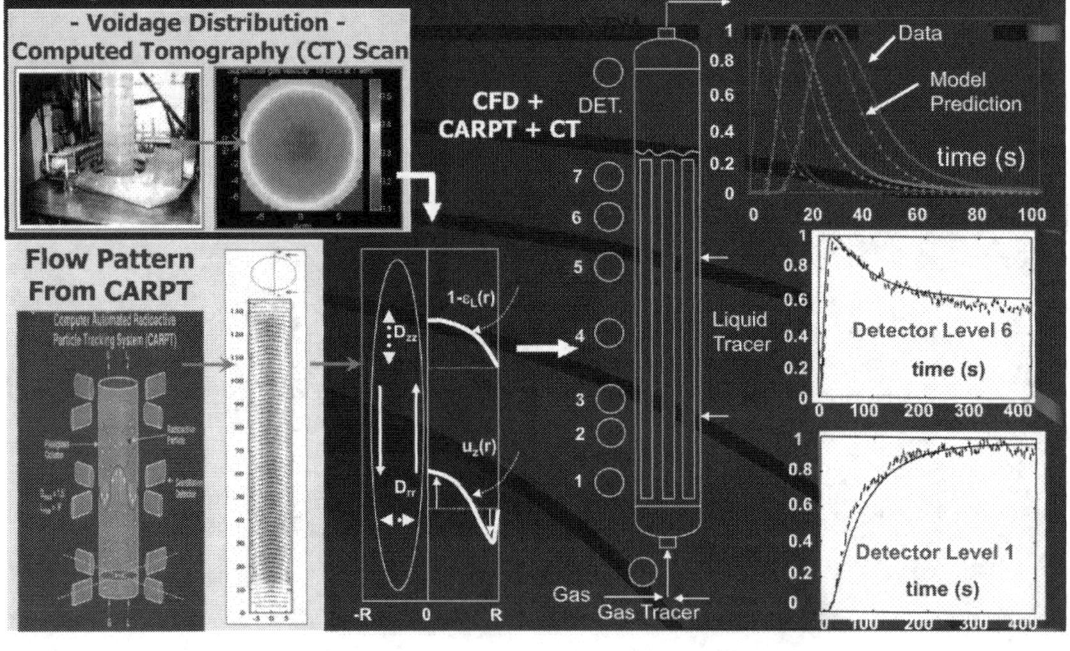

Future Developments

- Improved catalyst design via molecular simulation
- More robust kinetics models that connect composition to rates
- Increasing use of CFD for analysis of reactor hydrodynamics
- Commercial-scale reactor analysis using more advanced experimental and flow modeling tools
- More advanced process sensors for obtaining real-time data

Compositional Modeling Will Connect Analytical Chemistry to Process Models

- *Molecular Characterization of Refinery Streams Combined with Reaction Kinetics*
- *Process Models Based on > 5000 Species and Reactions*
- *Predicts Product Compositions, Yields, & Properties*

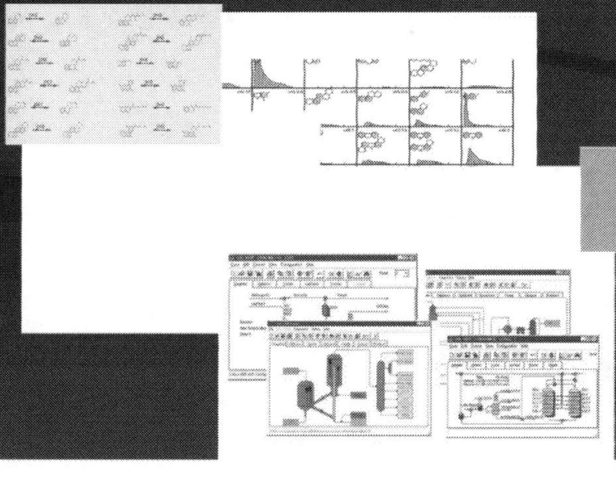

APPLICATIONS
- Crude Selection
- Refinery Yield Opt.
- Lube Performance
- Toxicological Char.
- Chemicals Mfg.
- Research Guidance

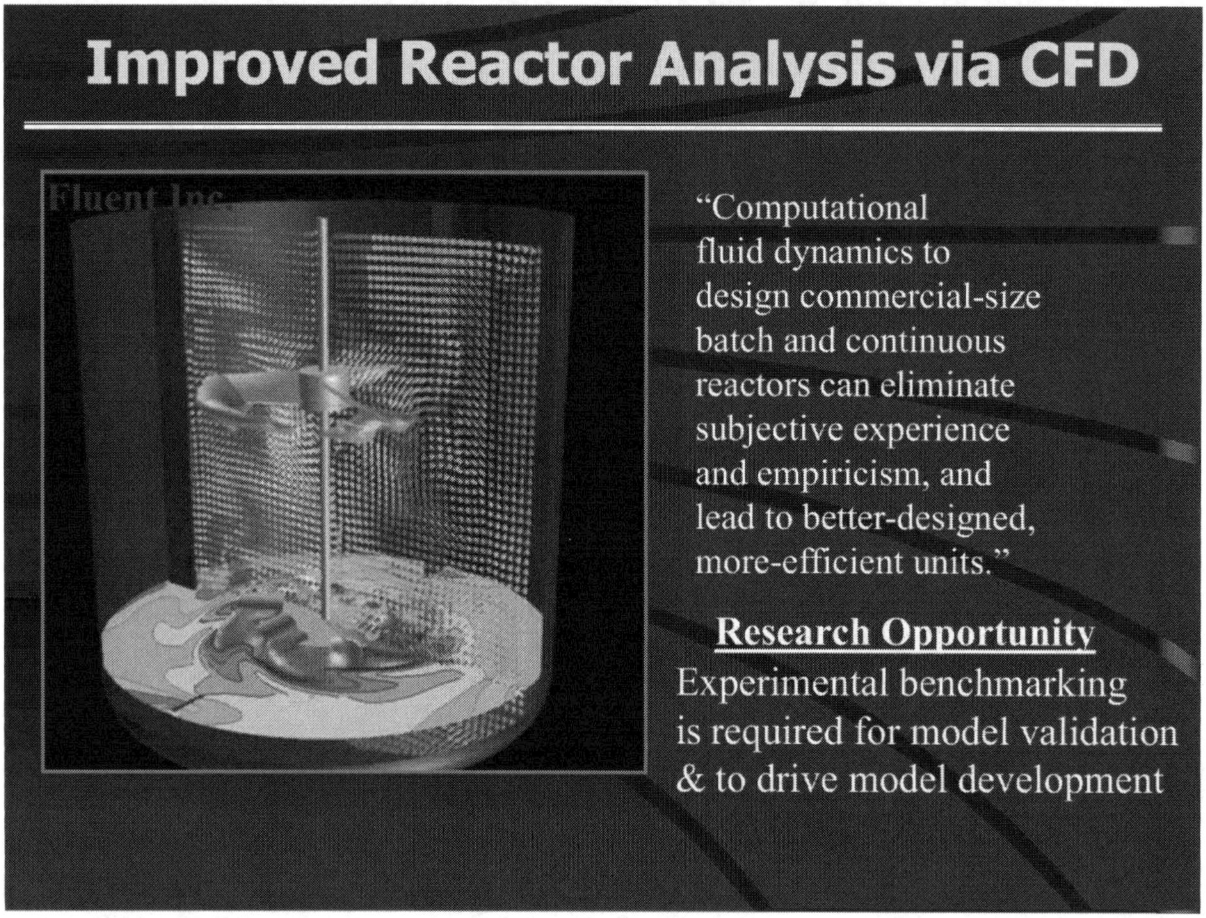
Improved Reactor Analysis via CFD
Fluent Inc.
"Computational fluid dynamics to design commercial-size batch and continuous reactors can eliminate subjective experience and empiricism, and lead to better-designed, more-efficient units."
Research Opportunity
Experimental benchmarking is required for model validation & to drive model development

MISTRAS
GROUP, INC.

Optimizing the Location of Condition Monitoring Locations (CML's)

F. Russ Davis CSP
AIMS Division Manager
Mistras Services

Corrosion Monitoring Location

API 510 defines CMLs as "Designated areas on pressure vessels where periodic examinations are conducted."

API 570 defines CMLs as "Designated areas on piping systems where periodic examinations are conducted."

Over the years CML's have been assigned on the presumption that MORE is better.

- A CML taken at the wrong place provides no useful information

Corrosion Monitoring Location

A CML is a location where monitoring is performed to evaluate the condition of the equipment.

CML condition monitoring NDE methodology
- Ultrasonic thickness measurements
- Profile X-ray
- Ultrasonic shearwave

Results from monitoring are used to predict a life-cycle of equipment
- Replacement/repair
- Next inspection/monitoring dates
- Schedule shutdown/turnarounds

Monitoring is used to detect damage such as:
- Corrosion/erosion – loss of wall thickness
- Cracking

Corrosion Monitoring Location

API 570 Piping Inspection Code states that:

Each piping system shall be monitored by taking thickness measurements at CML's. Piping circuits with high potential consequences if failure should occur and those subject to higher corrosion rates or localized corrosion will have more CML's and will be monitored more frequently. CML'S should be established for areas with continuing CUI (corrosion under insulation). Corrosion at soil to air interfaces, or other locations of potential localized corrosion as well as areas with general, uniform corrosion. CML's should be distributed appropriately throughout each piping circuit.

CML Placement

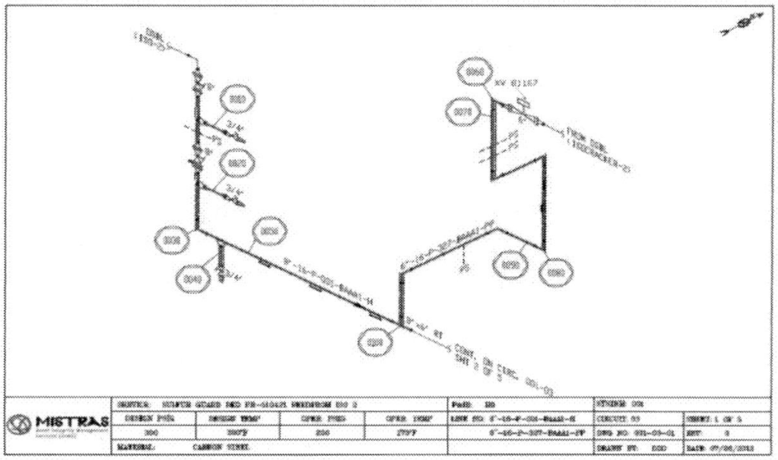

CML Placement

CML Placement Analysis

- Qualitative
- Quantitative

- Determine the appropriate number and location of CML's in the most effective location and utilize the appropriate inspection method
- Use a combination of RBI (risk based inspection) methodology and evaluated data considering expected damage mechanisms and effectiveness of inspection techniques
- RBI considers the risks associate with loss of containment
- Inspectors knowledge and experience used to establish CML numbers and locations

CML Placement

- Considerations when placing CML's
  - Accessibility
    - UT thickness reading
    - Profile x-ray
    - UT shearwave inspection
  - Probability of degradation
    - Evaluation of all potential damage mechanisms
  - Use of appropriate inspection technique
    - UT Thickness reading not effective in detecting cracking

One Source for Asset Protection Solutions

MISTRAS GROUP, INC.

CML Placement

Piping:

- Injection point circuits established as individual circuit
  - Monitoring upstream and downstream of injection points
  - Based on pipe wall thickness at the location of expected impingement
- Service specific and localized corrosion
  - Anticipate localized corrosion locations
    - Typically requires more CML's than generalized corrosion
  - More CML's should be considered when process is toxic or highly flammable
  - More CML's when higher corrosion rates have been detected
  - Configuration complexity
    - Fittings, branch connections, deadlegs require additional CML's

One Source for Asset Protection Solutions

MISTRAS GROUP, INC.

CML Placement

Piping

- Fewer CML's selected when:
  - Low potential for safety and environmental impact
  - Non-corrosive environment
  - Long, straight-run piping system

CML Placement

Vessels

The number and location of CML's shall be sufficient to be representative of the corrosion activity occurring within the vessel.

- Based on damage mechanism and corrosive environment, relevant history and shared knowledge from other sites and similar industry experience

- Sufficient CML's allocated in the following areas:
  - Each shell course
  - Each vessel head
  - Each nozzle or manway

CML Placement

Towers

- CML's assigned to the individual sections
  - Liquid phase (bottom, below the bottom tray)
  - Mixed phase area (tray area)
  - Vapor phase (above the top tray)

Exchangers

- Divided into 2 area with CML's assigned to each
  - Shell side
  - Channel side

Drums

- Drums typically consist of a liquid and vapor section or a top phase and bottom phase
  - CML's assigned to each section and at the interphase area

CML Placement

- CML calculation

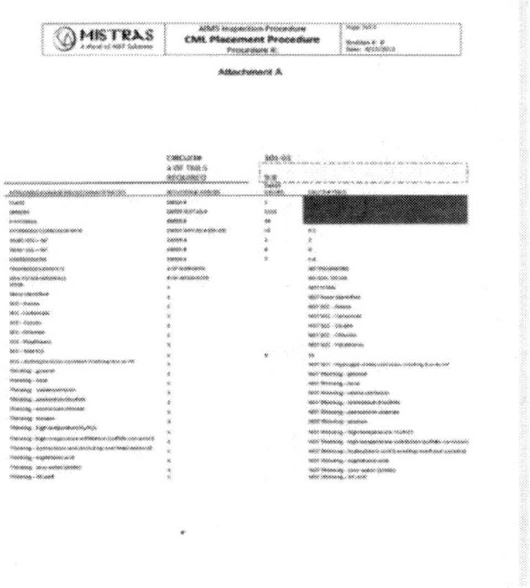
MISTRAS

CML Placement Procedure

Attachment A

One Source for Asset Protection Solutions

MISTRAS GROUP, INC.

Summary

- Condition monitoring locations (CML's) are used to determine the current fitness for service of piping and vessels and to predict future servicability
- CML's should be assigned so that effective testing and monitoring can produce data to evaluate the equipment
- The quantity of CML's should be selected based on consequence of release and the configuration of the equipment
- Excessive numbers of CML's should be avoided
- CML amounts and locations should be assigned using a formalized and accepted practice.

One Source for Asset Protection Solutions

MISTRAS GROUP, INC.

BAYER

Science For A Better Life

"A Containment Building and Scrubber Design for Reduction of Risk in a Process with a Highly Hazardous Chemical"

For presentation at the Regional AIChE Process Technology Conference,
Process Safety Management Session
League City, TX, Oct. 4, 2012

Mark W. Thiel, Deepak Gupta, Spotswood Miller* and John Gill

*Bayer MaterialScience

Overview

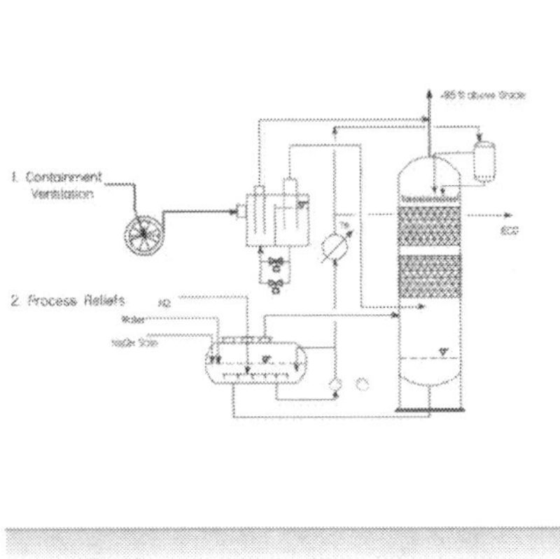

- Phosgene uses and hazards
- Safety specific principles – primary and secondary
- Highest level of safe design practices
- Containment enclosure design
- Technical design of caustic scrubber

Phosgene use

Production route to isocyanates

- Di-amine + 2 $COCl_2$ ⟶ Di-isocyanate + 4 HCl

  Example: Di-phenylmethlylene diamine (MDA) + 2 $COCl_2$ ⟶ Di-phenylmethylene diisocyanate (MDI) + 4 HCl

  - Di-isocyanate + polyol ⟶ polyurethane

- Polyurethane final product use examples

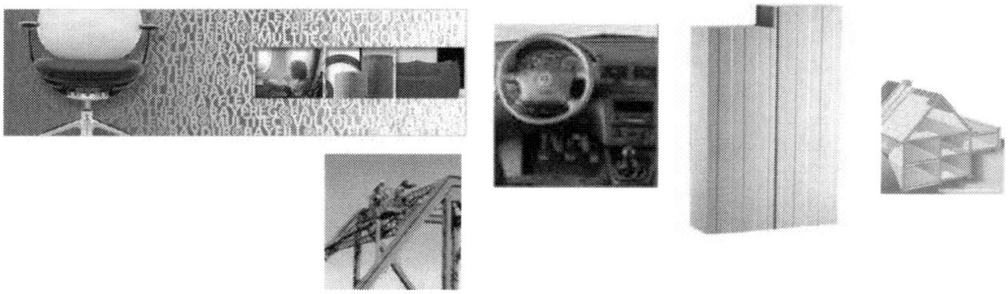

Phosgene hazards - basics

Safety/Hygiene Considerations

TLV phosgene	0.1 ppmv	ACGIH
STEL	0.5 ppmv	EU-SCOEL

Adverse Health Effects

Concentration	Effect
0.4 ppm	Perception of odor
1.5 ppm	Recognition of odor
>3.0 ppm	Irritation of nose, throat, eyes, and bronchi

Dose	Effect
<50.0 ppm-min	No clinical effect, no lung damage
50-150 ppm-min	Minor (subclinical) lung reactions
>150.0 ppm	Clinical lung edema to be expected
>300 ppm-min	Life threatening

Source: Bayer MaterialScience in-house training module
Public Sources: Health effects data: ACGIH, NIOSH, National Toxicology Information Program, Bethesda, MD Medical Library

Safety specific principles – primary and secondary

Global practices governed by corporate Standards:

Global Phosgene Steering Group

Phosgene Manual
Reports
Online, including
Safety
Assessments, Incidents, Documents and Communication

Manual
Logout

Industry data, standards and recommendations:

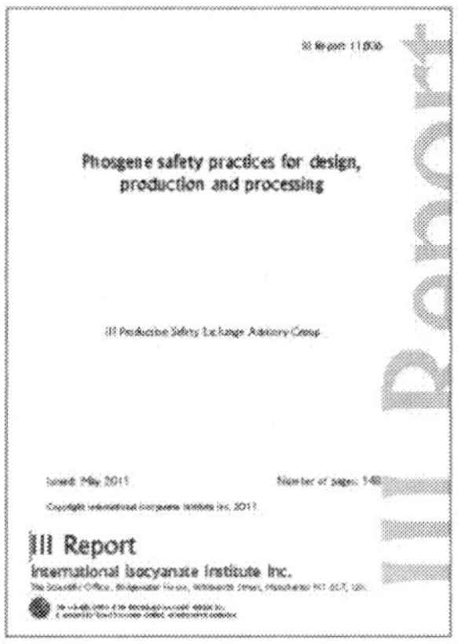

Safety specific principles – primary and secondary

Personnel safety measures

- Phosgene detection badges
- Personal 5 min. air escape bottles
- Site alarm systems, Evacuation procedures and practice

Example Operation safety measures

- Perimeter phosgene detectors and warning lights, PA anoucements
- Separate Master tags for "phosgene action" maintenance
- PPE for prescribed work
- Central breathing air systems for each unit area
- Emergency dump tank capacity to handle liquid volume of largest process vessel

Safety specific principles – primary and secondary

Process design – Highest level of safe design features

- No storage of pure phosgene – produce and use at the process
- In-process only inventory of phosgene is diluted with solvent
- Minimize number of flanges
- Sealless pumps (canned or magnetic drive)
- ASME Code vessels fabricated for lethal service with 100% of welds X-ray inspected
- Destruction of phosgene continuously in vent systems prior to final emission mitigation devices; In-process decomposition reactors convert phosgene to HCl and CO2.

Safety specific principles – primary and secondary

Steam - ammonia curtain

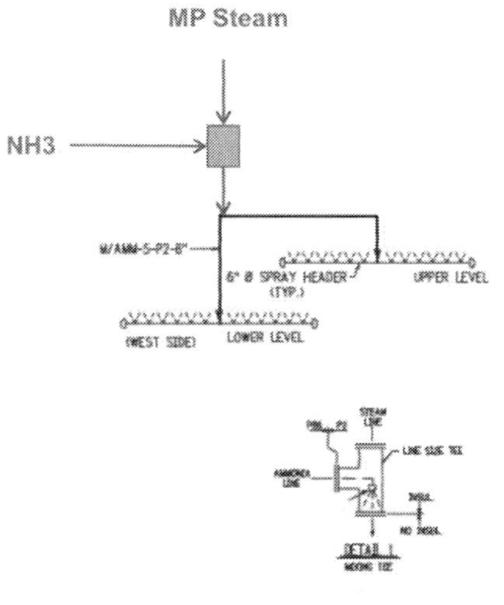

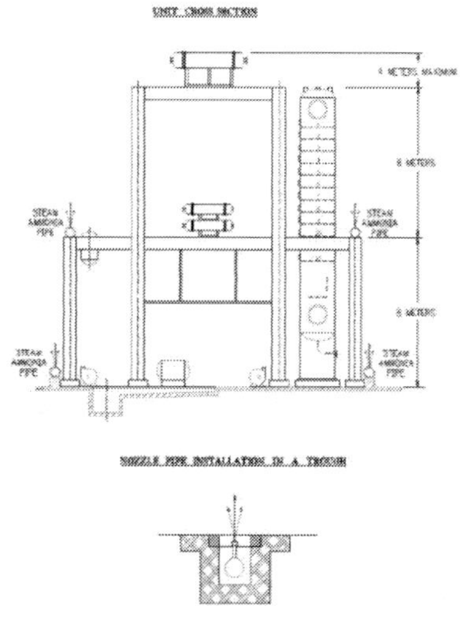

Safety specific principles – primary and secondary

Fully jacketed equipment and piping

Damage containment: Double-shell technology

- The compression-proof, double-sheathing of piping, apparatus and flange connections – also monitored by an inert gas* – is a well-developed safety technology, commonly used for unit sections which may contain liquid phosgene.

* Inert gas, which is continuously analytically monitored for phosgene

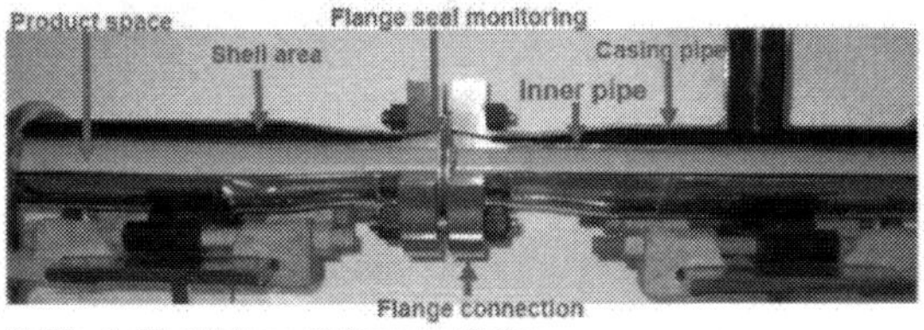

Double-shell techhology with flange monitoring

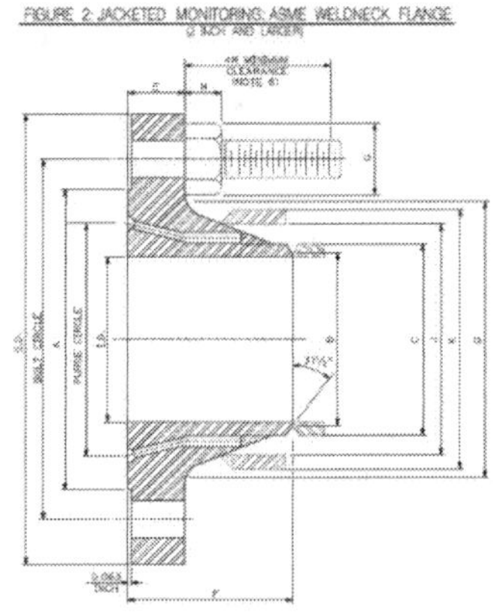

Safety specific principles – primary and secondary

Containment with Caustic scrubber

Considerations for implementation

- Size and complexity of equipment and piping the in phosgene containing area (> 1% phosgene in process streams)
- Dimension of unit (larger unit requires larger steam-ammonia curtain and the amount of ammonia will increase)

Containment enclosure design

Containment building-architectural design considerations

- Ventilation and HVAC – normal and emergency
- Layout and entry and egress methods and design – open 1st floor large doors for pumps maintenance
- Occupational rules and procedures – normal operation and maintenance
  - PPE requirements for both normal entry and maintenance work
- Motivation - Estimates of concentration in localized space near leakage point
- Wall construction for negative pressurization and joint sealing

Containment enclosure design

Ventilation air flow distribution

Normal: 10 changes/hour

Emergency: 5 changes/hour

Individual leakages from piping or equipment

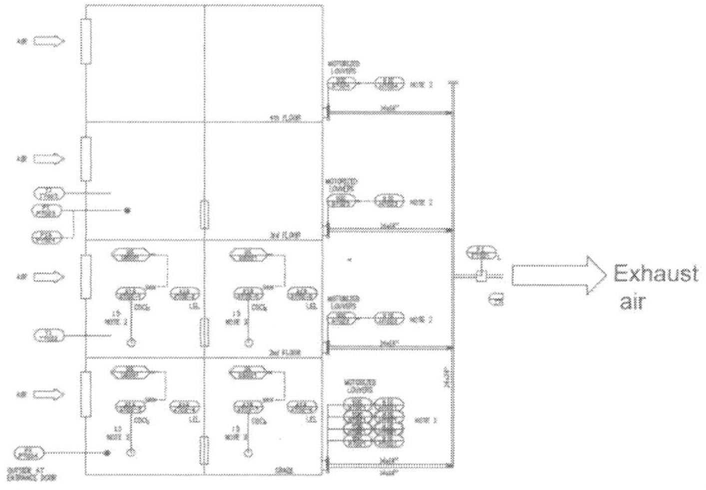

Technical design of caustic scrubber

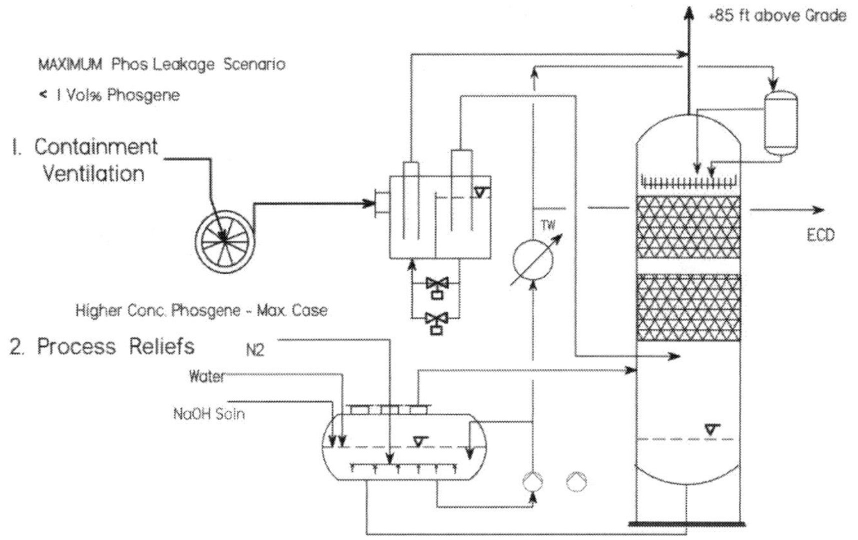

Technical design of caustic scrubber

Mass transport and neutralization chemistry

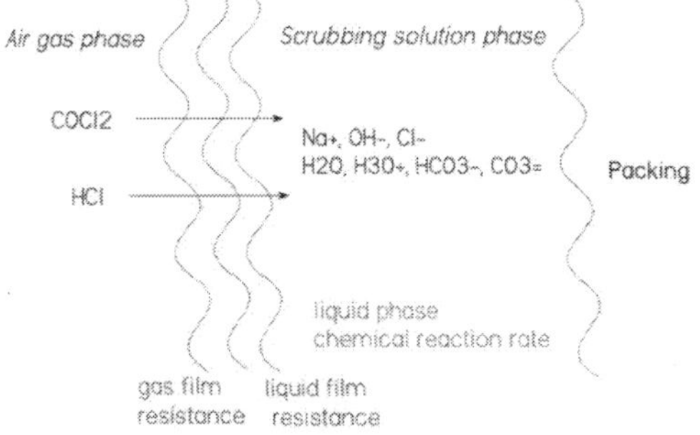

$COCl_2(g)$ + 4 NaOH ---> Na_2CO_3 + 2 NaCl + 2 H_2O

HCl(g) + NaOH ---> NaCl + H_2O

Technical design of caustic scrubber

HTUs were determined experimentally in a Pilot column in Germany: 100mm dia. X 2.4 m

<u>Gas-phase controlled mass transport model</u>

$Z = NTU \times HTU$

$$NTU = (y_{in} - y_{out}) \Big/ \left[(y - y^*) / \ln(y - y^*) \right] \Bigg|_{y_{out}}^{y_{in}}$$

Assume rapid absorption of gas solute at liquid surface, $y^* = 0$, and simplify :

$NTU = \ln(y_{in} / y_{out})$

$HTU = Z / NTU$

Technical design of caustic scrubber

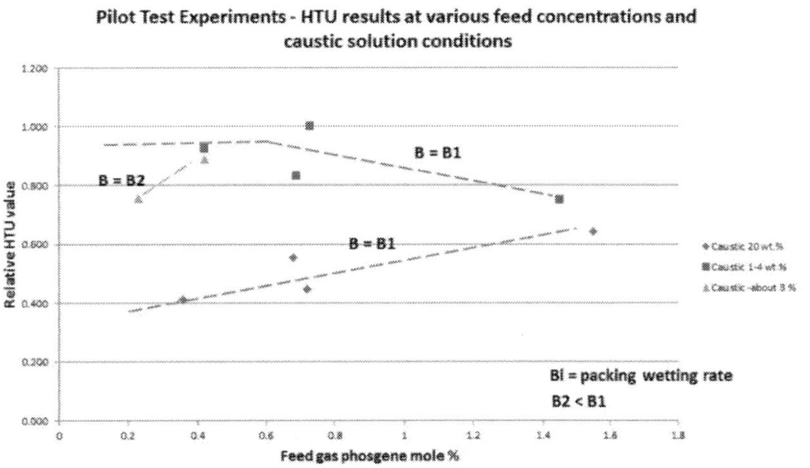

Source:
K. Hallenberger & D. Wastian, "Laborversuche zur Phosgen Vernichtung mit NaOH", 22 Feb. 2005

Technical design of caustic scrubber

Scale up HTU to full scale containment scrubber - method

- NTU calculated for ≤ 0.200 ppmv (200 ppbv) phosgene at outlet stack
- Calculate HTU for selected scale up packing from ratio of packing areas
- Adjust scaled up HTU for reduced wall effects compared to pilot column
- Add safety factor to HTU
- Z = NTU X HTU and determine number of beds (redistribution requirement)
- Diameter determined for reasonable F-factor and considering wetting density required

How does simulation approach with Aspentech^{TM} rate-based model compare with the above rigorous scale-up ?

Technical design of caustic scrubber

Aspen simulation model is with 30 cells and pilot column dimensions

- Containment case – representative feed:
- Results:

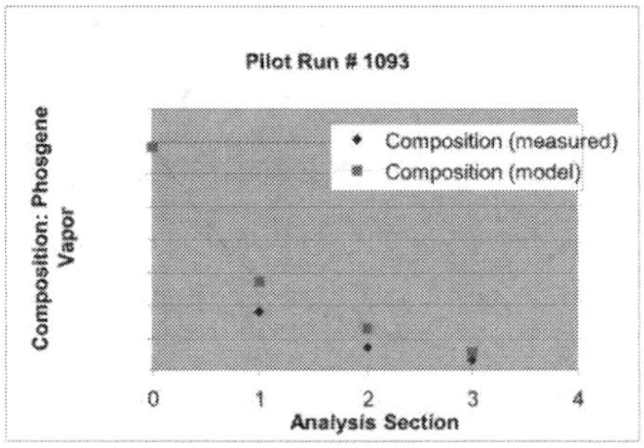

- Liquid resistance Ratesep model with 70% effective interfacial area input

Source (measured data):
Rieder, Hallenberger & Schubert, Bayer internal presentation, 20 Sept. 2005

Technical design of caustic scrubber

Aspentech™ rate-based model study result summary

- Input:
  - Liquid Scrubbing Stream
    20 % Caustic @ 45 C
  - HCl and COCl2 neutralization chemistries included with Δh_{reac} 's
- Results:
  - [Phosgene] @ Top Vapor: < 0.200 PPMV
  - HCl out top appears somewhat high
  - Bottom temperature rise is not excessive, 63 C max. liquid
  - Temp. is cycling segment (stage) to segment - anomaly in the model ?

Page 19 • Bayer Technology Services

Technical design of caustic scrubber

Containment Case: Packing Selection

Select 250Y - 250m2/m3 area structured packing, rather than 40 mm third generation random packing

Advantages / Cost impacts – capital cost savings

- Lower overall column and stack height by a significant amount , 33 % of height
- Lower delta P for packing, preliminary motor size reduced to 100 hp from 125 hp.
- Trade off for higher packing cost TBD
- Greater design safety margin
- Better liquid distribution through beds. One re-distributor versus two.
- Do not expect fouling to be an issue nor believe replacement of packing to be likely; reliable N2 purge to be implemented

Page 20 • Bayer Technology Services

Technical design of caustic scrubber

Scrubber System Design - Press Relief Case

- **Relief stream feed basis calculated from dynamic simulation with Superchems™ software**
- **Caustic Hold Tank sized for relief cases**
  - **Make-up addition of new fresh 20% caustic**
- **Scrubber packing height <u>not</u> limited by the relief case**
- **Sparger into Tank provides initial absorption-reaction of phosgene. ≥ 50 % removal expected**
- **Temp. rise of circulating scrubbing caustic-salt solution requires recycle cooler**

Technical design of caustic scrubber

Fire case at process vessel releases relief flow from PSE to "Catch Tank" where press. buildup occurs and gives flow basis for relief to Caustic scrubber

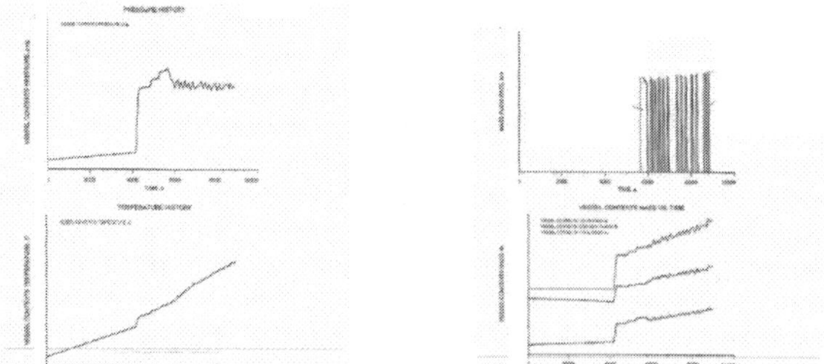

Technical design of caustic scrubber

Containment Caustic Scrubber under Construction – 2007, Asia

Technical design of caustic scrubber

Baytown, TX U.S. Gulf coast
FEL Capital Cost Estimate for Project :
Scrubber system and site infrastructure, engineering, contingency

$ 9,500,0000

with Escalation to 2012

Forward-Looking Statements

This presentation may contain forward-looking statements based on current assumptions and forecasts made by Bayer Group or subgroup management.

Various known and unknown risks, uncertainties and other factors could lead to material differences between the actual future results, financial situation, development or performance of the company and the estimates given here. These factors include those discussed in Bayer's public reports which are available on the Bayer website at www.bayer.com.

The company assumes no liability whatsoever to update these forward-looking statements or to conform them to future events or developments.

Science For A Better Life

Thank you!

Contacts: Mark Thiel, Principal Engineer, PE
mark.thiel@bayer.com

Michael Mueller, Customer and Product Manager
Phone : 281-383- 6644
michael.mueller3@bayer.com

Back -Up

Technical design of caustic scrubber

Containment Design

Dimensions						
Length	18.288	m	60	ft		
Width	12.8016	m	42	ft		
Height	23.7744	m	78	ft		
Volume	5,566	m3	196,560	ft3		
Normal Ventilation (Approx.)	10	Changes per hour				
Emergency Ventilation	**5**	Changes per hour				
Ventilation Air to Scrubber	**27,830**	**m^3/h Emergency flow to Scrubber**			16380	ACFM
	Variable speed ventilation Blower !					

Automatic switching time
for Water Seal Diverter Tank
Estimated to be ≤ 10 secs.

Technical design of caustic scrubber

Leakage Calculation Method*

$A = \text{Constant} \times (DN)^{2.2}$

DN = diameter of pipe, mm
A = Cross sectional area of leakage gap, mm^2, pseudo "cresent shaped"
Constant = proprietary value developed by Bayer in Germany*

For leak of liquid under pressure to atmosphere, where P_g is the gauge pressure in the pipe:
The energy equation (Bernoulli) for liquids applies and:

$v = \sqrt{(2 \times P_g) / \rho}$ where, ρ = density of the liquid

For leak of gas through the narrow gap:
Sonic, choked flow applies in determining the flow velocity, ie.

$v_{critical} = \sqrt{(k \times R / M \times T)}$ where, $k = Cp / Cv$

Then, the volumetric leak flow through the gap is calculated by:

$V = \mu \times A \times v$ where, μ = assumed coefficient of discharge

* Bayer internal methodology and spreadsheet from Bayer Technology Services, Germany

Reference: R. Kauer and K. Strohmeier, DETERMINATION OF LEAKAGE GAP AND LEAKAGE MASS FLOW OF FLANGE JOINTS SUBJECTED TO EXTERNAL BENDING MOMENTS, TU Munich, Germany

Page 29 • BTS 4:3 Template 2010 • June 2011 Bayer Technology Services

Application of Inherently Safer Technology Rules in New Jersey and Contra Costa County, California

Mannan, M. S., Texas A&M University
Originally presented on: 10/4/2012 12:50:00 - 13:15:00

Inherent safety technology (IST) has been developed with the goal to reduce or eliminate hazards based on the use of intrinsic properties of chemicals or conditions. After the concept was adopted by Trevor Kletz in the field of chemical industry, CCPS published the book "Inherently Safer Chemical Processes - A Life Cycle Approach," which developed a roadmap for the application of this technology. Multiple discussions on IST can be found in journal publications and patents and have provided good examples to apply IST in different industries. Since the 9-11 terrorism disaster, multiple federal bills have been proposed to mandate or consider Inherently Safer Design for hazardous chemicals security. However, arguments still exist about the potential economic impacts of mandatory IST initiatives on facilities.

The Inherently Safer Technology (IST) Review Rule was adopted under the Toxic Catastrophe Prevention Act (TCPA) program of the state of New Jersey in May 2008. The rule mandates IST reviews of all facilities covered by the TCPA by evaluating the four IST principles: minimization, substitution, moderation, and simplification. The rule requires reviewing and completing a checklist developed under the direction of the New Jersey Domestic Security Preparedness Task Force. Contra Costa County, California is another region in the US that has put an emphasis on inherently safer technologies. Contra Costa's Industrial Safety Ordinance was effective in January 1999, requiring facilities to consider human factors, security vulnerability analysis, and inherently safer systems. The Department of Environmental Protection recommends three methods that can be employed for IST analysis: (1) reviewing and completing a checklist containing a number of practical inherent safety considerations, (2) avoiding a particular hazard at a part of the process by employing a particular inherently safety strategy and (3) integrating IST into the facility's PHA study.

This paper will first investigate the IST rules adopted by New Jersey and Contra Costa County, the methodologies that are applied and the rules' application development. Then an industrial case study will be conducted based on the IST rules to study the effectiveness of the rules and to provide insight for legislation in the future.

New Leader Review Process

Dow Chemical

October 4, 2012

DOW RESTRICTED - For internal use only

New Leader Review - Dow Chemical

- **"A review by a new Plant Manufacturing Leader or Process Owner within 90 days of assuming the job.** The review shall cover their basic understanding of the reactive chemicals/process hazard potential associated with the operation. Any management of change issues that potentially involve reactive chemicals occurring prior to completion of this review must also be approved by a designated process knowledgeable contact. In certain situations there may be a question about who in the facility should actually have the "New Process Owner Reactive Chemicals Review". **In general, it should be that person that has authority for approving the highest level of Management of Change for the facility.**"

DOW RESTRICTED - For internal use only

New Leader Review - Dow Chemical

- This review requirement is for just a Reactive Chemicals review although it is recommended that the **new leader review as many of the broader Process Safety hazards associated with the facility as possible.** If the new leader chooses to follow the entire fixed facility RC/PHA protocol then the review will satisfy the periodic review requirements for the fixed facility version for the RC/PHA, and the clock for the next review will be reset.

DOW RESTRICTED - For internal use only

New Leader Review - Dow Chemical

The following are some key points to consider relative to this review:

- **The review is to be "demonstration" of knowledge of Reactive Chemicals by the new leader to a committee of knowledgeable experts typically from the site or business reactive chemicals committee.**

- **The review is not intended to be a training session put on by someone else to train the new leader. It is expected that the training of the new leader will have already occurred prior to this review.**

DOW RESTRICTED - For internal use only

Texas Operations NLR/RCPHA Committee

Texas Ops Reactive Chemicals Committee Members

Last Updated: 31-Mar-11

Champion: Gary Hockstra

Sponsor: Laura Ambrose

Co-Chairs: Harold Johnstone, Eric Atkins

PS Contact	RC Discipline Contact	Process Engineering	EH&S	Tech Center	Manufacturing	Research
Eric Atkins	Kevin Darst	Jon Bennett	Mary Schwartz	Dave Boerner	Scott Daigle	Ken Marshall
John Champion	Steve Horsch	Devon Hull	Brian Moore	Kevin Coulter	Gregor Layman	David West
Bill Efaw	Harold Johnstone	Tony Janeshek			Cliff Moore	
Sharon Fortin	Chad Mashuga	Konanur Manjunath			Dan Roberts	
Mike Gohlke	Rob Bellair	John Morgan				
David Martin		Jeff Pauls				
Marty Martin		Wendy Smades				
Brady McCollum						
Peder McHattie						
Jeff Melching						
Greg Nesmith						
Phil Partridge						
Scott Simon						
Karen Study						
Luiz Zacharias						

NLR Review

Pre-Review Work: A Must!!!

- Meet with local PS & RC Contacts to plan & prepare NLR/PHA Package
  - Overview of Process
  - Discuss & Review RC Concerns
  - Develop Worst-Case Scenarios
  - Obtain any missing/needed data
  - Complete RC/PHA Package

NLR Review
Pre-Review Work: A Must!!!

- Meet with local Process Safety Contact to prepare NLR Package
  - Review F&EI Summary
  - Review CEI Summary
  - Review HazOp/LOPA Status, as applicable
  - Review considerations from last audit
  - Review remainder of NLR questionnaire
  - Complete NLR Field verification

NLR Review

Example of Standardized Worst Case Scenarios

#	TECHNOLOGY	PROCESS AREA	UNCONTROLLED REACTION CHEMISTRY	NUMBER OF TIMES THIS SCENARIO HAS OCCURRED WITHIN THE TECHNOLOGY
1	Cellulosics	METHOCEL® powder handling	Combustion of METHOCEL powder (deflagration, dust explosion).	6 documented
2	Cellulosics	Dryer Fire	Exothermic thermal decomposition of METHOCEL®. Large agglomerates of METHOCEL can become lodged in the dryer and experience long heat histories. At high dryer temperatures, a slow decomposition reaction will begin and accelerate as the piece of METHOCEL	11 documented on file server
3	Cellulosics	METHOCEL® Pulp Grinders – Fire	Exothermic thermal decomposition of cellulose. Pulp may settle and accumulate on a small ledge behind the first bed knife, an apparent quiescent zone, in the MRP grinders. At high localized temperatures, a slow exothermic decomposition reaction can begi	6 documented
4	Cellulosics	METHOCEL® powder handling	Combustion of METHOCEL powder (deflagration, dust explosion).	7 documented
5	Cellulosics	Dryer Fire	Exothermic thermal decomposition of METHOCEL®. Large agglomerates of METHOCEL can become lodged in the dryer and experience long heat histories. At high dryer temperatures, a slow decomposition reaction will begin and accelerate as the piece of METHOCEL	12 documented on file server
6	Cellulosics	METHOCEL® powder handling	Combustion of METHOCEL powder (deflagration, dust explosion).	7 documented

New Leader Review - Dow Chemical

The following are some key points to consider relative to this review:

- The Chair of the review committee may choose to hold an abbreviated review with a subset of the normal review committee. This most often would occur in situations where the new leader has already spent a long time working in the technology and it is felt that they have already established a sufficient knowledge of the Reactive Chemicals issues.

- **Successful completion of the review shall be documented in the plant records and entered in the Process Safety PHA database.**

New Leader Review - Dow Chemical

V. REVIEW PRESENTATION FORMAT

- Present plant/process/chemistry overview.
- Summarize previous incidents and corrective actions taken by both plant and technology.
- Review the most significant Reactive Chemicals and Process Hazard worst case scenarios and key lines of defense.
- Review significant process changes completed since last review, including process related MOCs and any planned changes. (Existing Facilities)
- Discuss Reactive Chemicals/Process Safety Training and awareness programs. (Existing Facilities)
- Discuss how Reactive Chemicals and Process Safety Preventative Actions are included in plant Operating Discipline and Policies. (Existing Facilities)
- Review the RC/PHA Questionnaire by exception only.
- A walk through inspection of the facility.
- Review Summary including Action Item Review and Timing agreement.

NLR Review
RC/PHA Questionnaire

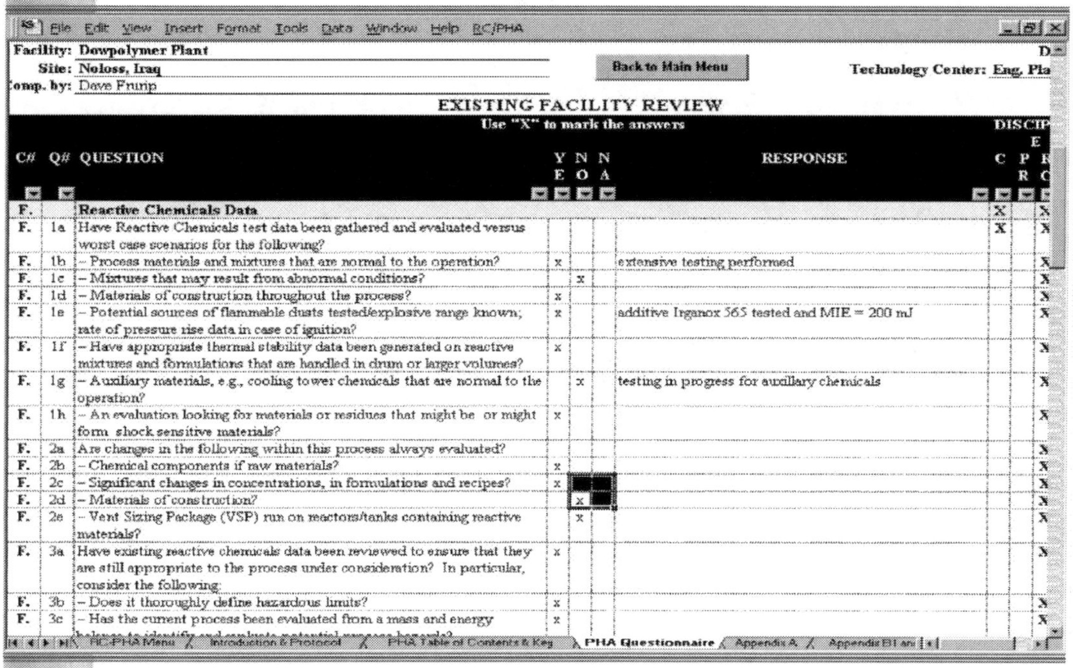

Facility: Dowpolymer Plant
Site: Noloss, Iraq
Comp. by: Dave Frurip
Technology Center: Eng. Pla

EXISTING FACILITY REVIEW

Use "X" to mark the answers

C#	Q#	QUESTION	YE	NO	NA	RESPONSE	C	PR	E RC
F.		**Reactive Chemicals Data**					X		X
F.	1a	Have Reactive Chemicals test data been gathered and evaluated versus worst case scenarios for the following?					X		X
F.	1b	– Process materials and mixtures that are normal to the operation?	x			extensive testing performed			X
F.	1c	– Mixtures that may result from abnormal conditions?		x					X
F.	1d	– Materials of construction throughout the process?	x						X
F.	1e	– Potential sources of flammable dusts tested/explosive range known; rate of pressure rise data in case of ignition?	x			additive Irganox 565 tested and MIE = 200 mJ			X
F.	1f	– Have appropriate thermal stability data been generated on reactive mixtures and formulations that are handled in drum or larger volumes?	x						X
F.	1g	– Auxiliary materials, e.g., cooling tower chemicals that are normal to the operation?		x		testing in progress for auxillary chemicals			X
F.	1h	– An evaluation looking for materials or residues that might be or might form shock sensitive materials?	x						X
F.	2a	Are changes in the following within this process always evaluated?							X
F.	2b	– Chemical components if raw materials?	x						X
F.	2c	– Significant changes in concentrations, in formulations and recipes?	x						X
F.	2d	– Materials of construction?		x					X
F.	2e	– Vent Sizing Package (VSP) run on reactors/tanks containing reactive materials?		x					X
F.	3a	Have existing reactive chemicals data been reviewed to ensure that they are still appropriate to the process under consideration? In particular, consider the following:	x						X
F.	3b	– Does it thoroughly define hazardous limits?	x						X
F.	3c	– Has the current process been evaluated from a mass and energy	x						X

Overview of Operations and Key Process Hazards

•Demonstrate knowledge of key chemistry practiced and major unit operations in the plant – it's very useful to have an Index Flow sheet and Plant Layout to use in this discussion.

•Demonstrate general knowledge of hazards associated with these unit operations including fire and explosion potentials as well as community exposure issues. (Toxicity, Flammability, Reactivity) - the CEI and F&EI were useful as a basis to focus on highest hazards
- understand the main raw materials and hazards associated with each;
- understand potential consequences associated with worst case scenarios and how they might occur. What are the key mitigation equipment in the plant such as dikes, sprinkler systems, ventilation systems, etc?

•Demonstrate knowledge of key learning experiences / incidents in this technology as relevant to the chemicals and hazards being addressed in current plant. What is the plant incident history and what has happened to similar plants in the technology?

Training and Communication of Key Hazards

•Does the facility have "Cardinal Rules" or a "Top 10 List? If so, what are they and how are they shared?

•How does the plant compare to business MET-2, if applicable?

•Demonstrate knowledge of the plant reactive chemicals training program. Has the New Leader completed any plant Reactive Chemical testing?

•Do plant operations understand and use compatibility charts. Are they available in the control room?

•Are there any unique alloy equipment and why?

•Demonstrate general knowledge of reactive chemicals testing (recognize that a DSC onset temperature alone does not represent a safe operating limit, etc), what data exists, who their global business RC Thermal focal point is and when to engage them, etc.

Safety Standards and LPPs

•Is the New Leader familiar with the Emergency Plan and taken the appropriate testing?

•Is the New Leader familiar with Locked (Administrative) Valve Policy or Safety System Impairment Standard?

•What is the status of LPP compliance

•What is the status of LOPA / DowGEP / SIS / Critical Instruments?

•What is the status of previous Burner Management reviews

•What is the status of the Mechanical Integrity (GMISS) program?
•What is the status of previous RC-PHA review action items?
•What is the status of active Variances or Compliance Plans?
•What are recent major changes and status of follow-up actions?

DOW RESTRICTED - For internal use only

New Leader Review - Dow Chemical

V. REVIEW PRESENTATION FORMAT

- Present plant/process/chemistry overview.
- Summarize previous incidents and corrective actions taken by both plant and technology.
- Review the most significant Reactive Chemicals and Process Hazard worst case scenarios and key lines of defense.
- Review significant process changes completed since last review, including process related MOCs and any planned changes. (Existing Facilities)
- Discuss Reactive Chemicals/Process Safety Training and awareness programs. (Existing Facilities)
- Discuss how Reactive Chemicals and Process Safety Preventative Actions are included in plant Operating Discipline and Policies. (Existing Facilities)
- Review the RC/PHA Questionnaire by exception only.
- A walk through inspection of the facility.
- Review Summary including Action Item Review and Timing agreement.

New Leader Review - Dow Chemical

The following are some key points to consider relative to this review:

- If the new leader does not convince the committee that they have sufficient Reactive Chemicals knowledge, the committee may ask for a remedial follow up review meeting.

- The review should follow the same general presentation format for all new leaders -- 9 review topic areas as outlined previously.

Layer of Protection Analysis (LOPA) Basics

Randy Freeman
S&PP Consulting

1

Randy Freeman

- 35+ years of industrial process safety experience
- 25 years experience in design and application of instrument systems to prevent accidents
- Author of over 40 technical papers
- Committee member or technical reviewer of approximately 20 CCPS guideline books
- Member ISA S84 Committee
- AIChE Fellow, Emeritus Member Loss Prevention Symposium Planning Committee and Emeritus Member and Fellow of CCPS
- Past Chairman of AIChE Safety and Health Division
- BS, MS, PhD Chemical Engineer - Univ of MO - Rolla
- Recipient 2011 Walton-Miller Award

4 Oct 2012

2

Typical Events of Concern

- Fire
- Explosion
- Toxic material release
- Significant environment impact
- Release to a flare system
- Vessel Overpressure
- Runaway reaction

4 Oct 2012 3

LOPA

- Layer of Protection Analysis
- Semi quantitative method for the evaluation of safety risks
- Widely used in Industry
- Becoming accepted if not required by regulators (OSHA. EPA. Etc.)

4 Oct 2012 4

LOPA Concept

- Define a Scenario that we are worried about
- Look at a single cause for scenario
- Look at various protective measures to stop the scenario from occurring
- Evaluate the likelihood of the protective measures failing
- Evaluate the Risk against Company Requirements
- Define action plan to reduce the risk to a tolerable level

4 Oct 2012 5

Layers of Defense

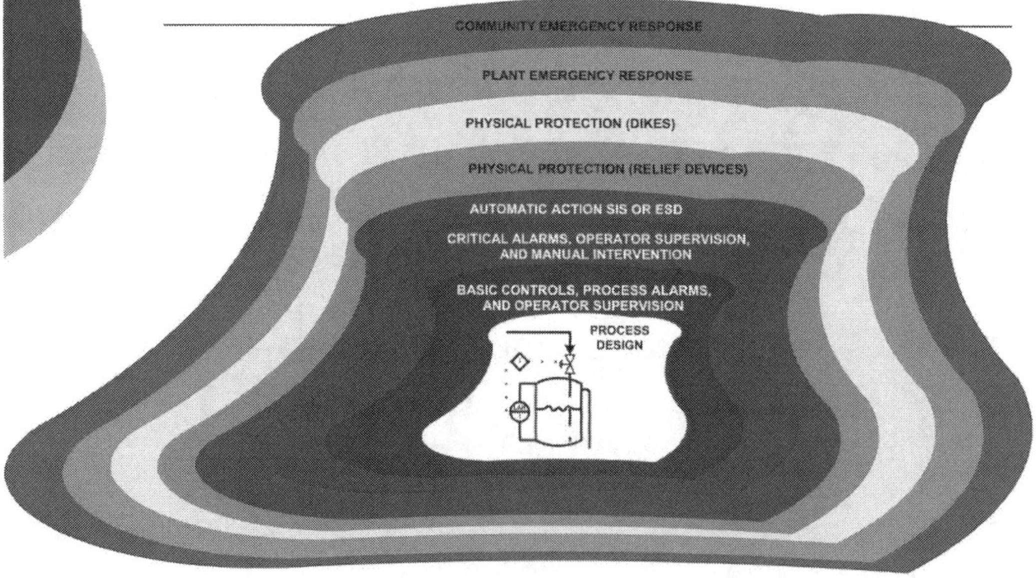

4 Oct 2012 6

What is an IPL?

- An IPL is something that will prevent the train from reaching the disaster

4 Oct 2012 7

What is an IPL?

1. Capable of Preventing Event of Concern
2. Independent of other Protective Layers
3. Auditable or Verifiable

4 Oct 2012 8

Swiss Cheese View of Layers

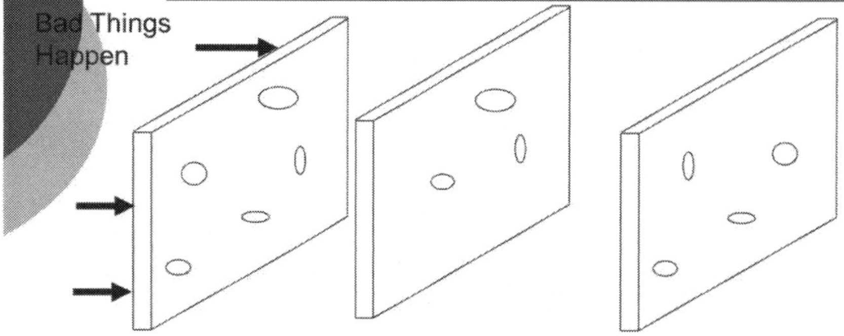

Protective Layers 1, 2, 3

Holes in Cheese are weaknesses in each layer
GOAL: Keep the holes from alignment

4 Oct 2012 9

Swiss Cheese View of Layers

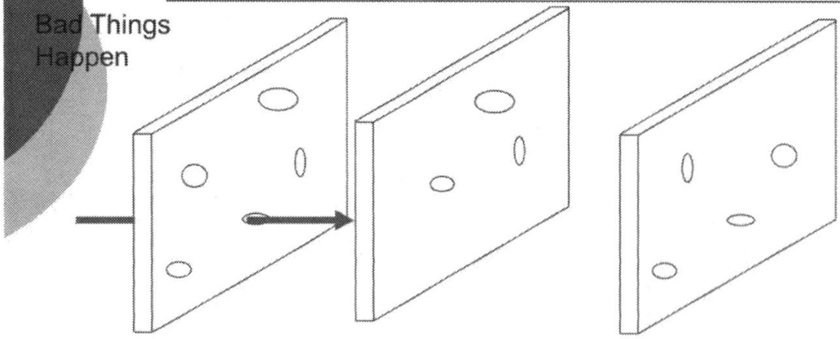

Protective Layers 1, 2, 3

Holes in Cheese are weaknesses in each layer
GOAL: Keep the holes from alignment

4 Oct 2012 10

Swiss Cheese View of Layers

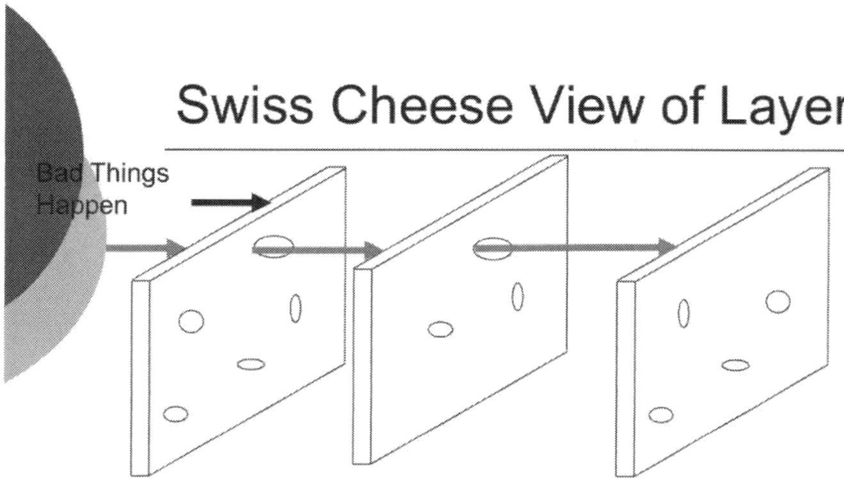

Protective Layers 1, 2, 3

Holes in Cheese are weaknesses in each layer
GOAL: Keep the holes from alignment

4 Oct 2012 11

Swiss Cheese View of Layers

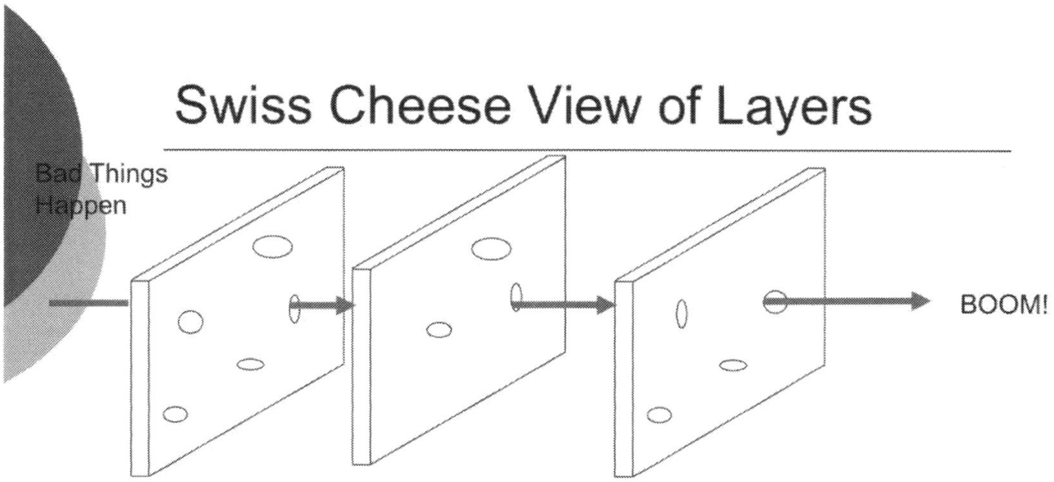

Protective Layers 1, 2, 3

Holes in Cheese are weaknesses in each layer
GOAL: Keep the holes from alignment

4 Oct 2012 12

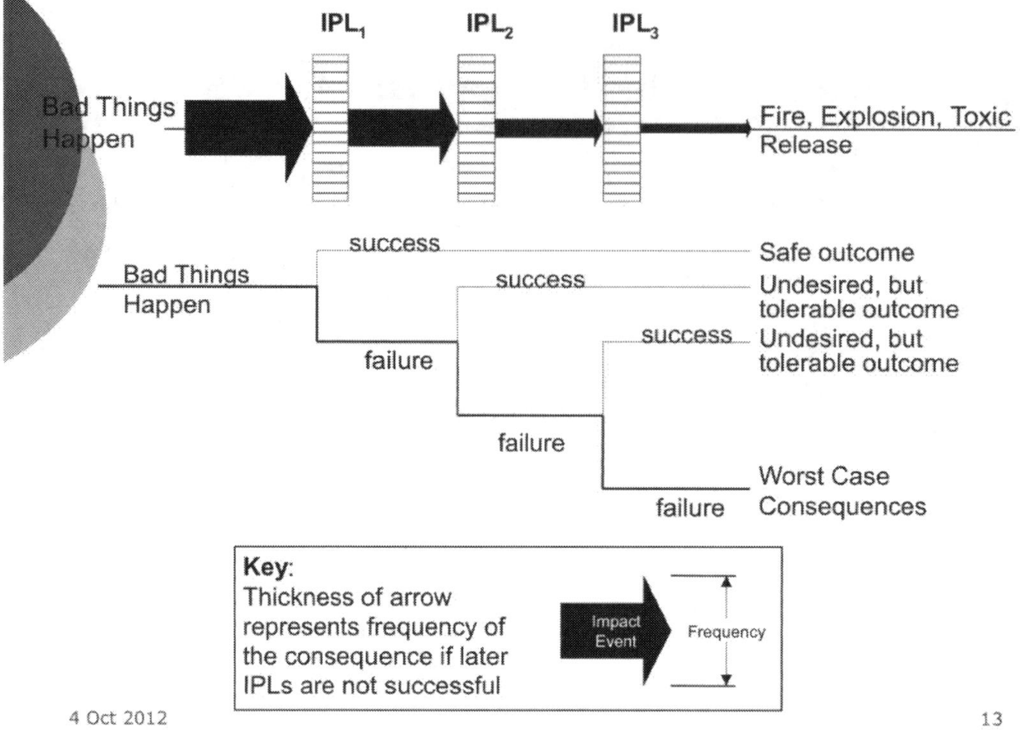

Typical LOPA Team Participants

Team Participant	Function
LOPA Team Leader	Lead and document the LOPA sessions. Primarily a facilitator of the team meetings
Process Engineer	Provides process chemistry knowledge to the LOPA team
Control Systems Engineer	Provides control system knowledge to the LOPA team
Production operator	Provides hands-on operations knowledge of how the system operates. For a new facility, the production operator may be chosen from a sister production unit or from a similar process system. A senior operator is normally assigned to this function.
Production supervision	Provides management and operating policy input to the team
Safety advisor	Provides knowledge of plant safety policies and risk toleration
Maintenance	Provides knowledge on how the system will be maintained

Independent Protective Layers Must Be

- Capable of preventing the undesired event of concern
- Be independent of other safeguards counted as protective layers
- Be auditable

4 Oct 2012 15

Possible IPLs

- Relief Valve
- Basic Process Control System
- Safety Instrumented Function Implemented in a Safety Instrumented System - INTERLOCK

4 Oct 2012 16

Generic LOPA Credits

Protective Layer	LOPA Credit (PFoD)
Basic Process Control System	1×10^{-1}
Procedural control with more than10 minutes to complete the task	1×10^{-1}
Procedural Control with more than 40 minutes to complete the task	1×10^{-2}
Active mechanical safeguard (relief valve, rupture disk, etc.)	1×10^{0} to 1×10^{-3} depending upon process conditions and history of the device

4 Oct 2012

17

Generic LOPA Credits

Protective Layer	LOPA Credit (PFoD)
Passive mechanical safeguard (dike)	1×10^{-2}
Safety interlock (Safety Instrumented Function implemented in an Safety Instrumented System) at a Safety Integrity Level 1	1×10^{-1}
Safety interlock (Safety Instrumented Function implemented in an Safety Instrumented System) at a Safety Integrity Level 2	1×10^{-2}
Safety interlock (Safety Instrumented Function implemented in an Safety Instrumented System) at a Safety Integrity Level 3	1×10^{-3}

4 Oct 2012

18

People as IPLs???

Can we credit human actions in the prevention of an incident??

19

Model of a Protective Layer

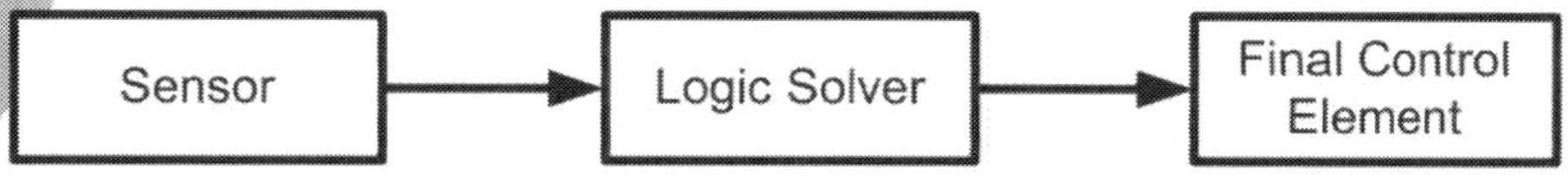

4 Oct 2012

20

Comparison of Different Types

Type of Safeguard	Sensor	Logic Solver	Final Control Element
Interlock in Basic Process Control System	Field sensor – temperature, pressure, etc.	Basic Process Control System computer	Remotely operated control valve, block valve, motor starter switch
Interlock in Safety Instrumented System	Field sensor – temperature, pressure, etc.	Safety Instrumented System Interlock	Remotely operated control valve, block valve, motor starter switch
Operator action procedural control – control room	Field sensor – temperature, pressure, etc.	Human action based on response to annunciated alarm from BPCS or SIS	Remotely operated control valve, block valve, motor starter switch
Operator action procedural control – Field Operator	Field sensor – temperature, pressure, etc.	Human action based on response to local alarm, sensor reading or observation.	Remotely operated control valve, block valve, motor starter switch.

4 Oct 2012 21

Definition of a Procedural Control

Procedural control – A layer of protection based on the combination of a field sensor, human operator and final control element (switch, valve, relay) governed by a **written procedure.**

4 Oct 2012 22

Elements of a Procedural Control

1. Written procedures that define the task or action to be taken
2. A clear indication that the task must be performed
3. Training on how to perform the task
4. Adequate tools and materials to perform the task
5. Proper and available PPE to perform the task
6. Adequate time to perform the task
7. Time to diagnosis abnormal problems with the task
8. Clear indication of proper performance of the task
9. Ability to verify that the task was performed (auditability)

4 Oct 2012 23

Time Available to Prevent an Accident

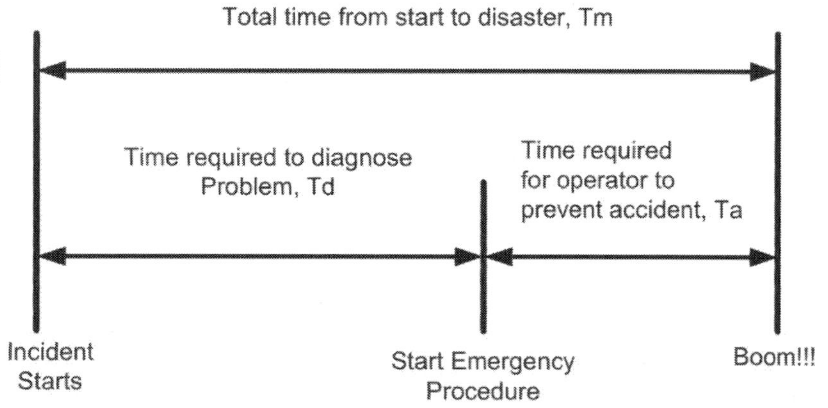

For Safe Design: Tm > Td + Ta

4 Oct 2012 24

Probability of Failure to Diagnosis

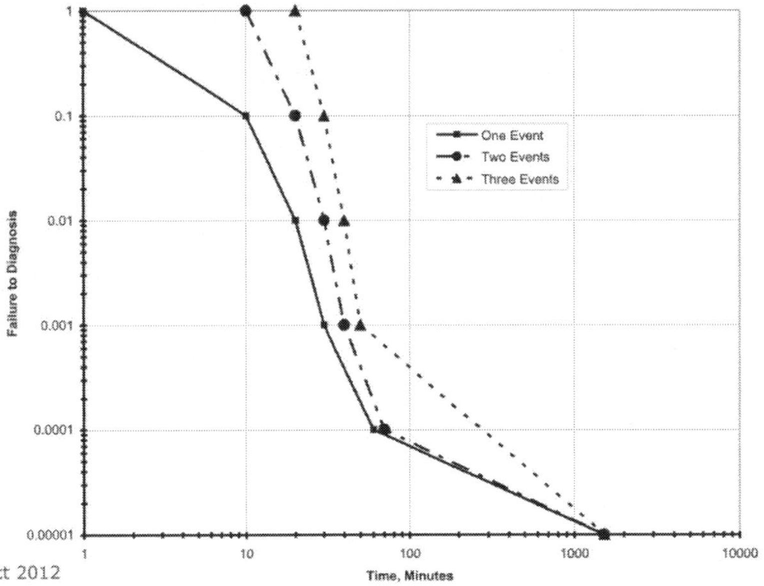

Checklist for Procedural Control

1	Is the procedure capable to stopping the bad event of concern from occurring?
2	Is the procedure auditable? Is the trail of written procedure, operator training, re-training and certification of operator understanding complete?
3	Is there a written procedure? Does the procedure clearly define the steps that need to be completed in the proper sequence to successfully prevent the undesired event of concern from occurring?
4	Have the operators been trained in the use of the procedure? Re-training as needed?
5	Is a hard copy (paper) of the procedure available for the immediate use by the operator? Is this procedure clearly identifiable as an emergency procedure (for example, printed on a different color of paper)?

Checklist for Procedural Control

6	Is the procedure based on a written check list?
7	Is the procedure a two person (done-by checked-by) written checklist?
8	Is the failure to complete the procedure clearly and immediately presented to the person doing the procedure? Can they tell that they have made a mistake in doing the procedure?
9	Is a record completed that indicates the procedure was completed?

4 Oct 2012 27

Checklist for Procedural Control

10	How much time will it take for the operator to take action to stop the bad event of concern, once the need for action is determined?
11	How long does it take to complete the procedure once started?
12	Is there any travel time to the site where action is required?
13	How much time is available from the initiating event until the undesired event occurs?
14	Are any special tools required to perform the procedure?
15	Are any special tools available for immediate use?

4 Oct 2012 28

Checklist for Procedural Control

16	Are special protective equipment (gloves, respirators, air packs, etc.) needed at the work site? Are these items immediately available and how long will it take to gather them?
17	Is the need for the completion of the procedure clearly annunciated?
18	Is there an alarm on a control panel to indicate the initiating event has occurred? How is the alarm implemented? Using the BPCS or the SIS?
19	Is the control panel normally manned?
20	If there is an alarm on a control panel, is the alarm activated using the basic process control system (BPCS)?

4 Oct 2012 29

Checklist for Procedural Control

21	Is the alarm activated using a SIS? If yes, what is the SIL rating of the alarm?
22	Is this procedure independent from other procedures or actions taken and claimed as an IPL during the LOPA analysis?

4 Oct 2012 30

Model for Procedural Control

For a 10 step procedure:

HEP = 0.003 for each step

$P_s = (P_i)^n$

$P_i = 1 - HEP$

$P_f = 1 - P_s = 1 - (1 - HEP)^n$

$P_f = 1 - P_s = 1 - (1 - 0.003)^{10}$

$P_f = 0.03$ or rounding up

$P_f = 0.1$ ⇦

4 Oct 2012 31

Suggested LOPA Credits for Procedural Controls

Description of Procedural Control	LOPA credit	PFOD
Single human operator with less than 10 minutes to diagnosis and take action on a clearly annunciated alarm event.	0	1E0
Single human operator with 10 minutes to 40 minutes to diagnosis and take action on a clearly annunciated alarm event	1	1E-1
Single human operator with 40 minutes to 8 hours to diagnosis and take action on a clearly annunciated alarm event activated by a SIS.	2	1E-2

4 Oct 2012 32

Suggested LOPA Credits for Procedural Controls (Cont)

Description of Procedural Control	LOPA credits	PFOD
Single human completing a simple routine task with between 10 minutes and 40 minutes to complete the task	1	1E-1
Two operators acting independently completing a short (10 items or less) written checklist procedure (done-by/check-by) with between 10 minutes and 40 minutes to complete the task	2	1E-2
Two independent operations groups completing a task with one group independently checking the work of the other group (operations checking maintenance) with adequate time (greater than 10 minutes) to detect any issues	2	1E-2

4 Oct 2012

33

Risk Evaluation

34

Uses of LOPA

- Develop List of Critical Safety Devices, Procedures and Systems
- Screen Scenarios for those requiring detailed CPQRA
- Develop Safety Integrity Level (SIL) targets for new safety instrumented systems
- Rank potential hazards to allow for efficient risk reduction expenditures
- **MUST** Have Risk Criteria to do LOPA

4 Oct 2012 35

LOPA Risk Matrix

Severity -> Likelihood (event/yr)	5	4	3	2	1
1.E+00	I	I	I	M	T
1.E-01	I	I	M	T	T
1.E-02	I	I	M	T	T
1.E-03	I	M	T	T	T
1.E-04	M	M	T	T	T
1.E-05	T	T	T	T	T
1.E-06	T	T	T	T	T

4 Oct 2012 36

Qualitative Consequence Categorization

Category 1/2	
Personnel	Minor or no injury, no lost time
Community	No injury, hazard, or annoyance to public
Environment	Recordable event with no agency notification or permit violation
Facility	Minimal equipment damage at an estimated cost of less than $100,000 and with no loss of production
Category 3	
Personnel	Single injury, not severe, possible lost time
Community	Odor or noise annoyance complaint from the public
Environment	Release that results in agency notification or permit violation
Facility	Some equipment damage at an estimated cost greater than $100,000 and with minimal loss of production
Category 4	
Personnel	One or more severe injuries
Community	One or more minor injuries
Environment	Significant release with serious offsite impact
Facility	Major damage to process area(s) at an estimated cost greater than $1,000,000 or some loss of production
Category 5	
Personnel	Fatality or permanently disabling injury
Community	One or more severe injuries
Environment	Significant release with serious offsite impact and more likely than not to cause immediate or long-term health effects
Facility	Major or total destruction of process area(s) at an estimated cost greater than $10,000,000 or a significant loss of production

Typical LOPA Team Members

*LOPA Team Leader	Leads and documents the LOPA sessions. Primarily a facilitator of the team meetings and ensures LOPA rules are understood and followed by the team
*Process Engineer/ Responsible Engineer	Provides technical knowledge of process equipment, basic chemistry and accepted engineering practices to the LOPA team
*Instrumentation/Automation Engineer	Provides instrumentation and automation knowledge to the LOPA team
*Process Operator	Provides hands-on operations knowledge of how the system operates. For a new facility, the process operator may be chosen from a sister production unit or from a similar process system. A senior operator is normally assigned to this function.
*Operations Supervision	Provides management and operating policy input to the team
*Safety/Process Safety Professional	Provides knowledge of plant safety policies and risk toleration
Maintenance	Provides knowledge on how the system will be maintained

LOPA Output

- The LOPA team leader is responsible for completion of a final report of the efforts of the team
- Report will define those safety measures that were counted as IPLs. These become critical safety items and must be managed as such.
- Report will define those scenarios where addition risk reduction measures are needed

4 Oct 2012 39

LOPA Take Away Points

- Simplified Risk Analysis Tool
- One Cause One Consequence
- **Very Conservative**!
- If LOPA Study says system OK, confident that it is OK
- If LOPA Study says system NOT OK, two choices:
  - Accept result and take action to correct
  - Do a more advanced study (CPQRA)

4 Oct 2012 40

QUESTIONS

????

4 Oct 2012 41

LOPA References

1. Center for Chemical Process Safety, *Layer of Protection Analysis*, ISBN: 0-8169-0811-7, American Institute of Chemical Engineers, New York, 2001
2. ANSI/ISA-S84.00.01-2004 Part 3 (IEC 61511-1 Mod) "Functional Safety: Safety Instrumented Systems for the Process Industry Sector – Part 3: Guidance for the Determination of the Required Safety Integrity Levels - Informative"
3. R.A. Freeman, "Using Layer of Protection Analysis to Define Safety Integrity Level Requirements," Presented at the 2006 Spring National AIChE Meeting in Orlando, Florida. Submitted for Publication in Process Safety Progress.

4 Oct 2012 42

EXAMPLES

These examples are a work of fiction. Equipment, locations and chemicals are either the product of the authors imagination or are used fictitiously. Any resemblance to an actual chemical plant or process is entirely coincidental.

4 Oct 2012 43

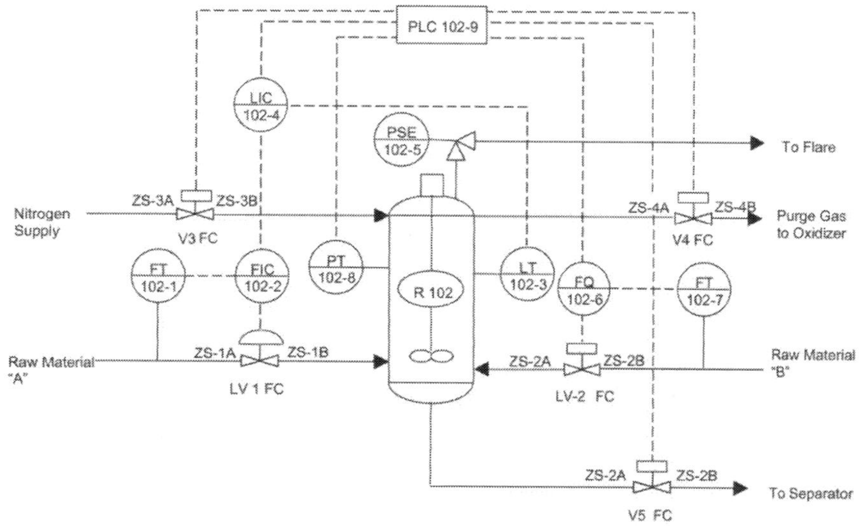

Reactor R-102 P&ID

4 Oc

Steps in a LOPA Study

1. Define the system to be studied (Scope)
2. Define the reason for the study (Why or what question are you trying to answer)
3. Organize the Team
4. Define Scenarios
5. Define Possible Protective Layers
6. Define Initiating Events
7. Determine Consequences
8. Evaluate Risk
9. Develop Recommendations for Improvement

4 Oct 2012 45

Example 1 – ESD Activation

During the review of a batch process system, the question of emergency shutdown of raw material addition was discussed. The high level alarm is activated using the BPCS. The operators are trained in the use of the emergency shutdown button on the control panel which will stop all flows to the reactor. The alarm will be activated 15 minutes prior to overflowing the reactor. A high level shutoff will be automatically tripped by the BPCS prior to the high level alarm.

4 Oct 2012 46

Example 2 – Rewire ESD

The LOPA team proposes that the high level alarm will be routed from a sensor into the SIS logic solver associated with the batch process. An alarm will be activated on the SIS control panel and an alarm signal will be sent to the batch process alarm panel. A second valve in raw material "A" supply line to the reactor will be tied into the emergency shutdown button on the control panel.

4 Oct 2012 47

Example 3 – How to Double Charge Reactor

1. The operations group on Shift A must fail to empty the reactor
2. The maintenance group on Shift B must fail to clean out the reactor
3. The operations group returning on Shift A must fail to recognize that the controls indicate that the reactor is still full and attempt process a new batch.
4. The operations group on Shift A must bypass the high level BPCS permissive set to prevent addition of raw materials on high level in the reactor. This would require the intervention of control engineering group to reprogram the BPCS logic solver.
5. The operations group on Shift A must then charge raw material into the reactor and ignore the alarms installed in Examples 1 and 2 above.

4 Oct 2012 48

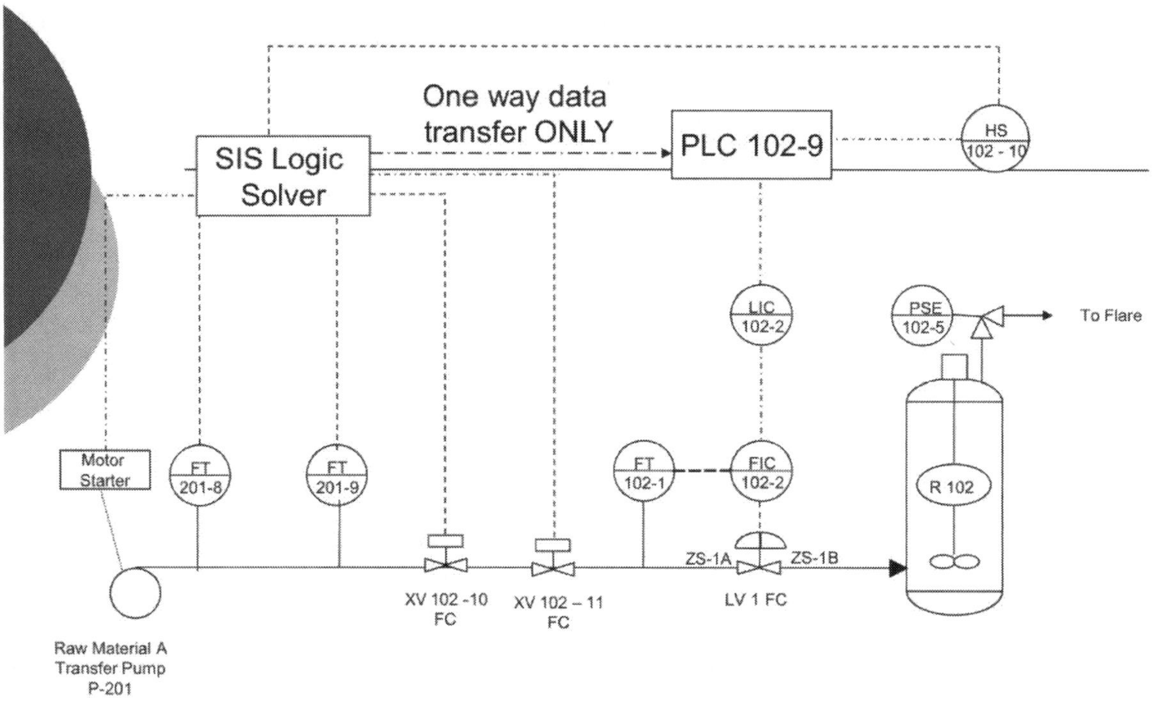

Revised Raw Material "A" System

4 Oct 2012

49

Effects of Process Variables on Dynamic Relief Load Estimates for a Depropanizer and Debutanizer

John Wilkins
Dustin Smith P.E.

Smith & Burgess
Process Safety Consulting
5700 NORTHWEST CENTRAL DR. SUITE 301
HOUSTON, TX 77092
SMITHBURGESS.COM

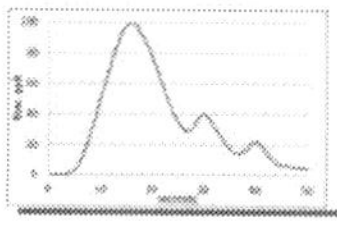

DYNAMIC RELIEF
PROCESS VARIABLES EFFECTS

AGENDA

- Introduction
- Methodology
  - Setting up the dynamic simulation
  - Modeling relief
- Results
  - Depropanizer
  - Debutanizer
- Conclusions
- Recommendations

Smith & Burgess
Process Safety Consulting

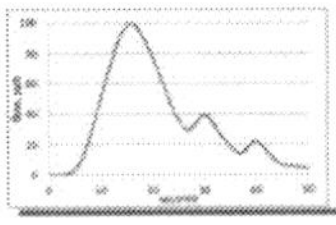

DYNAMIC RELIEF
PROCESS VARIABLES EFFECTS

INTRODUCTION

JOHN WILKINS
STAFF PROCESS CONSULTANT
SMITH & BURGESS
PROCESS SAFETY CONSULTING

DUSTIN SMITH P.E.
PRINCIPAL ENGINEER
SMITH & BURGESS
PROCESS SAFETY CONSULTING

Smith & Burgess
Process Safety Consulting

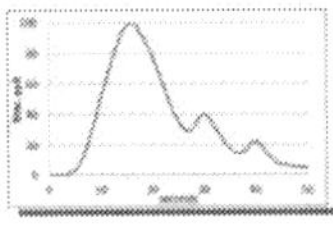

DYNAMIC RELIEF
PROCESS VARIABLES EFFECTS

BACKGROUND

- Dynamic analysis allows an engineer to view the transient response of a system during an overpressure scenario.
- Peak relief rate through a pressure relief device can be predicted using dynamic software, but it can be affected by process variables.
- A sensitivity analysis can show the effects of process variables on the response of the relief device.

Smith & Burgess
Process Safety Consulting

BACKGROUND [CONT.]

The following notes are from API 521 5th Ed. Sec 5.22:

- It can be necessary to perform sensitivity analyses with respect to control response in order to identify appropriate control response.
- In general, no credit is taken for automatic control action unless it tends to increase the relief load.
- If dynamic simulation is used for column-relief-system design, it is necessary to ensure that the model is conservative with respect to calculating the maximum relief load.
- These assumptions shall be checked by sensitivity analyses to assess their impact on the column-relief load.

METHODOLOGY

- Three operating conditions were varied
  - Column liquid level
  - Feed temperature
  - Column pressure

- Different columns were analyzed
  - Depropanizer (Column temp range: 104-210°F)
    - (4' diameter, 21,600 lb/hr feed)
    - (8' diameter, 32,400 lb/hr feed)
  - Debutanizer (Column temp range: 179-384°F)
    - (10' diameter, 623,000 lb/hr feed)
    - (15' diameter, 716,000 lb/hr feed)

DYNAMIC RELIEF
PROCESS VARIABLES EFFECTS

METHODOLOGY [CONT.]

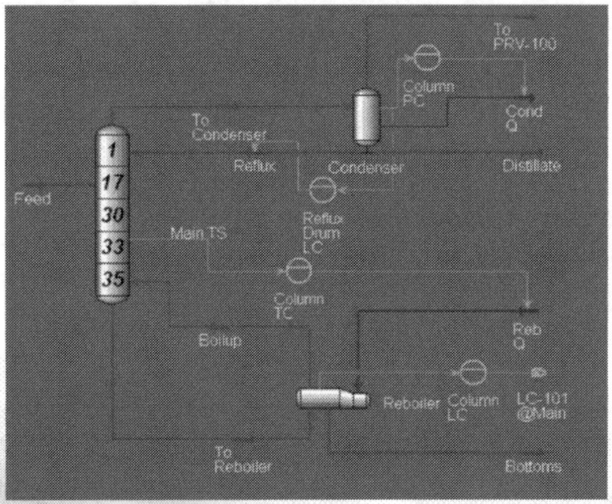

Column PFD for the depropanizer

1. Steady state columns were created in HYSYS
2. Dynamic mode was initiated
3. PID controllers were created
   - Column Pressure
   - Column Temperature
   - Reflux Drum Liquid Level
   - Column Liquid Level
   - Product Flow Rate
4. Controllers were set to achieve desired conditions

Smith & Burgess
Process Safety Consulting

DYNAMIC RELIEF
PROCESS VARIABLES EFFECTS

METHODOLOGY [CONT.]

5. Program ran in dynamics until the system was stable
6. A PRV of adequate size was added to the vapor overhead
7. Boilup was obtained by closing product, feed, and cooling streams
8. Relief data was collected using stripcharts

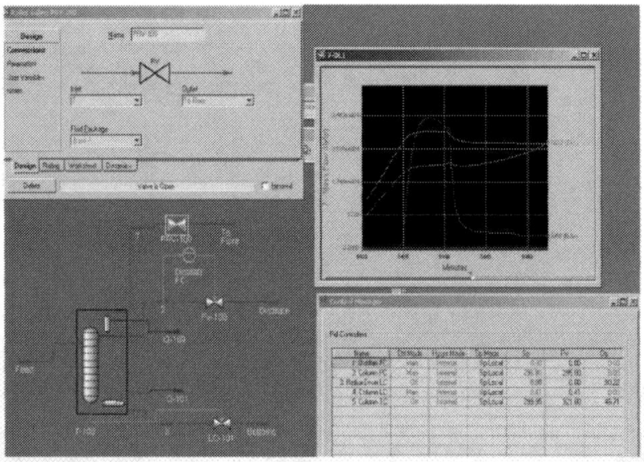

Collecting data for the depropanizer

Smith & Burgess
Process Safety Consulting

DYNAMIC RELIEF
PROCESS VARIABLES EFFECTS

METHODOLOGY [CONT.]

- Sensitivity was quantified by calculating the peak ratio (PR) of the maximum and minimum curves.

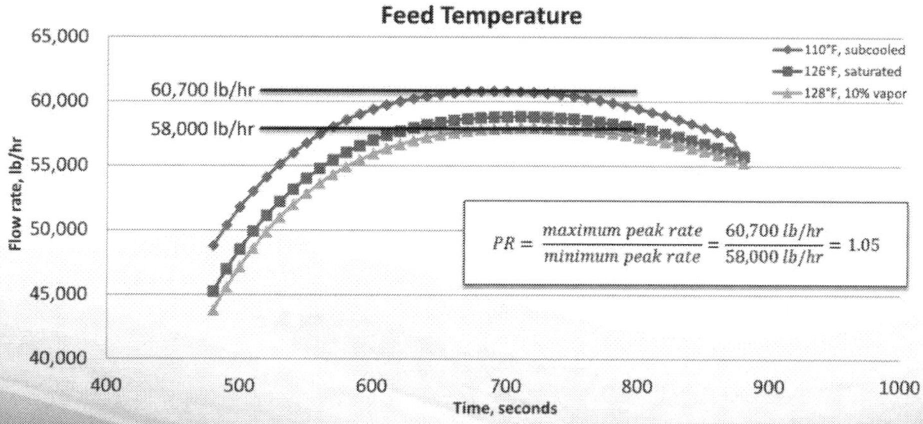

$$PR = \frac{maximum\ peak\ rate}{minimum\ peak\ rate} = \frac{60{,}700\ lb/hr}{58{,}000\ lb/hr} = 1.05$$

Smith & Burgess
Process Safety Consulting

DYNAMIC RELIEF
PROCESS VARIABLES EFFECTS

RESULTS

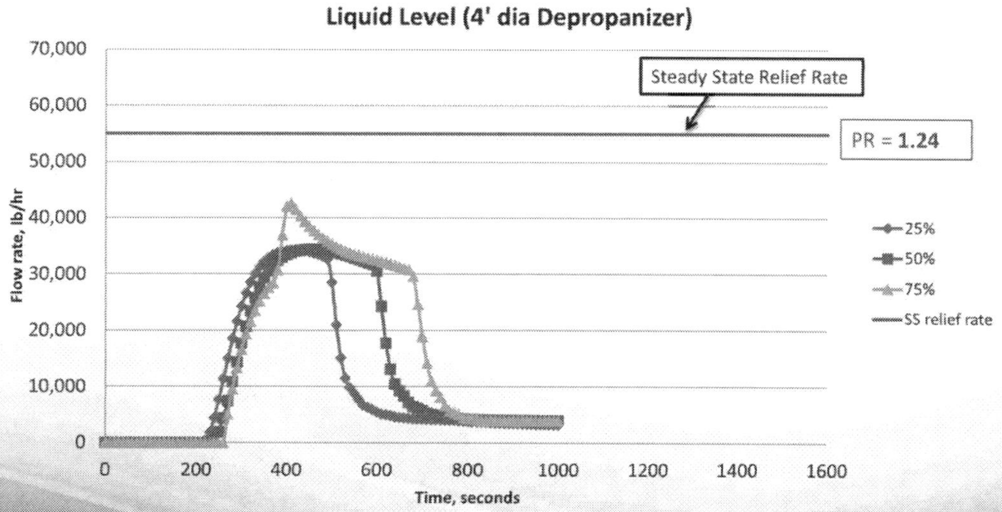

Smith & Burgess
Process Safety Consulting

DYNAMIC RELIEF
PROCESS VARIABLES EFFECTS

RESULTS [CONT.]

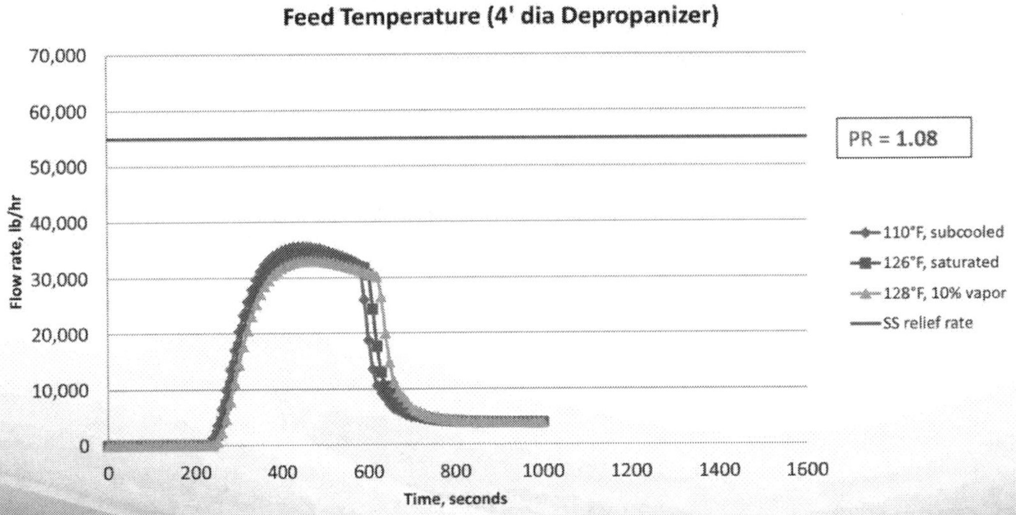

Smith & Burgess
Process Safety Consulting

DYNAMIC RELIEF
PROCESS VARIABLES EFFECTS

RESULTS [CONT.]

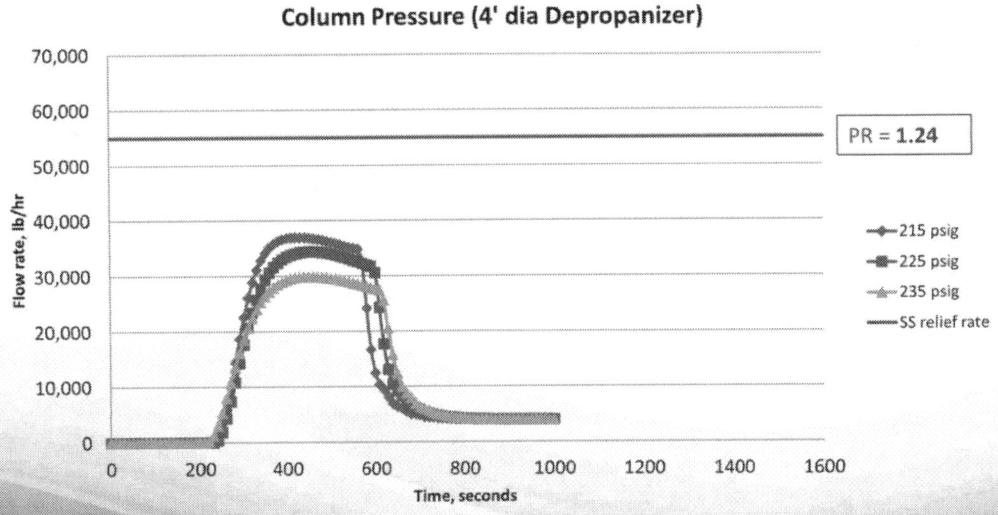

Smith & Burgess
Process Safety Consulting

DYNAMIC RELIEF
PROCESS VARIABLES EFFECTS

RESULTS [CONT.]

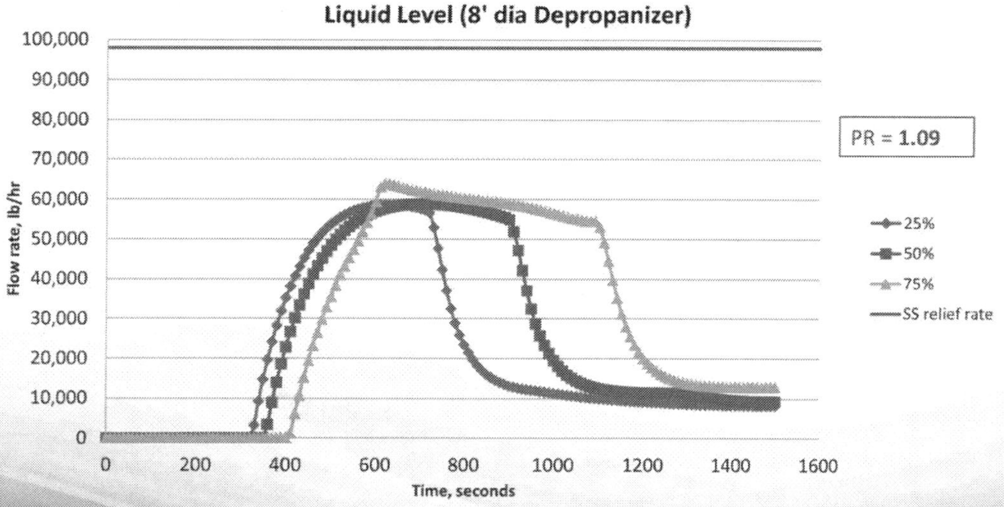

Smith & Burgess
Process Safety Consulting

DYNAMIC RELIEF
PROCESS VARIABLES EFFECTS

RESULTS [CONT.]

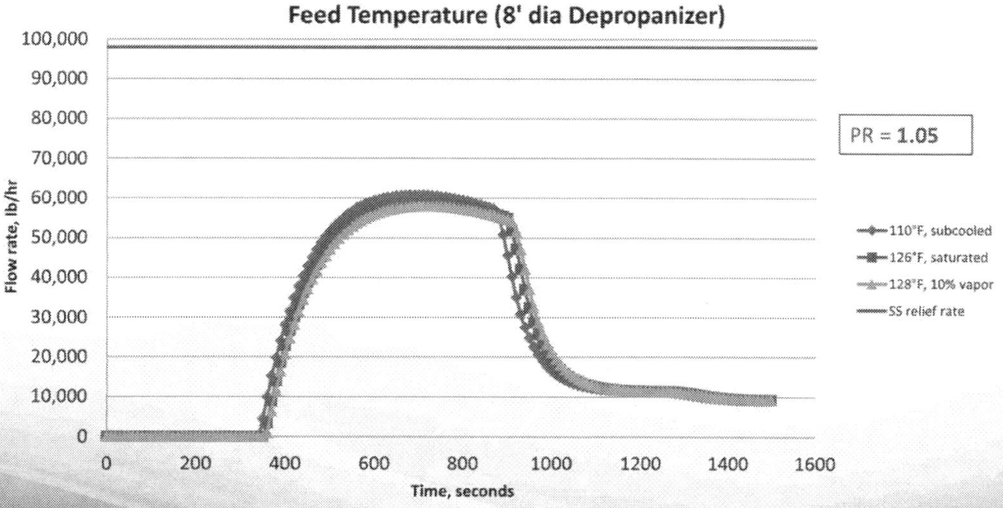

Smith & Burgess
Process Safety Consulting

DYNAMIC RELIEF
PROCESS VARIABLES EFFECTS

RESULTS [CONT.]

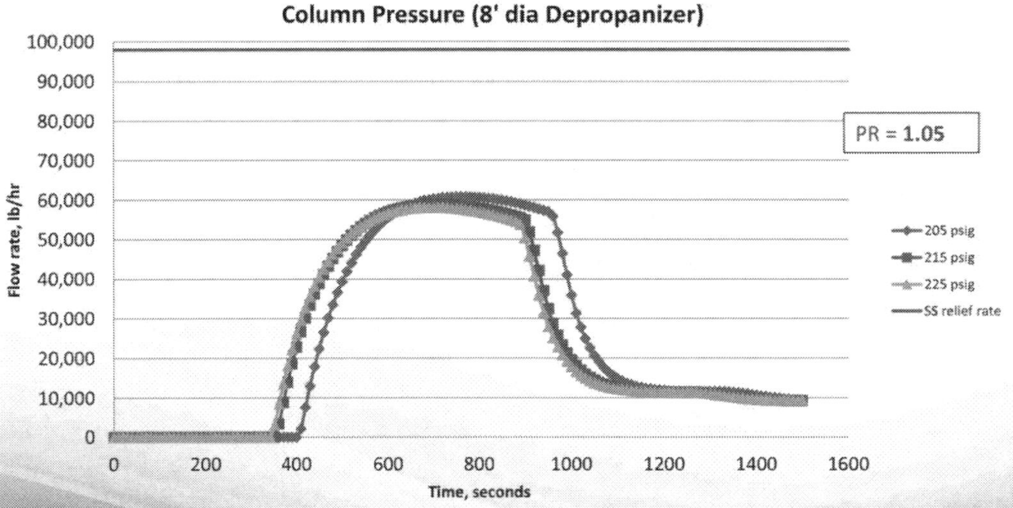

Smith & Burgess
Process Safety Consulting

DYNAMIC RELIEF
PROCESS VARIABLES EFFECTS

RESULTS [CONT.]

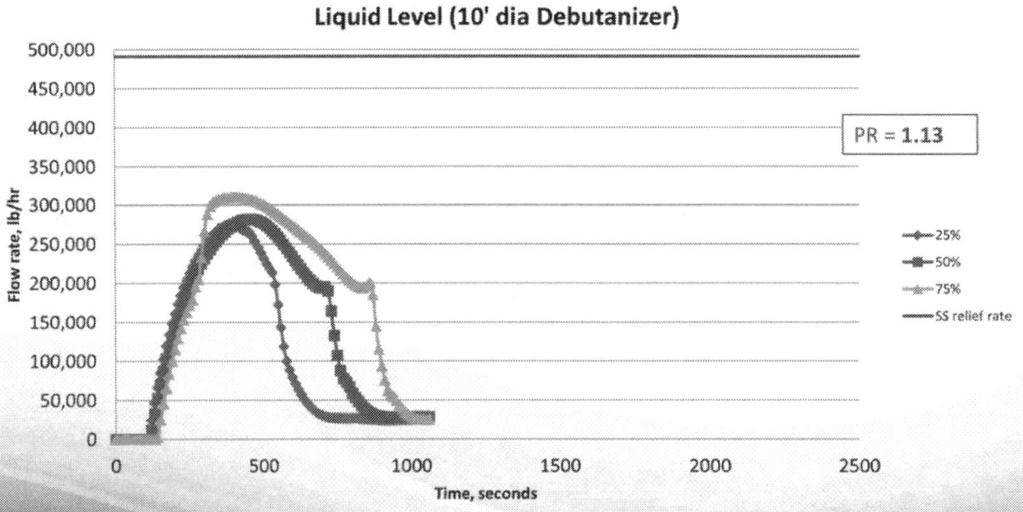

Smith & Burgess
Process Safety Consulting

DYNAMIC RELIEF
PROCESS VARIABLES EFFECTS

RESULTS [CONT.]

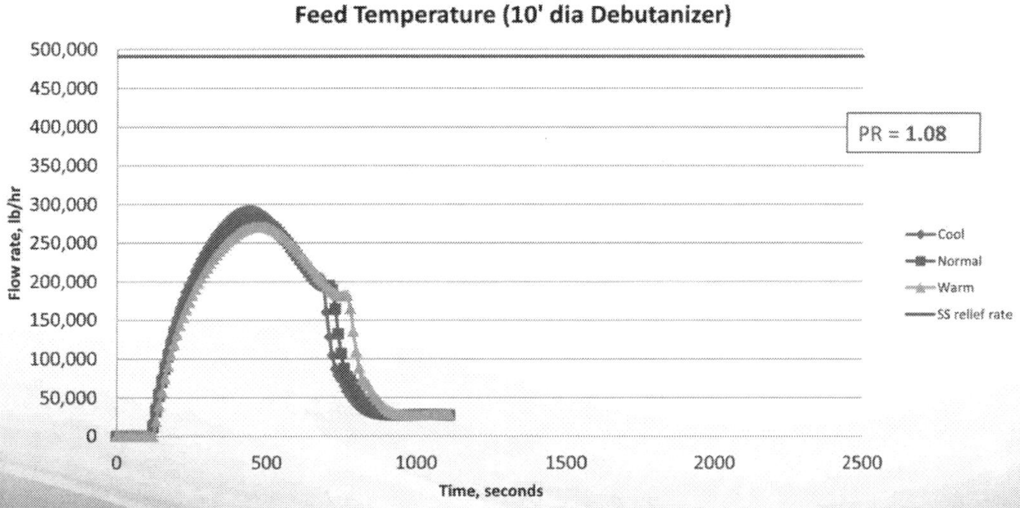

Smith & Burgess
Process Safety Consulting

DYNAMIC RELIEF
PROCESS VARIABLES EFFECTS

RESULTS [CONT.]

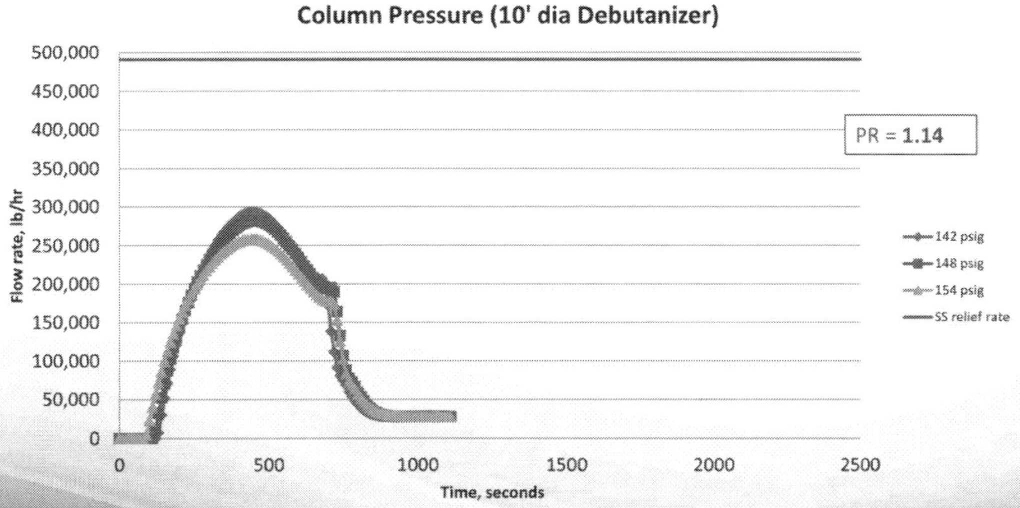

Smith & Burgess
Process Safety Consulting

DYNAMIC RELIEF
PROCESS VARIABLES EFFECTS

RESULTS [CONT.]

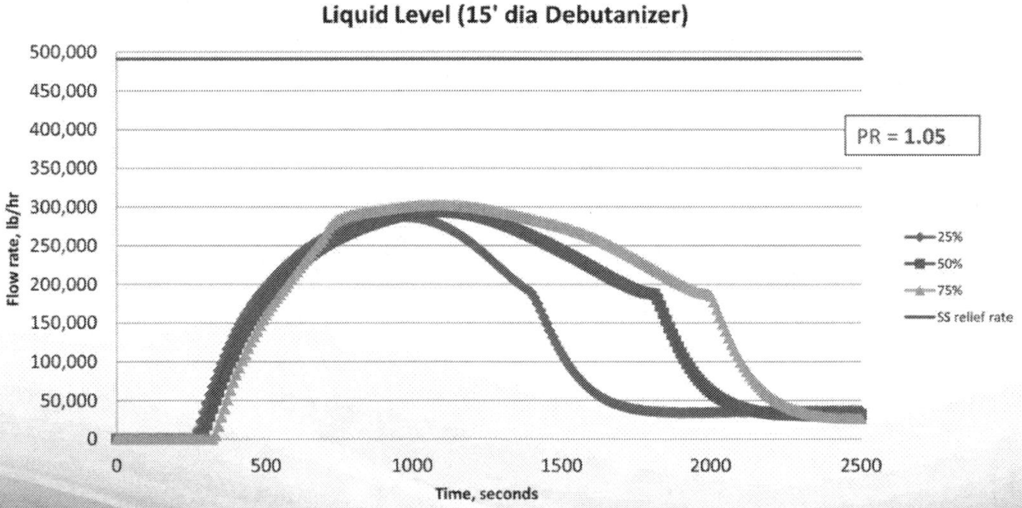

Smith & Burgess
Process Safety Consulting

DYNAMIC RELIEF
PROCESS VARIABLES EFFECTS

RESULTS [CONT.]

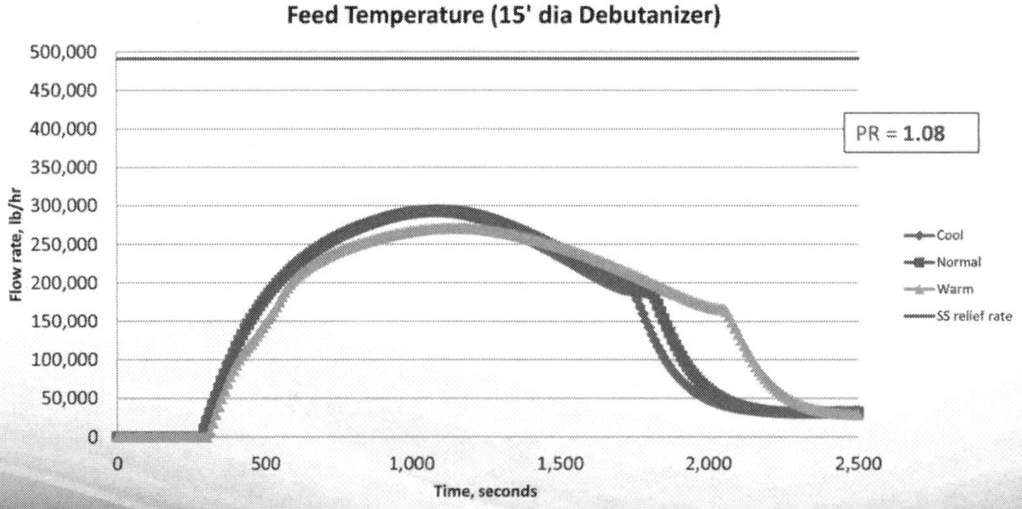

Smith & Burgess
Process Safety Consulting

DYNAMIC RELIEF
PROCESS VARIABLES EFFECTS

RESULTS [CONT.]

Column Pressure (15' dia Debutanizer)

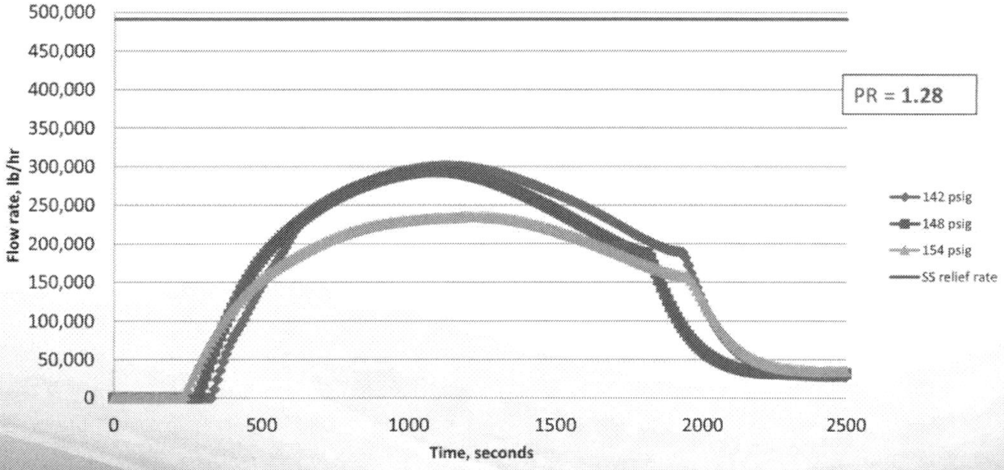

Smith & Burgess
Process Safety Consulting

DYNAMIC RELIEF
PROCESS VARIABLES EFFECTS

CONCLUSIONS

1. Some process variables had a significant effect on the peak flow rate.
   - Liquid Level (highest PR of 1.24)
   - Column Pressure (highest PR of 1.28)
2. Feed temperature had a minimal effect.
3. Larger columns seemed to be less sensitive.
4. Columns with more narrow boiling ranges appear to be more sensitive.
5. Process variables also affected the time to reach peak relief and how long the PRV stays open.

Smith & Burgess
Process Safety Consulting

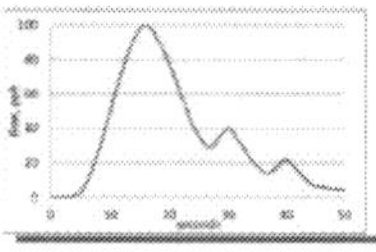

DYNAMIC RELIEF
PROCESS VARIABLES EFFECTS

RECOMMENDATIONS

1. A sensitivity analysis involving process conditions should be performed for pressure relief valves that are sized via dynamic simulation.
2. The sensitivity analysis should be as complete as possible.
3. This analysis could be used for sizing undersized relief devices.

Smith & Burgess
Process Safety Consulting

Superior Performance by Design

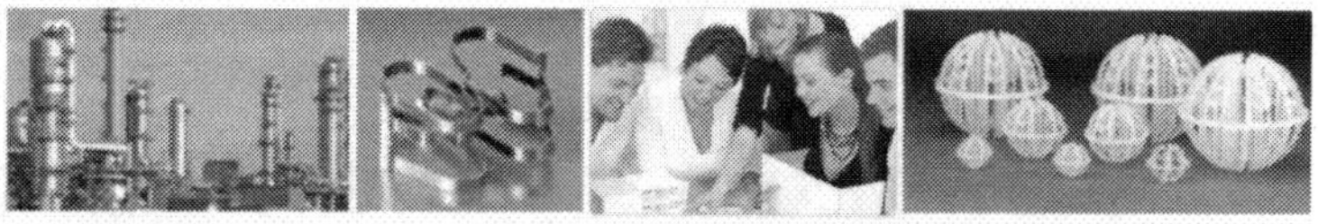

PRODUCTS & SERVICES | DISTRIBUTOR TESTING | APPLICATIONS | DOWNLOADS | ABOUT US | CONTACT US | HOME

Packing Trays & Column Internals

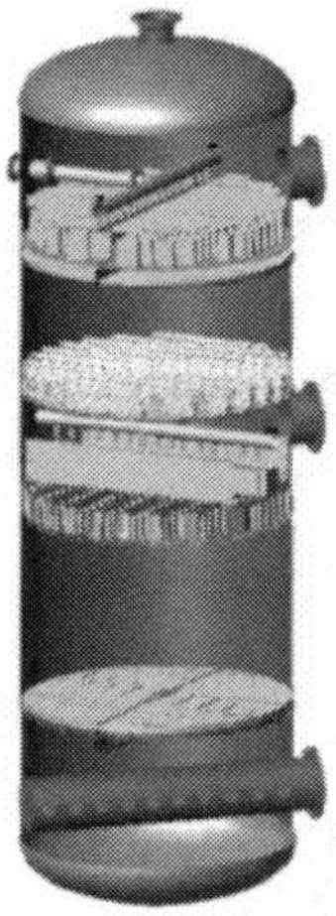

Raschig Jaeger Technologies

Superior Performance by Design

M. Schultes

Agenda

1. **Raschig Super-Pak (New innovative Packing Structure)**
2. **Liquid Distributor Test Center (Germany)**
3. **Raschig Super-Ring (Recent Tests and Application)**

Raschig Jaeger Technologies Structured Packing Portfolio

Raschig Super-Pak

MAX Pak

Raschig-Pak

Raschig Super-Pak

9 sizes available

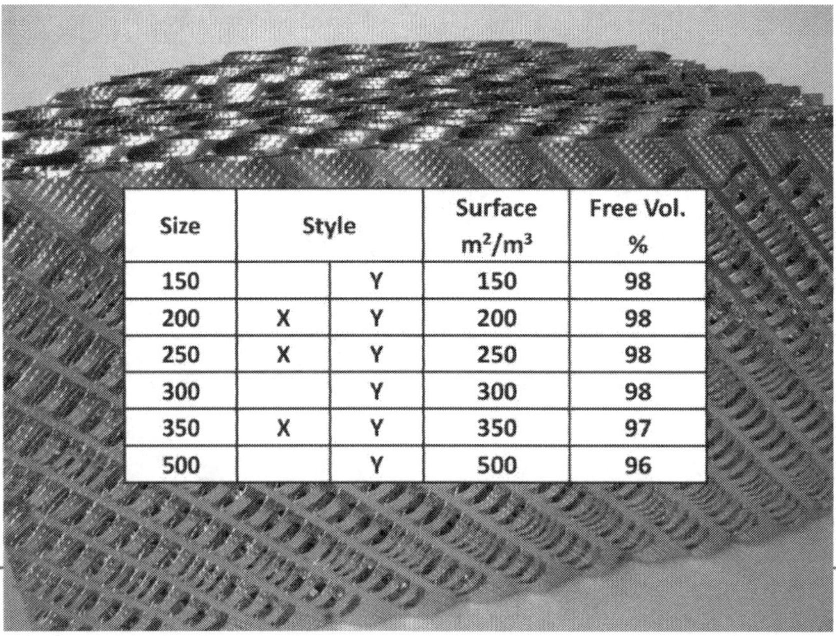

Size	Style		Surface m^2/m^3	Free Vol. %
150		Y	150	98
200	X	Y	200	98
250	X	Y	250	98
300		Y	300	98
350	X	Y	350	97
500		Y	500	96

FRI

Results for

Raschig Super-Pak ®

C6/C7-system at 1.62 bar (24 psia)

FRI Test (2009)

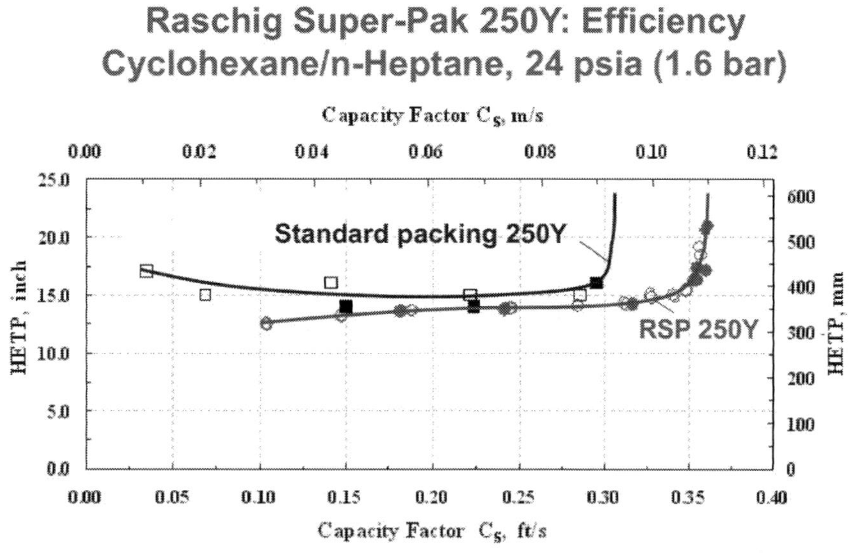

FRI Test (2009)

RASCHIG JAEGER TECHNOLOGIES

Raschig Super-Pak 250Y: Pressure Drop
Cyclohexane/n-Heptane, 24 psia (1.6 bar)

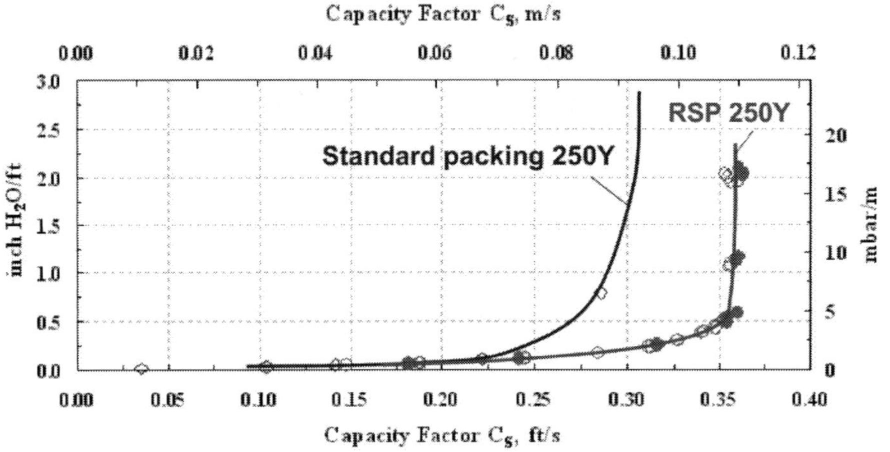

RASCHIG JAEGER TECHNOLOGIES

SRP

comparison of RSP 150Y, 200Y, 200X p = 24 psia (1.6 bar)

SRP test of RSP 150Y, 200Y, 200X

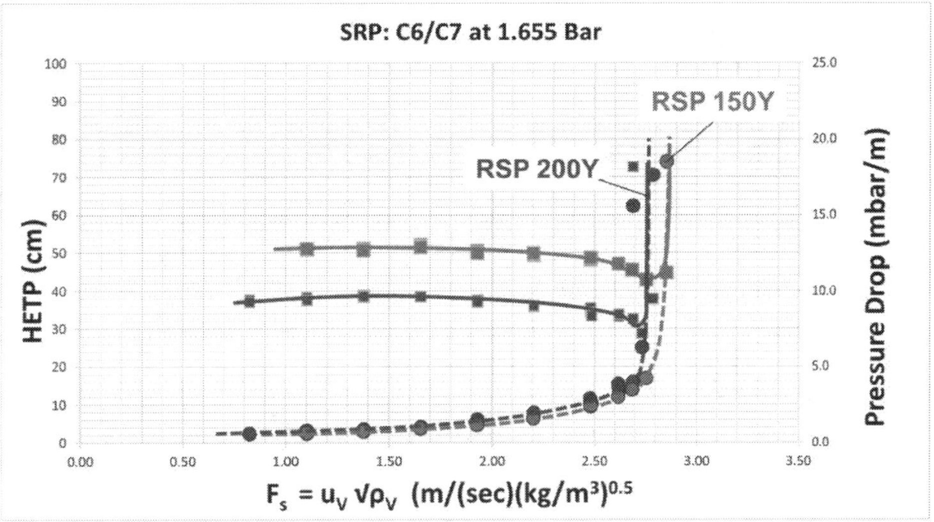

SRP test of RSP 150Y, 200Y, 200X

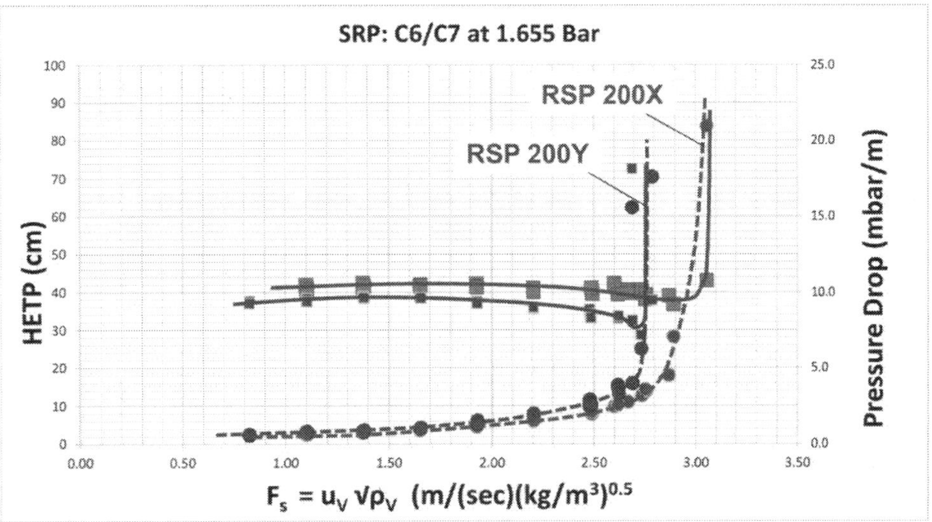

Raschig Super-Pak®

Grasroot Project: Rectification Tower in USA

Rectification Section:

2 Beds Raschig Super-Pak 250Y

D = 10 ft (3 m) ; H = 2 x 18.5 ft (5.64 m)

Stripping Section: 1-pass valve trays

Column provided excellent performance in regards to efficiency, pressure drop and capacity

Raschig Super-Pak®

Revamp of MDEA-Absorber in Europe

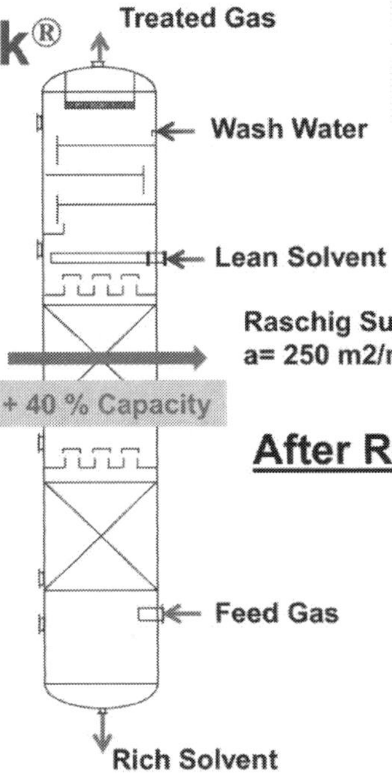

D = 1600 mm = 5ft
H = 2 x 7 m = 2 x 23 ft

Standard Structured Packing
a = 250Y m2/m3

Raschig Super-Pak 250Y
a= 250 m2/m3

Before Revamp

After Revamp

p = 28 bar = 410 psia

Target:
Increase capacity

Raschig Super-Pak®

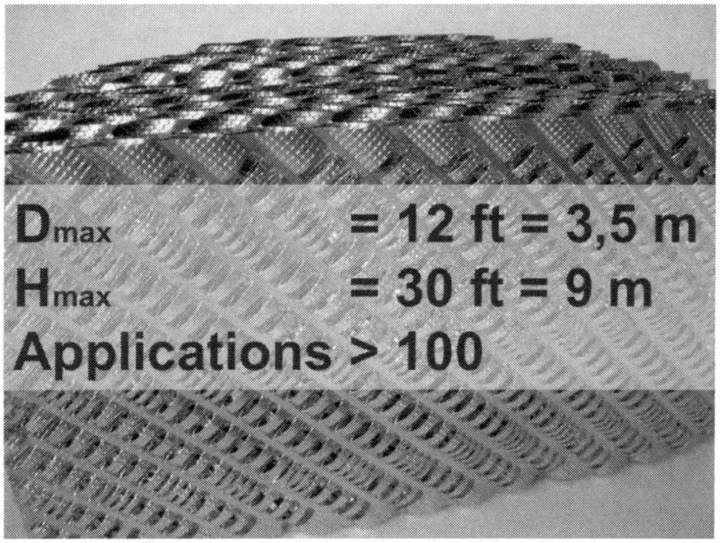

Liquid Distributor Test Facility

D = 23 ft = 6.9 m

Liquid Distributor Test Facility

D = 23 ft = 6.9 m

Raschig Super-Ring®

Flash Gallery with Distributor

D = 20 ft = 6.0m

Liquid Distributor Test Facility

D = 25 ft = 7.7m; L = 12,300 gpm = 2,800 m3/h

Raschig Jaeger
Random Packing Portfolio

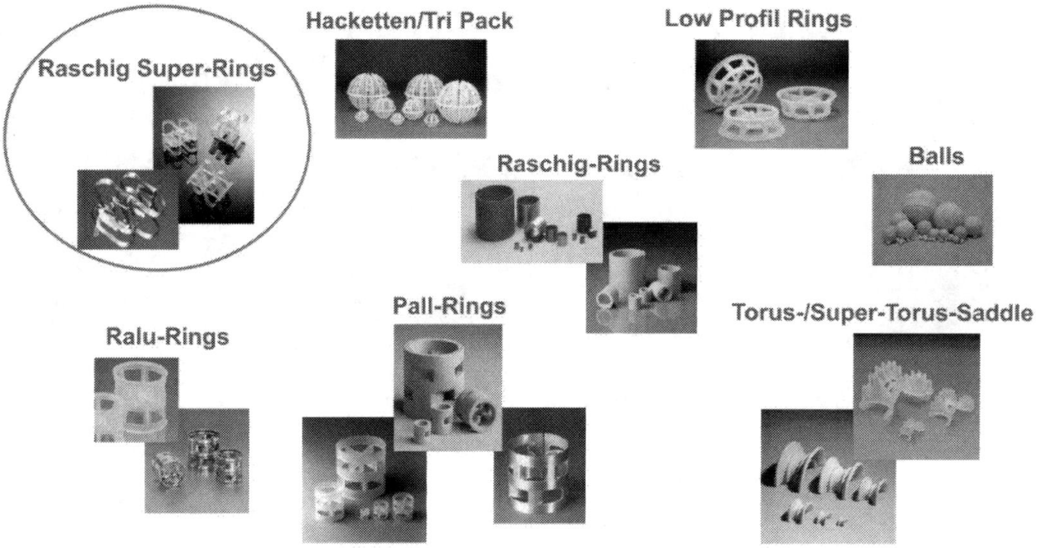

Raschig Super-Ring®

9 sizes available

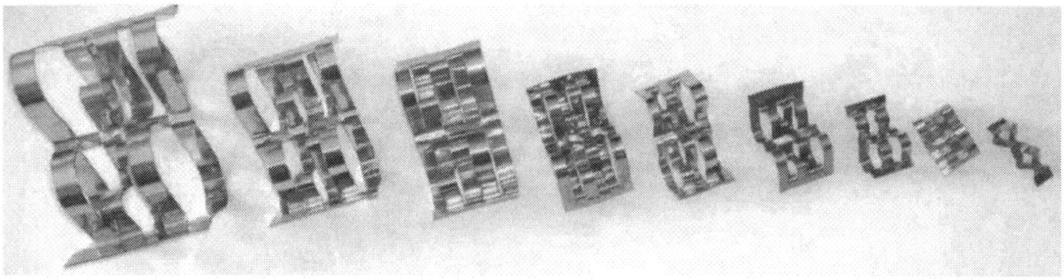

3.0 # 2.0 # 1.5 # 1.0 # 0.7 # 0.6 # 0.5 # 0.3 # 0.1

Raschig Super-Ring® No. 0.7

FRI Test Data and Comparison

Test systems:

-C6/C7 at 4.5 psia (0.31 bar)

-C6/C7 at 23.5 psia (1.62 bar)

-o/p-xylene at 14.7 psia (1 bar)

FRI Comparison Data

25, 50, 90mm Pall Rings and RSR #0.7

- 1" Pall Ring : a = 207 m^2/m^3
- 2" Pall Ring : a = 102 m^2/m^3
- 3.5" Pall Ring : a = 56 m^2/m^3
- RSR #0.7 : a = 175 m^2/m^3

FRI test

Efficiency
Cyclohexane/n-Heptane, 24 psia (1.62 bar)

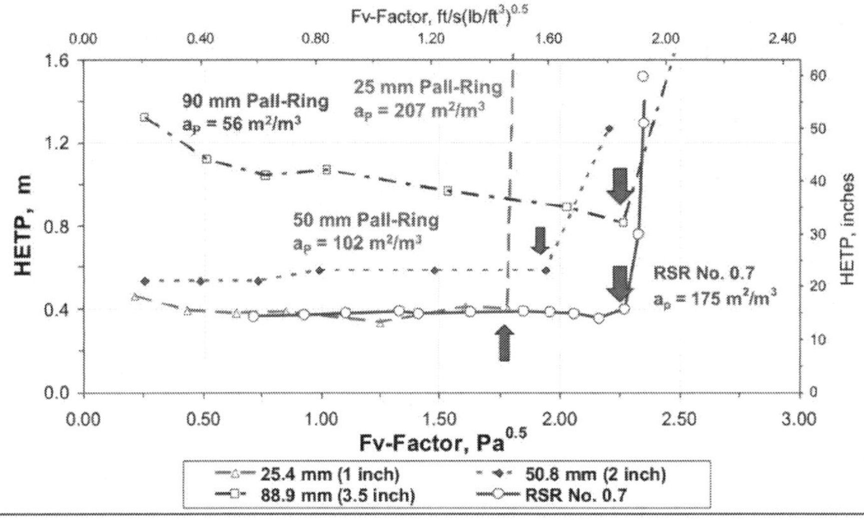

FRI test

Pressure Drop
Cyclohexane/n-Heptane, 24 psia (1.62 bar)

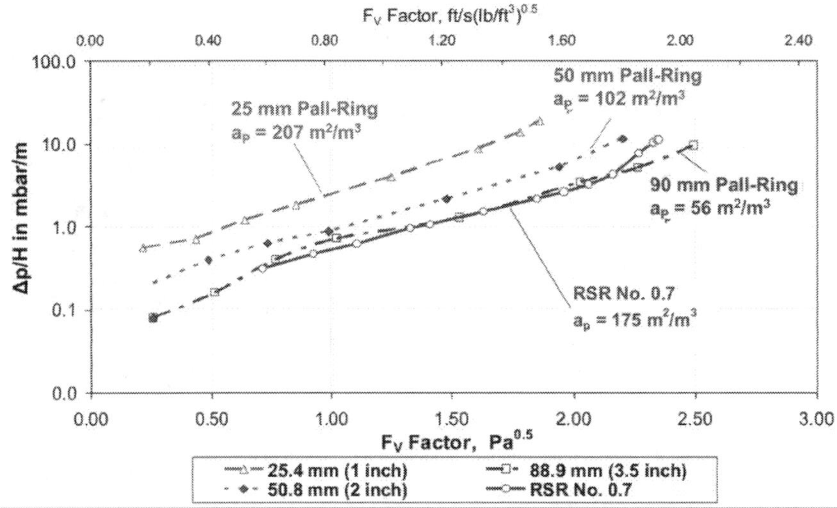

Raschig Super-Ring®

New Applications

Offshore – BOG Recondenser

Floating Regasification Unit

Raschig Super-Ring®

New Applications

Offshore - BOG Recondenser

Floating Regasification Unit

Raschig Super-Ring

Offshore BOG Recondenser

Floating Storage and Regasification Unit

Raschig Super-Ring®

Revamp of Ethylene Oxide Plant in Germany (Catacarb – Process)

- 2 x CO_2 Absorber
Diameter: Ø = 8.5 ft = 2,6m
Ø = 11.0 ft = 3,4m
2 pass valve trays → Raschig Super Ring#1.5

- 2 x CO_2 Regenerator
Diameter: Ø = 12.0 ft = 3,7m
2 pass valve trays → Raschig Super Ring#0.3

Targets/Results:
- Significant capacity increase
- No sensitivity to fouling any more
- Same separation efficiency

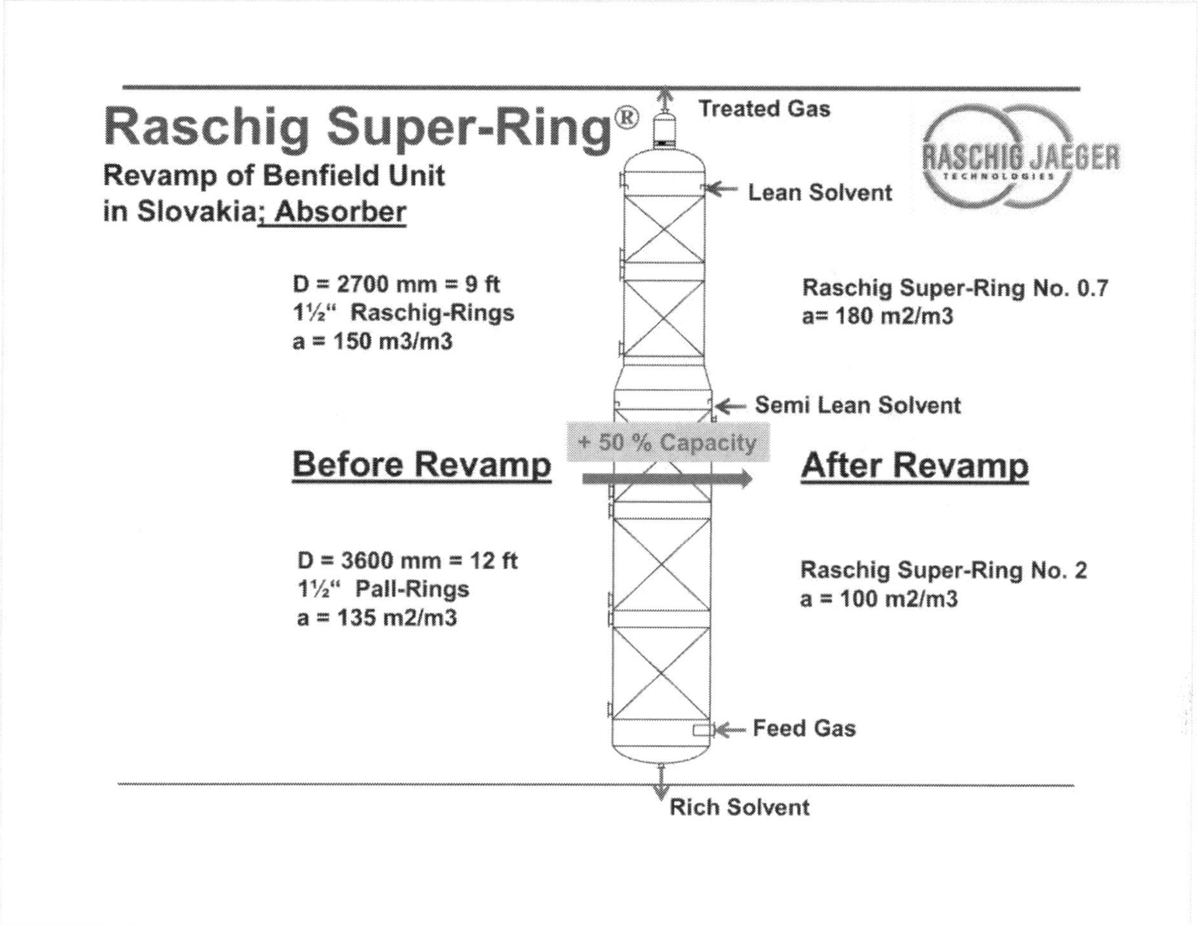
Raschig Super-Ring®
Revamp of Benfield Unit
in Slovakia; Absorber
Treated Gas
RASCHIG JAEGER
TECHNOLOGIES
Lean Solvent
D = 2700 mm = 9 ft
1½" Raschig-Rings
a = 150 m3/m3
Raschig Super-Ring No. 0.7
a= 180 m2/m3
Semi Lean Solvent
+ 50 % Capacity
Before Revamp
After Revamp
D = 3600 mm = 12 ft
1½" Pall-Rings
a = 135 m2/m3
Raschig Super-Ring No. 2
a = 100 m2/m3
Feed Gas
Rich Solvent

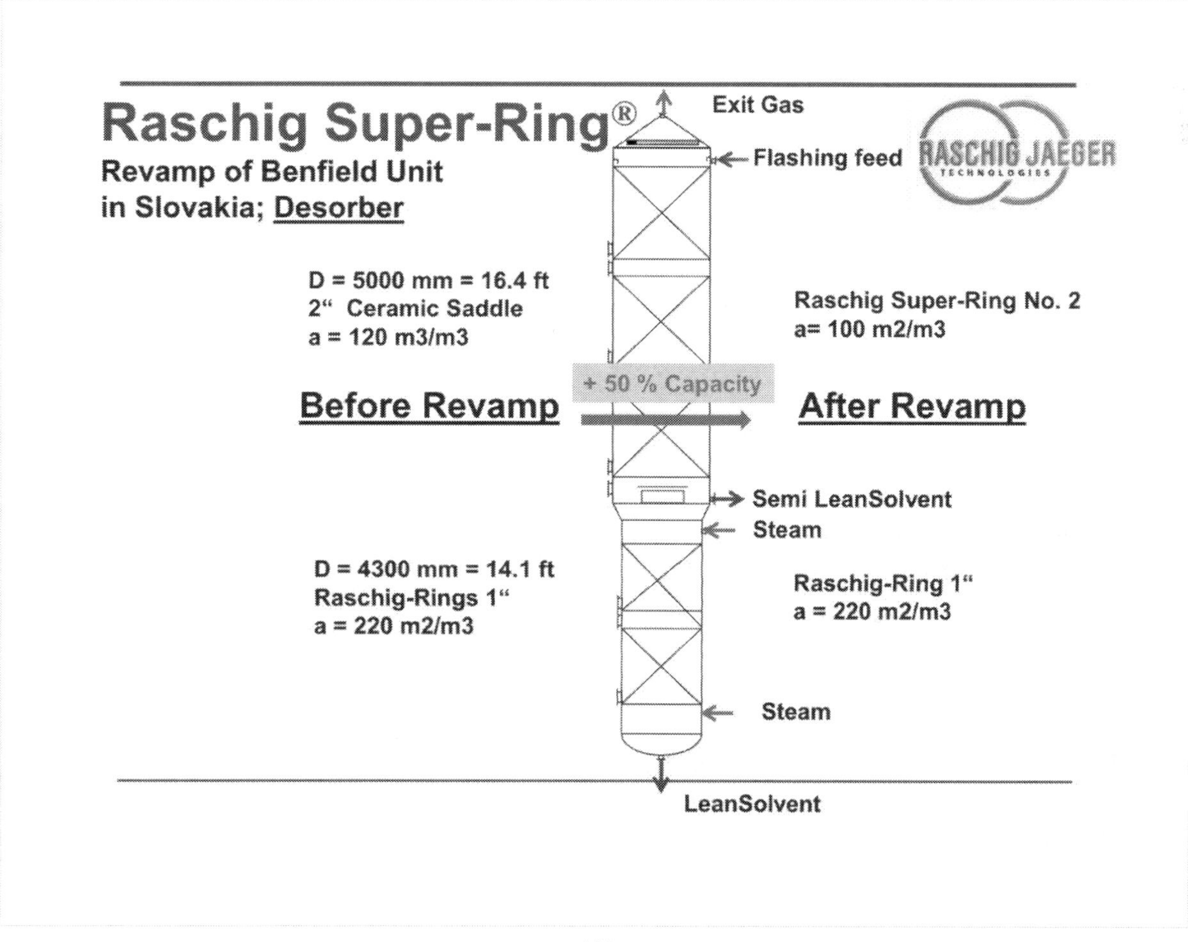
Raschig Super-Ring®
Revamp of Benfield Unit
in Slovakia; Desorber
Exit Gas
Flashing feed
RASCHIG JAEGER
TECHNOLOGIES
D = 5000 mm = 16.4 ft
2" Ceramic Saddle
a = 120 m3/m3
Raschig Super-Ring No. 2
a= 100 m2/m3
+ 50 % Capacity
Before Revamp
After Revamp
Semi LeanSolvent
Steam
D = 4300 mm = 14.1 ft
Raschig-Rings 1"
a = 220 m2/m3
Raschig-Ring 1"
a = 220 m2/m3
Steam
LeanSolvent

Raschig Super-Ring®

Revamp of Benfield Unit in Slovakia

-> Raschig Super-Ring provides significant more capacity
-> Raschig Super-Ring provides more efficiency
-> Raschig Super-Ring creates significant less pressure drop

Raschig Super-Ring®

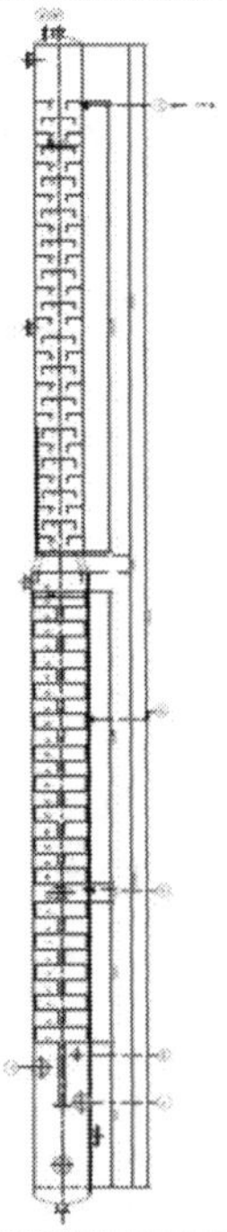

Revamp of Aromatic Rectification in Germany

2-Pass Valve Trays to Raschig Super-Ring No. 0.1

Target: Efficiency increase

New: Raschig Super-Ring® No. 0.1

Smallest Raschig Super-Ring for Miniplant and Industrial Scale:

Specific surface: 450 m2/m3
Void fraction: 95%

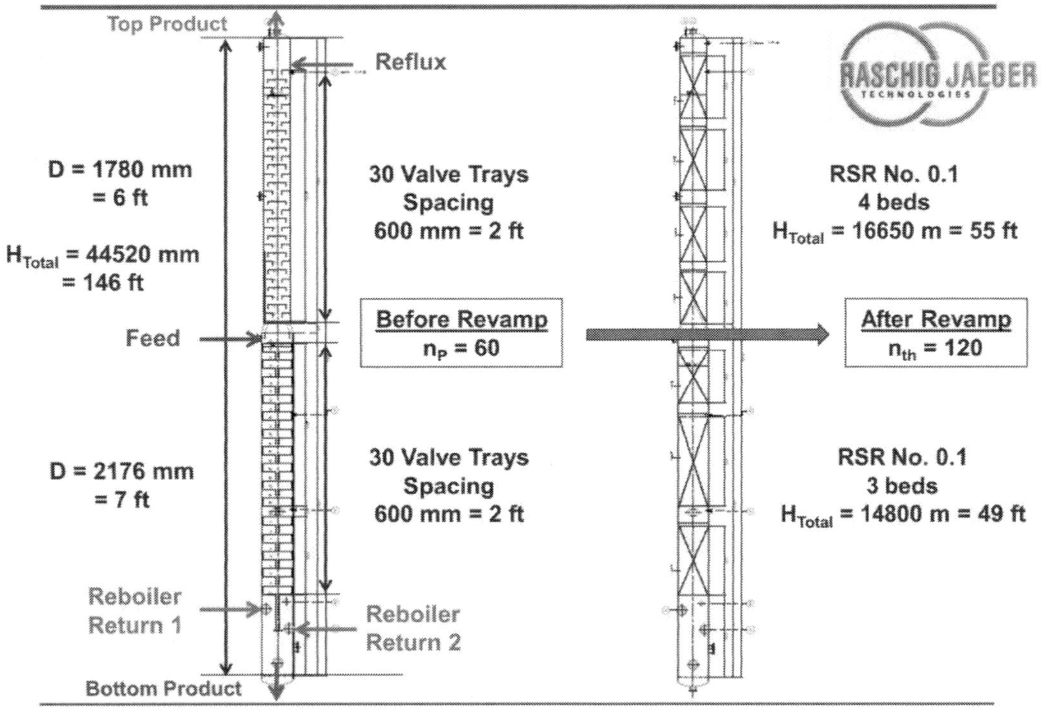

Raschig Super-Ring®

RASCHIG JAEGER TECHNOLOGIES

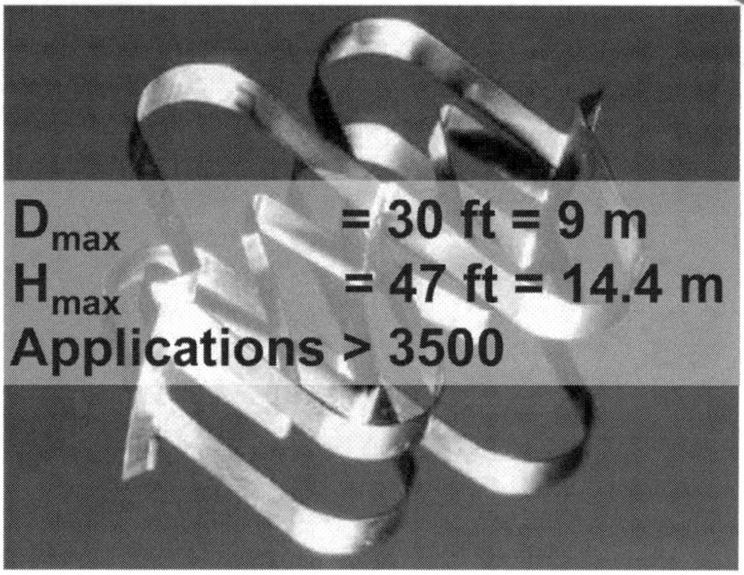

Superior Performance by Design

RASCHIG JAEGER TECHNOLOGIES

PRODUCTS & SERVICES | DISTRIBUTOR TESTING | APPLICATIONS | DOWNLOADS | ABOUT US | CONTACT US | HOME

Packing Trays & Column Internals

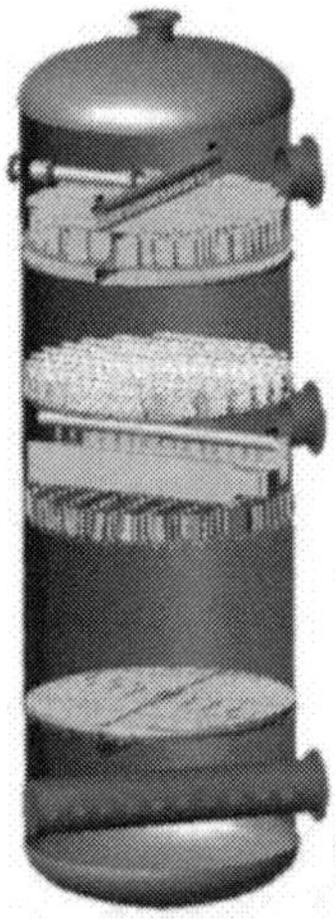

Raschig-Jaeger Technologies

Superoir Performance by Design

Any question?

Tower Scanning as a Quantitative Tool

Pless, L., Tracerco
Nieuwoudt, I., Koch-Glitsch

Originally presented on: 10/4/2012 12:50:00 - 13:15:00

Tower scanning is well established in the process industries as a qualitative tool to help troubleshoot separation towers. In this presentation it will be shown that through careful and innovative application of gamma scanning and neutron backscatter techniques, quantitative information about tray performance can be extracted. This will be illustrated by comparing data taken under very controlled conditions on a hydrocarbon tower equipped with windows, with the data obtained from scans of the tray downcomer and active area. Three sets of trays designed to flood under jet flood, downcomer backup or choke flood mechanisms were tested and studied. The experimental observations, tray model predictions, and scan data for each of these sets of trays will be compared. Based on this information it can be concluded that tower scanning can be used to generate quantitative information about tray performance.

Tray Capacity and Low Surface Tensions

Summers, D. R., Sulzer Chemtech USA, Inc.
Originally presented on: 10/4/2012 13:15:00 - 13:40:00

Tray vapor capacity correlations today do not include a surface tension physical property parameter. Low liquid surface tensions in a distillation column can lead to tiny droplet sizes, high liquid entrainment and limited vapor capacity. Several authors have examined surface tension in the past but today's leading vapor capacity correlations have omitted this important physical property parameter. This paper will examine very low surface tension data and then present a new tray vapor capacity correlation that uses surface tension as one of the primary correlating parameters. The intent of this paper is to show that tray designs of very high pressure trayed distillation towers, such as demethanizers and deethanizers, can be made without the need for "system factors" with this new correlation. It is the author's opinion that the system factor should be used strictly for those tray applications that are limited in capacity due to foaming and not be used as a "fudge factor" to account for capacity correlation deficiencies.

Improving packed towers performance: Severe service grid & Random packing

Nieuwoudt, I., Koch-Glitsch
Quotson, P., Koch-Glitsch, LP

Originally presented on: 10/4/2012 13:40:00 - 14:05:00

FLEXIGRID® severe service grid has been used extensively in challenging applications such as vacuum tower wash beds, FCC main fractionators and ethylene quench towers. In these applications there is a continuous push for more capacity, improved fouling resistance and improved droplet removal. This led to the development of the PROFLUX® severe service grid. The PROFLUX® grid delivers capacities comparable to that of structured packing, good droplet removal, and better fouling resistance than conventional grid products. During this presentation experimental data on this new product will be shared.INTALOX® ULTRA™ random packing was introduced to the market as a packing with a combination of efficiency and capacity that exceeds that of existing random packing product lines. The packing performance was verified in testing at F.R.I. INTALOX® ULTRA™ random packing has been installed in a large number of light hydrocarbon fractionators in gas fields. Data from industrial case studies will be presented.

Performance data for Shell HiFi high capacity trays

DeVilliers, W., Shell Global Solutions Inc.
Wilkinson, P., Shell Global Solutions
Vos, E., Shell Global Solutions
Originally presented on: 10/4/2012 14:05:00 - 14:30:00

Shell HiFi high capacity trays have recently been tested by an independent research consortium. This presentation will present results obtained at 12" and 24" tray spacing using the iso-butane/n-butane test system at 165, 300 and 400 psia system pressures. An overview of the main operating characteristics of HiFi trays will also be presented.

Asphaltene Flow Assurance: Modeling Phase Behavior and Deposition in Crude Oil Systems

Chapman, W. G., Rice University
Panuganti, S. R., Rice University
Chen, Y., Nanjing University
Al Hammadi, A.,
Punnapala, S., The Petroleum Institute
Vargas, F. M., The Petroleum Institute
Originally presented on: 10/4/2012 12:00:00 - 12:25:00

Simulation of flow assurance issues in the energy industry requires accurate modeling of fluid properties and phase behavior as well as deposition rate. This requires tuning of an appropriate equation of state and transport model to laboratory data. Successes in tuning the PC-SAFT equation of state to model asphaltene phase behavior are discussed. Predictions related to live oil depressurization, gas injection and oil based mud contamination are presented. Further, an asphaltene deposition tool (ADEPT) to predict deposition flux is described with an example application.

Nanoparticle Transport and Interfacial Behavior in EP Operations

Krishnamoorti, R., University of Houston

Originally presented on: 10/4/2012 12:25:00 - 12:50:00

Nanoparticles have been shown to be outstanding interfacial agents and in particular nanoparticles have been shown to drive the interfacial tension between oil and water to values below 15 mN/m at concentrations of the order of a few ppm. We have developed a comprehensive program to examine and understand the ability of nanoparticles to act as interfacial agent. Specifically, we have attempted to understand the role of particle chemistry (metal, metal oxide, inorganic, carbon) along with surface functionalization chemistry and that of the particle size and to a limited extent shape on the interfacial tension reduction achieved by such nanoparticles. Additionally, to understand the origins of these interfacial properties we are also using grazing incidence x-ray scattering from oil (alkane) water interfaces with particles with tailored surface properties to examine the interfacial structuring of the particles. Recent studies using "Janus" spherical gold nanoparticles (diameter 10 nm) with a hydrophobic polymer on one side and a hydrophilic polymer on the other have shown a significant enrichment of the surface concentration with the incipient formation of hexagonally ordered domains at the surface coinciding with the saturation of the interfacial tension reduction caused by the particles.

Asset Integrity - Risk Assessment with Chemical Engineering Principles

Alapati, R.,

Originally presented on: 10/4/2012 12:50:00 - 13:15:00

Chemical engineering principles are an absolute necessity in the oil and gas industry to be able to assess the asset integrity risk. This risk unmitigated could cause lost production revenue, environmental damage and threaten the people safety. For this reason, it is of extreme importance to examine, comprehend, design, predict, and mitigate these risks. To achieve this, principles of thermodynamics, heat and mass transfer, fluid dynamics and transport phenomena are utilized to attain simple and practical solutions to mitigate asset Integrity risk.

Thermodynamic principles are utilized in determining the Phase Behavior of fluids, which will predict the presence of water, a necessary ingredient for acid gas corrosion.. Heat and mass transfer are equally important concepts required to calculate the temperature and pressure profiles across the production system, which will play a role in the severity of asset integrity risk. Moreover, chemical engineers study fluid dynamics to understand the distribution and characteristics of phases present at any given point in the system. These parameters aid in the prediction of asset integrity related issues.

The combination of chemical engineering principles together with mathematics, chemistry, and physics has led to the development of many simulation models. These are constantly utilized upstream to mitigate asset integrity issues, either through the use of chemical or mechanical solutions. Chemical engineers are an essential part of upstream operations, it is the one profession that studies, analyzes, models and solves asset integrity problems to provide more reliable process system, avoid problematic operational conditions and optimize production.

Panel Discussion: The need to integrate Chemical Engineering along with Petroleum, Mechanical and Civil Engineering to develop/improve Upstream Oil and Gas Technologies and how it is relevant to various aspects of Upstream Engineering and Flow Assurance

Kondapi, P. B., FMC Technologies

Originally presented on: 10/4/2012 13:15:00 - 13:40:00

Panel Discussion:

"The need to integrate Chemical Engineering along with Petroleum, Mechanical and Civil Engineering to develop/improve Upstream Oil and Gas Technologies and how it is relevant to various aspects of Upstream Engineering and Flow Assurance"

Panelists:

- 1.David Jennings, PhD, Baker Hughes
- 2.Juan Luongo, PhD, Genesis Oil and Gas
- 3.Tommy Golczynski, Assured Flow Solutions
- 4.Phaneendra Kondapi, PhD, FMC Technologies
- 5.Norman Carnahan, PhD, Carnahan Corporation
- 6.Ramanan Krishnamoorti, PhD, University of Houston

Role of Chemical Engineers in Design and Maintenance of Deepwater Facilities: A Case Study

Nikhil B. Joshi

MOULINEX BUSINESS SERVICES LLC

1

Myth?

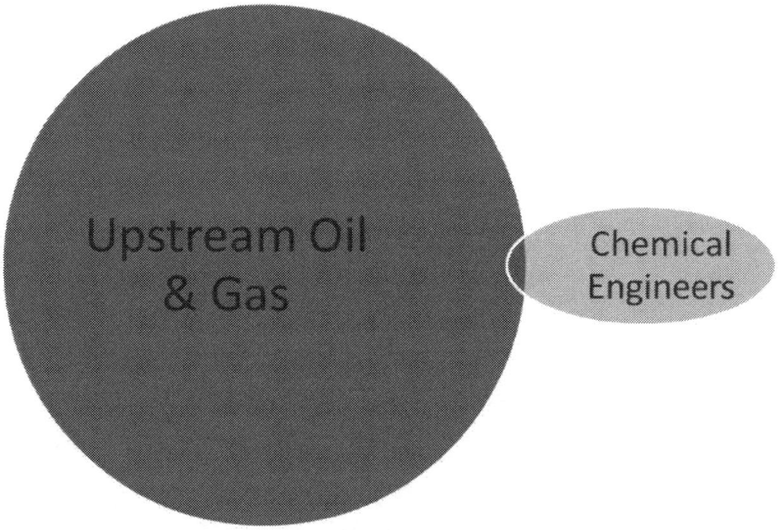

2

CHE: Employment Opportunities

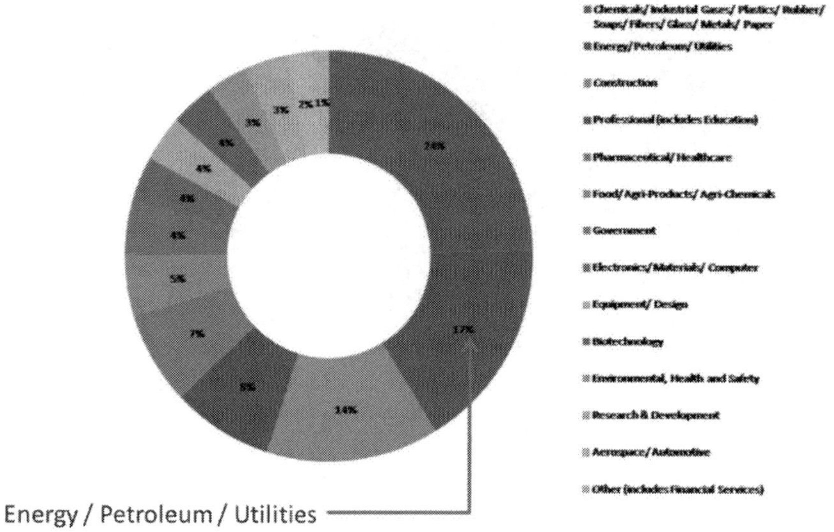

Source: Texas Higher Education Coordinating Board

3

CHE: Employment Opportunities (Bureau of Labor Statistics)

The industries employing the largest numbers of chemical engineers in 2010 were as follows:

Architectural, engineering, and related services	16%
Scientific research and development services	12
Basic chemical manufacturing	12
Resin, synthetic rubber, and artificial synthetic fibers and filaments manufacturing	7
Pharmaceutical and medicine manufacturing	5

No mention of upstream oil and gas industry!

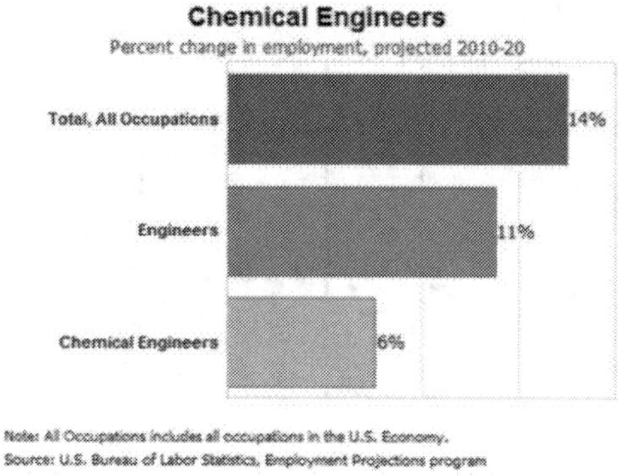

Note: All Occupations includes all occupations in the U.S. Economy.
Source: U.S. Bureau of Labor Statistics, Employment Projections program

4

Domestic Oil Production Growth Forecast

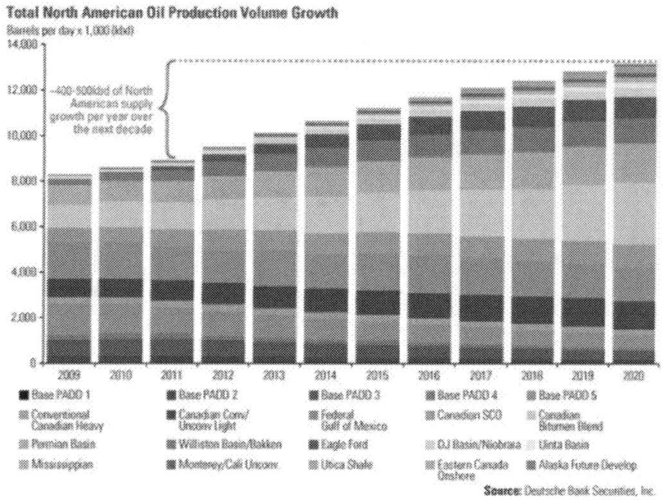

Claim!

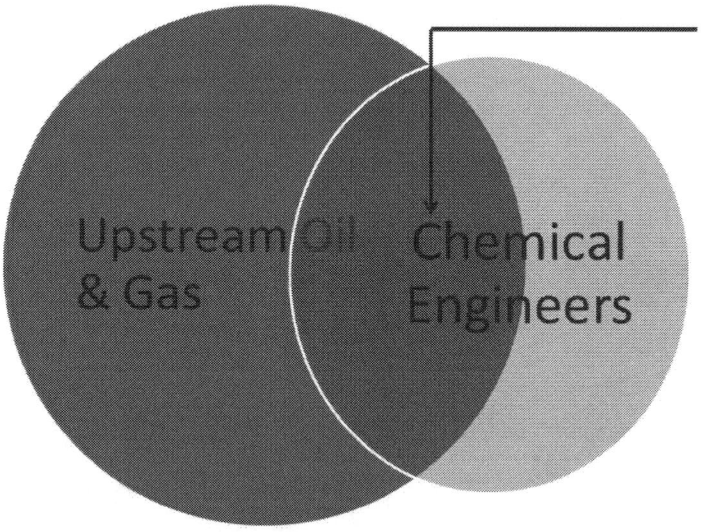

CHE contribute in the following domains:

- Reservoir
- Production Operations
- Completions
- Instrumentation and controls

Chemical Engineers can excel at over 50% of Upstream Oil and Gas Opportunities

CHE Student's Journey

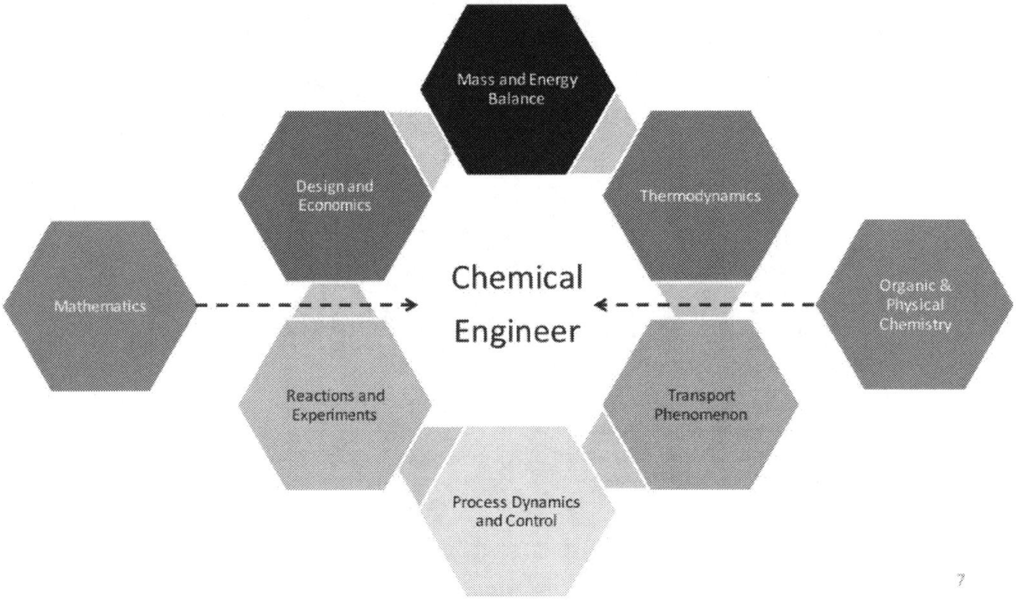

7

Deepwater Project Life Cycle

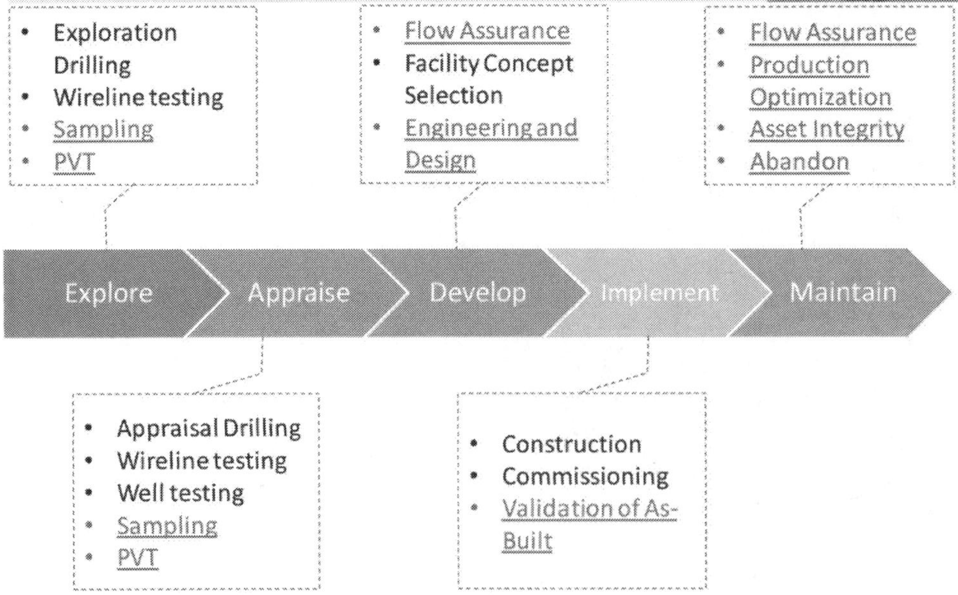

8

Deepwater Project Life Cycle

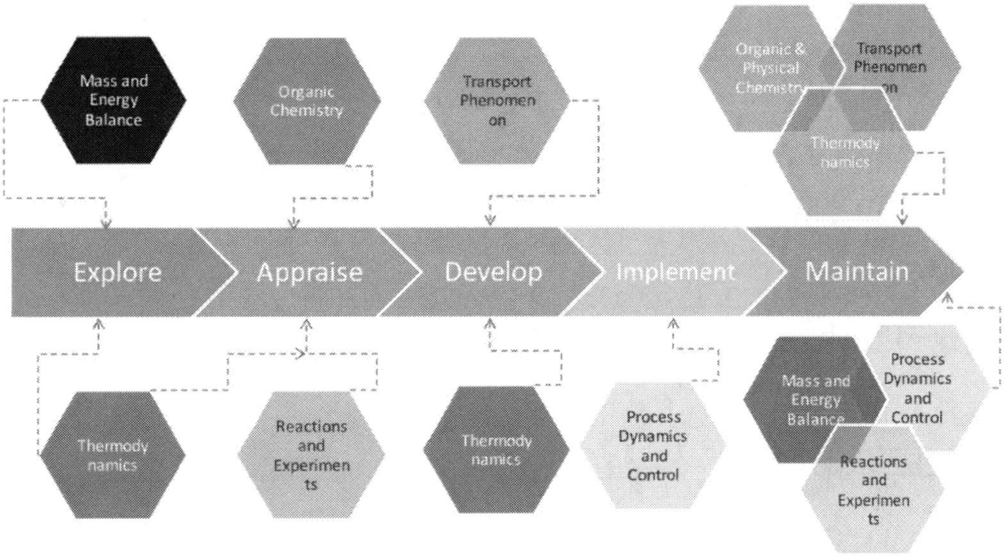

9

CHE Contributions

- Fluid phase behavior
- Reservoir characterization
  - Maximize recovery
- Process Engineering and Design
- Simulation
  - Reservoir simulation
  - Pressure and Rate Transient Analysis
  - Steady state
  - Transient
  - Process plant design
- Compliance
  - Safety and Environment Management Systems (SEMS)

10

Explore & Appraise: Contribution

- Organic Phase Behavior Testing (Devil is in the details)
  – Pressure Volume Temperature tests
  – Solids precipitation test (Hydrate, Paraffin, Asphaltene)
- Water Chemistry Testing
  – pH
  – Scale, Corrosion

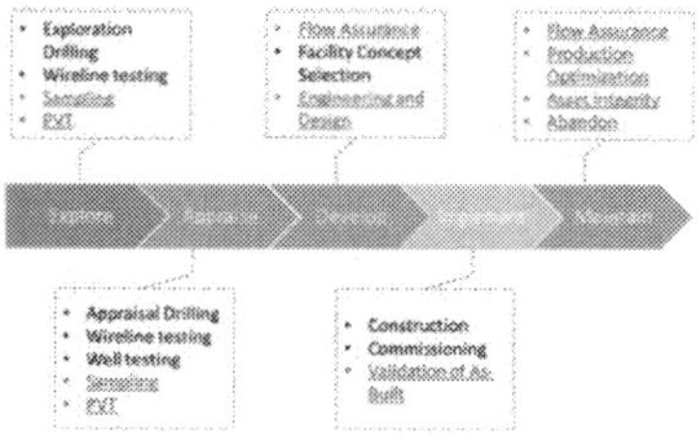

11

Explore & Appraise: Benefits

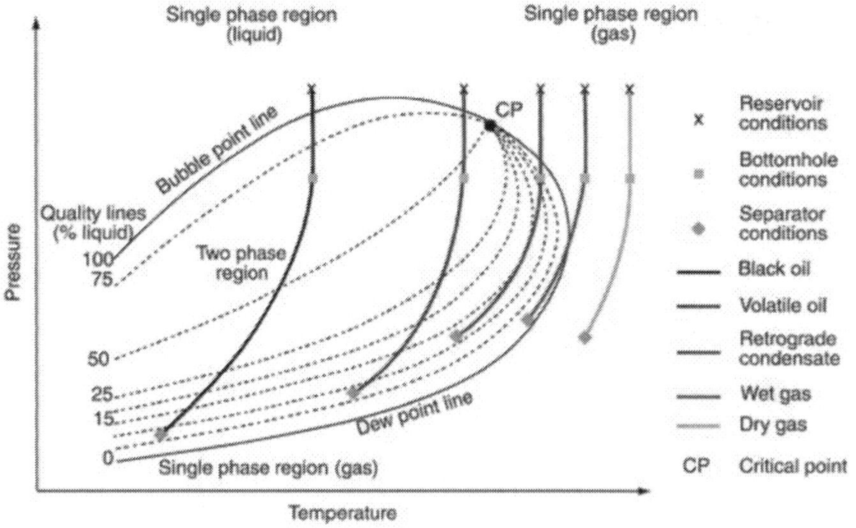

Source: Developments in Petroleum Science

12

Explore & Appraise:

Benefits

- Reserves Estimation
  - Formation volume factor
  - Density
  - Gas-Oil Ratio
- Facility Capacity Design & Well Counts
  - Deliverability of wells (viscosity)
  - Saturation pressure
  - Drawdown (pressure drop across completion)
- Water Flood Considerations
  - Corrosion and Scale
  - Souring

13

Develop & Implement:

Contribution

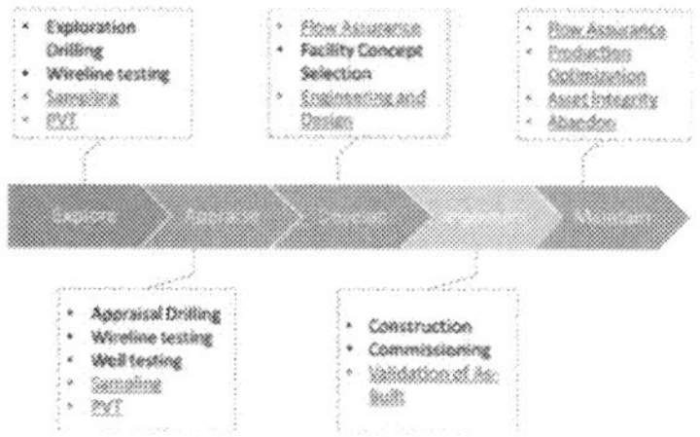

- Equation of state models
- Multi-phase simulations
  - Steady state
  - Dynamic
  - CFD
- Advanced lab testing
  - Define solid phase behavior boundaries
  - Chemical selection to mitigate solids precipitation
- Support control logic requirements (safe charts)
- Operating philosophy and system requirements
  - Consolidate system boundary conditions

14

Develop & Implement: Benefits

- Wellbore Design
  - Use of VIT (vacuum insulated tubing)
  - Location of subsurface safety valve
  - Chemical injection mandrel locations
  - Screen sizing
  - Compatibility of completion fluids with reservoir fluids

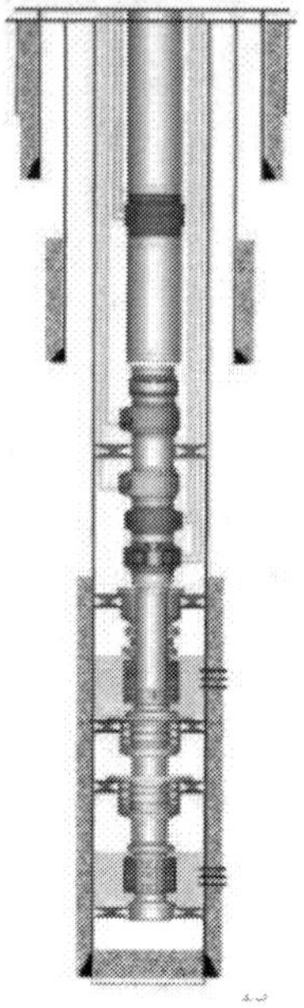

Develop & Implement: Benefits

- Subsea Infrastructure Design
  - Multi-phase flow simulations for pipeline design
  - Insulation requirements to mitigate solids precipitation/deposition
    - Hydrates
    - Paraffin
    - Asphaltene
  - Location and logic of subsea valve operation
  - Chemical injection system requirements
  - Define erosion and corrosion limits and mitigation plans
  - Define pigging requirements

16

Develop & Implement: Benefits

- Topside Facility Design
  - Chemical injection and storage requirements
  - Support processing design
    - Residence time of vessels
    - Separation of phases (gas, oil and aqueous)
    - Support to meet water overboard discharge requirements
    - Slug control design
    - Valve design and timing
  - Planned and un-planned shut-in requirements

17

Develop & Implement: Benefits

- Commissioning Support
  - De-watering of flowlines
  - Control system validation
- Export Facility Design
  - Export gas and oil pipeline sizing
  - Chemical injection requirements
  - Liquid drop out and sweep rate limits
  - Pigging requirements

18

Maintain: Contribution

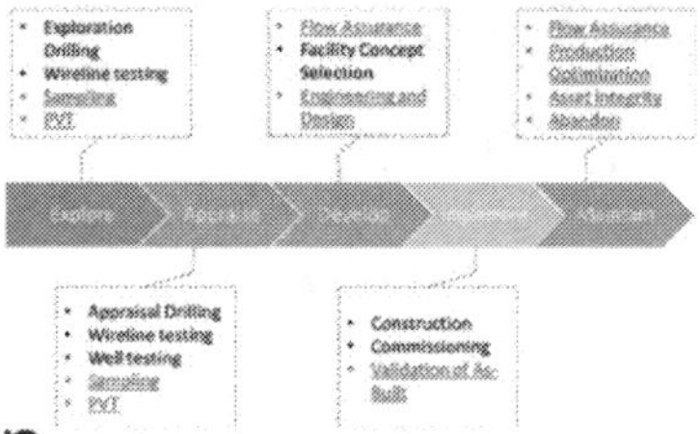

- Multi-phase simulations
  - Steady state
  - Dynamic
- Periodic fluid characterization
  - SARA
  - Composition using gas chromatography
- Chemical injection optimization
- Monitor corrosion, erosion, scale

19

Maintain: Benefits

- Support during planned & un-planned shut-in
- Optimize well routing and production
- Optimize OPEX (chemical costs)
- Maintain asset integrity
- Monitor effective separation of aqueous phase to meet HSE and regulatory requirements

20

Maintain: Benefits (Avoidance/Mitigation)

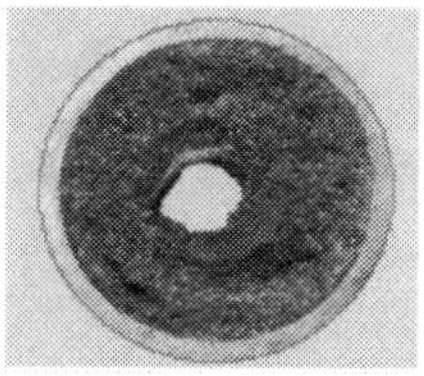

Paraffin Plug

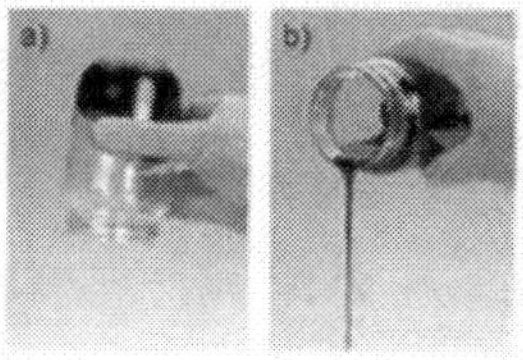

Viscous Gel

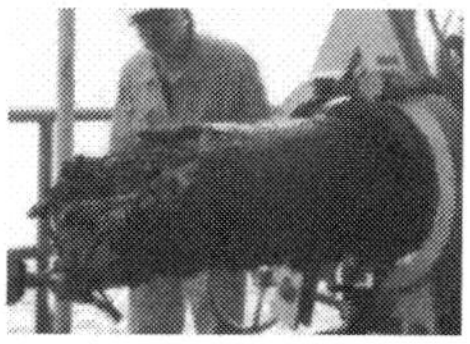

Asphaltene + Paraffin Plug

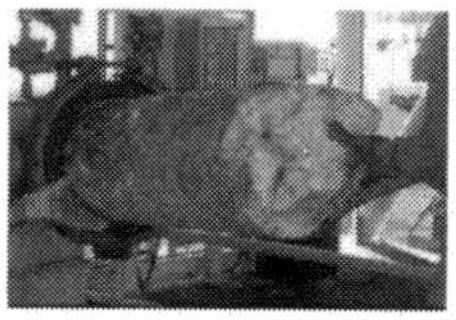

Hydrate Plug

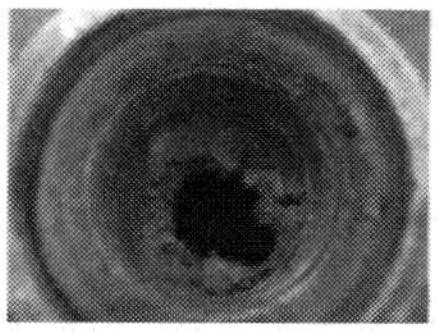

Corrosion

Overboard water quality

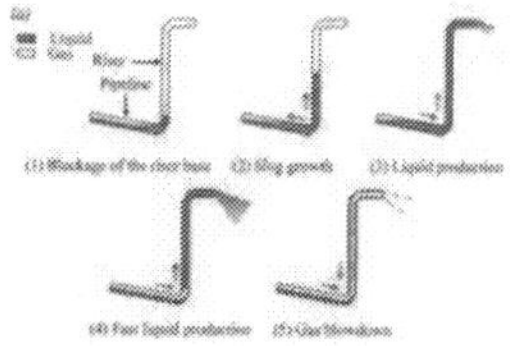

Slugging

21

Maintain: Benefits (Avoidance/Mitigation)

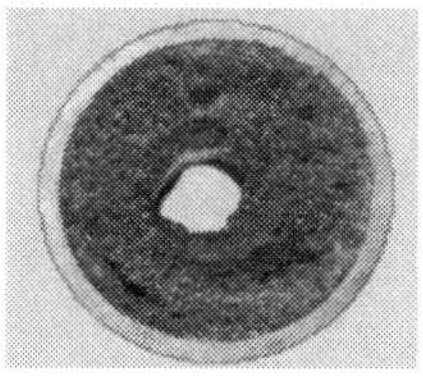

Paraffin Plug

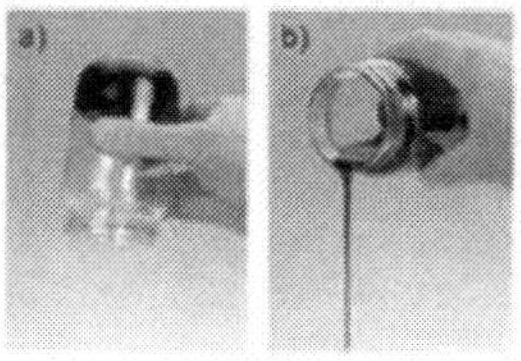

Viscous Gel

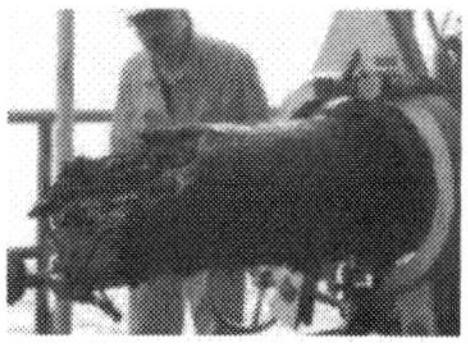

Asphaltene + Paraffin Plug

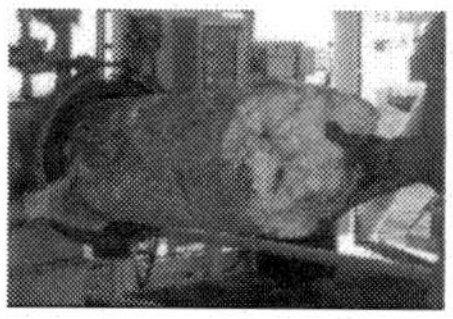

Hydrate Plug

Avoid expense of millions of dollars in remediation measures!

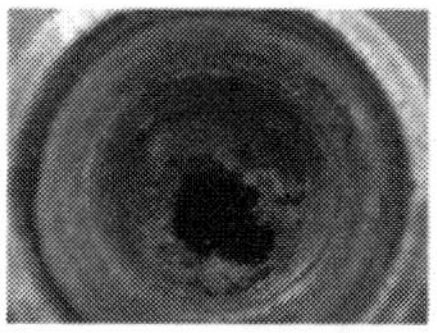

Corrosion

Overboard water quality

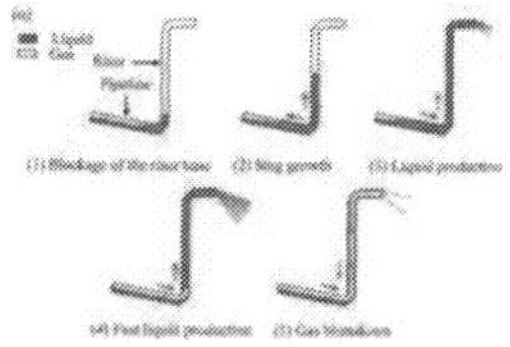

Slugging

22

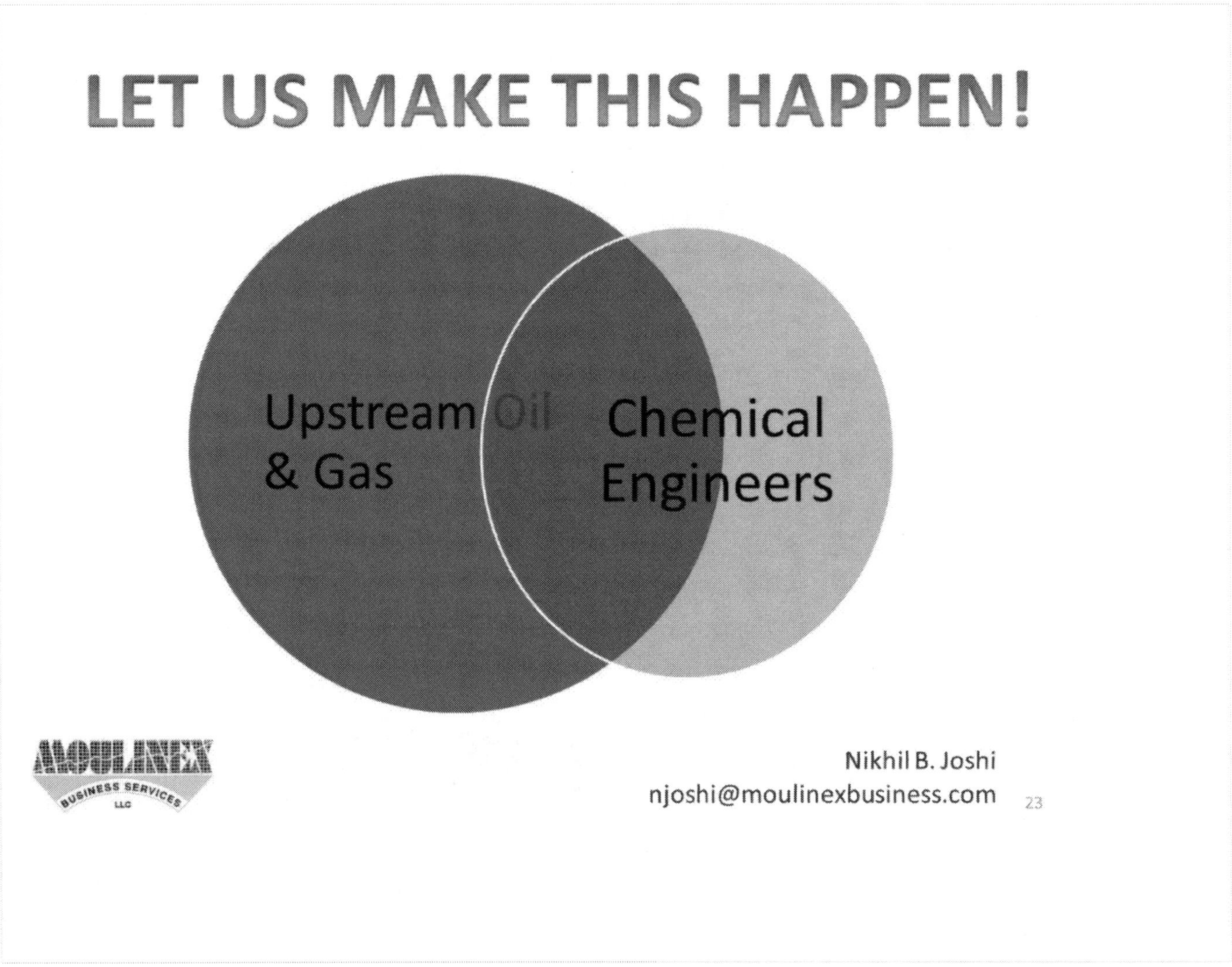
LET US MAKE THIS HAPPEN!
Upstream Oil & Gas
Chemical Engineers
MOULINEX BUSINESS SERVICES LLC
Nikhil B. Joshi
njoshi@moulinexbusiness.com
23

Texas Board *of* Professional Engineers
Engineering For A Better Texas

TBPE

Texas Board of Professional Engineers

Professional Practice Update / Ethics

C. W. Clark, P.E.
Director of Compliance & Enforcement
October 4, 2012

Engineering for a better Texas

TBPE

Agenda

- Overview of Professionalism and Ethics
- Laws/Rules/Scenarios
- Legislative News
- TBPE Focus on Excellence – what it means to you

Engineering for a better Texas

TBPE

Excellence

Outstanding or superior to an exceptionally or unusually high degree; among the best in its class

Engineering for a better Texas

TBPE

Engineering Excellence

- Professionalism
- Competency
- Laws and Rules
- Ethics

Engineering for a better Texas

TBPE

Professionalism – TBPE Board

"It is one of the primary obligations of the Texas Board of Professional Engineers to assure that Texas engineers are held to the highest standards of competency, integrity, and honesty. Public safety depends upon that standard and consistently achieving in our profession requires faithful and unwavering practice of honesty and integrity in all aspects of our lives not only when we sit at our desk."

Engineering for a better Texas

TBPE

Engineering Competence

- Core protection of health, safety, welfare
- Professional Engineers must not practice outside of their competence
- Competence is determined by experience, education and examination
- Competence is an expectation of the public

Engineering for a better Texas

TBPE

Engineering Ethics

- Ethical behavior – you know it when you don't see it
- Understanding laws, rules, and appropriate responses
- Scenarios to illustrate best choices

Engineering for a better Texas

TBPE

Law and Rules

- TBPE derives authority from the Texas Engineering Practice Act (TEPA)
- Board Rules interpret and implement the statute
- It is incumbent upon each professional engineer to know both

Engineering for a better Texas

TBPE

Scenarios

- QBS/PSPA – publicly funded projects
- Sealing Rules – proper use/control of
- Continuing Education - requirements
- Firm Registration – when to register
- Conflict of interest – when is it?

Engineering for a better Texas

TBPE

Scenario – QBS/PSPA

City of Arlen puts out bid for building the new Tom Landry Middle School, asking for engineering costs. There is an implication that if you don't provide costs that you will not be considered, even though you have discussed this with the Arlen city planner.

TBPE

What should a Texas PE do?

A. Show some civic pride and provide costs; after all, how else can the city save money?
B. Submit a bid with an estimate of the number of hours it will take for your team members to complete.
C. Speak with the city officials and explain the Professional Services Procurement Act
D. Submit without costs at the risk of losing the job

Engineering for a better Texas

TBPE

And the best answer is….

- C & D
- Engineers are required to follow the PSPA, regardless of the competition. Sharing the law with the building officials, or with the ISD purchasing division may help them understand the state laws as well.

Engineering for a better Texas

TBPE

Common PSPA Issues

- Professional Services Procurement Act (PSPA)/Qualifications-Based Selection (QBS)
- Legal requirements
- Exceptions
  - Privately funded projects
  - Federally funded projects (Brooks Act)

Engineering for a better Texas

TBPE

Scenario - CE

- You have run late with a project and forgotten to have your secretary renew your PE license and the deadline is tomorrow. You haven't completed all of the continuing education, but you have just been too busy because you are the lead on the project.

Engineering for a better Texas

TBPE

What should a Texas PE do?

A. Go online and renew, and try to take the CE as soon as you can.
B. Wait to renew until you have taken the CE which could be a few weeks until the project is done, and then renew
C. Take some self-study courses and hope you don't get audited.
D. Call the board and ask for an extension.
E. None of the above.

Engineering for a better Texas

TBPE

And the best answer is?

- E – None of the above – to avoid late fee, go to inactive status (still can't practice)
- If you don't renew, your license is delinquent and you should not be practicing. If you do renew and certify that you have completed CE but have not your license is subject to censure. The safest thing to do is to keep your continuing ED current.

Engineering for a better Texas

TBPE

Continuing Education

- 15 hours – includes self-study (5 hours)
- Random audits ongoing
- Keep documentation for 3 years
- Over 65 not exempt
- Suggested sanction is $2,500; separate violations for not responding to Board.

Engineering for a better Texas

TBPE

Common CE Issues

- Timely response to Board inquiries within 21 days
- Delinquent practice prohibited
- Firm registration required
- Updates to the board:
  - Name change
  - Address
  - Business affiliation

Engineering for a better Texas

TBPE

Scenario - Seals

- You are the lead engineer on a project on which you are overseeing all aspects of the design and construction. A family emergency has called you away from the job for a week, at the same time that there is a deadline for plan submission.

Engineering for a better Texas

TBPE

What should you do?

A. Tell the team members to copy your seal from the shared drive and use it whenever there is a need.

B. Tell them to submit the documents with your seal but you'll come in later to sign everything.

C. Mark the plans as "preliminary"

D. Negotiate a new deadline

E. Ask the architect to do it.

Engineering for a better Texas

TBPE

Best Answer

- C or D – Mark as preliminary or renegotiate the deadline
- Per Board Rule, 137.33(d): License holders shall take reasonable steps to ensure the security of their physical or electronic seals and electronic signatures; work in Texas requires a Texas seal or mark as preliminary.

Engineering for a better Texas

TBPE

Scenario - Firms

A licensed professional engineer has worked for 20 years performing engineering work for her employer, a petrochemical company. She will be retiring soon and wants to continue performing engineering work on her own. What actions must she perform to be in compliance with the Act and Board Rules?

Engineering for a better Texas

TBPE

What should she do?

A. Notify the Board of her change of employer within 30 days of the change.

B. Register the new engineering business with the Board AFTER she has been awarded her first engineering job.

C. Notify the Board of her change of employer; but, since she will be performing engineering services under her engineer license, she does not need to register as a business.

D. Register the new engineering business with the Board before she offers or provides any consulting engineering services.

Engineering for a better Texas

TBPE

Best Answer

- A & D
- **§137.77 Firm Registration Compliance**
  Offering engineering services as a business or consulting company requires registration with the board, even if sole practitioner.

Engineering for a better Texas

TBPE

Scenario – conflict of interest

You have been providing consulting services to a small suburban town on development of a parcel of land. Your firm decides to make a bid on the same project.

What should you do?

25

TBPE

What Should You Do?

A. Stop work/negotiations on the project and walk away after providing an explanation to the client.
B. Continue work/negotiations on the project since as a professional engineer, you are able to maintain complete objectivity and ethical responsibility.
C. Notify all involved parties in the situation in writing of concerns regarding a possible conflict of interest and continue working only after obtaining written confirmation from the parties that they have been notified.
D. None of the above.

26

TBPE

Best Answer

- C – Notify all parties in writing of your concerns
- **§ 137.57 (c) and (d):** A professional engineer must be objective and truthful, and shall disclose a possible conflict of interest in writing.

27

TBPE

Know the Law

- PEs are required to know the TEPA, Board Rules, applicable state laws and local building codes.
- Windstorm, Architectural Barriers/ADA
- If you have any questions about TEPA interpretation, let us know.
- Keep an eye on our emails and newsletter to make sure you know about changes.

Engineering for a better Texas

TBPE

Filing A Complaint

- Online, email, phone, mail – all are acceptable for initial contact
- A complaint form or detailed letter/email is needed to cover all the bases
- Provide specific instances of violation
- Provide evidence to show probable cause

Engineering for a better Texas

TBPE

Complaint Process

- If a complaint is filed against a PE
- A letter would be sent with the details
- The respondent – would respond in writing,
- Providing as much documentation as possible
- The person should be open, honest and forthcoming.

Engineering for a better Texas

TBPE

Sanctions

- Reprimands (Formal and Informal)
- Suspension (possible probation)
- Refuse to Renew
- Revocation
- $3,000 per violation per day

Engineering for a better Texas

TBPE

Violations are Published

- By law, all violations, except informal reprimands, must be published
- On website by Board Meeting Date
- Add to NCEES Enforcement Exchange
- In newsletter which is mailed at least annually

Engineering for a better Texas

TBPE

Preventing Complaints

- Communication
- Contract
- Calculations

Most importantly – know the law, and call if you have a question

Engineering for a better Texas

TBPE

Legislative News

Engineering for a better Texas

TBPE

2011 Legislative session

Bills passed

- HB 2067 – Petroleum Engineering
- HB 2284 – Engineer / Architect Overlap
- SB 652 – Sunset
- HB 3 – Windstorm (special session)

Engineering for a better Texas

TBPE

HB 2067 – Petroleum Engineering

- Allows very specific petroleum engineering practice in Texas by P.E.s from other states that allow reciprocal practice.

- Currently, no other states have a similar statute or rule.

Engineering for a better Texas

TBPE

House Bill 2284

- "Overlap Bill"
- Resolved questions of what is architecture or engineering
- Defined areas of practice within overlap
- Eliminated the Joint Advisory Committee

Engineering for a better Texas

TBPE

HB 2284 – Overlap Bill

- Made provision for engineers practicing architecture to continue to do so
- Window closed 12/31/11
- Texas Board of Architecture Examiners is now processing these applications
- Meetings ongoing; 80 applications received
- Engineer Exception list on www.tbae.state.tx.us

Engineering for a better Texas

TBPE

HB 2284 – Overlap Bill

- TBPE and TBAE staffs have worked very closely on communication and implementation
- Joint Task Force established to further clarify statute for up to two years.

Engineering for a better Texas

TBPE

House Bill 3 – Windstorm

- Created in the first called (special) session
- Addresses the Texas Windstorm Insurance Association
- Professional Engineers are affected if they practice Windstorm Inspection

Engineering for a better Texas

TBPE

House Bill 3 – Windstorm

- Windstorm Inspectors must apply to continue
- TBPE will create a roster of qualified engineers who must apply with TDI
- Texas Department of Insurance (TDI) and TBPE are collaborating on implementation

Engineering for a better Texas

TBPE

Windstorm Applications

- Online application available
- TDI will change their rules; current roster in effect until then
- http://www.engineers.texas.gov/windstorm.html

Engineering for a better Texas

TBPE

Senate Bill 652

- Sunset Review for 24 agencies/programs
- TBPE is included
- Self-evaluation report complete and on our website
- Opportunities for feedback from stakeholders
- 2013 Legislative Session will make changes to TBPE enabling statute
- Self Directed Semi Independent (SDSI)

Engineering for a better Texas

TBPE

Update on NCEES

- BS+30/Masters Degree Or Equivalent is a part of NCEES model law to be effective 2020
- CBT – Computer Based Testing – FE 2014

Engineering for a better Texas

TBPE

Excellence

Engineering for a better Texas

TBPE

Journey to Excellence

- Our goal in 2012 – excellence in all we do
- Implementing Malcolm Baldrige Quality Principles
- Pursuing Quality Texas Award for Performance Excellence

Engineering for a better Texas

TBPE

What this means to you…

Engineering for a better Texas

TBPE

Communication Improvements

- New logo – new look
- Newsletter – new layout
- Website redesign scheduled
- More frequent email updates
- Reaching out to the more stakeholders

Engineering for a better Texas

TBPE

Outreach – continuing excellence

Fiscal Year	Attendees	Presentations
2008	5,739	100
2009	6,928	120
2010	9,145	157
2011	8,954	153

- Webinars
- Includes K-12
- E-Week

Engineering for a better Texas

TBPE

Licensing Improvements

- ECHO – online renewals and payments
- Online Application
- Automated email increases responsiveness
- Returned calls and emails more rapidly

Engineering for a better Texas

TBPE

Compliance & Enforcement Improvements

- Focus on violation deterrent
- 10% increase in the number of disciplinary actions since last year
- Opened over 700 cases in FY2011
- State Office of Administrative Hearings
  - In last five months of 2011 filed nine cases
  - In previous two years only three filed

Engineering for a better Texas

TBPE

Compliance & Enforcement Improvements

- Increasing capacity to prosecute cases
- Hired new staff attorney
- Reducing complexity in correspondence
- Streamlining processes
- Online complaint filing

Engineering for a better Texas

TBPE

Compliance & Enforcement Improvements

- Becoming more focused on enforcement
- Reaching out to public and professionals
- Communicate directly with local officials
- Become more accessible to our licensees and the public

Engineering for a better Texas

TBPE

We want to hear from you

- How did we do today?
- How are we doing in general?

http://engineers.texas.gov/outreachsurvey
http://engineers.texas.gov/feedback

Engineering for a better Texas

TBPE

Contact TBPE

- 1917 S. Interstate 35
- Austin, TX 78741
- Phone: 512-440-7723
- Web: http://engineers.texas.gov
- Email: cw.clark@engineers.texas.gov

Engineering for a better Texas

Aromatics Configuration Case Studies

Agenda

- Integrated Refining / Petrochemical Facilities
- Modern Aromatics Complexes
- Three Configuration Case Studies
- Summary
- Q & A

Integrated Refining / Petrochemical Facilities

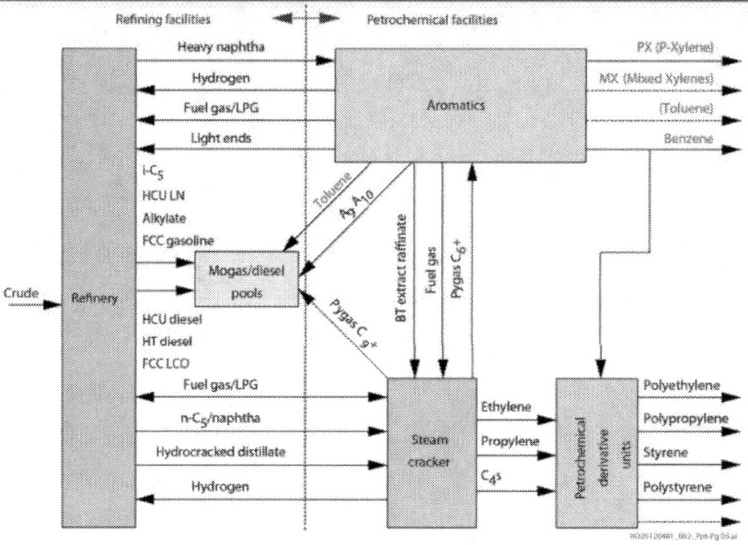

FLUOR

Modern Aromatics Complex

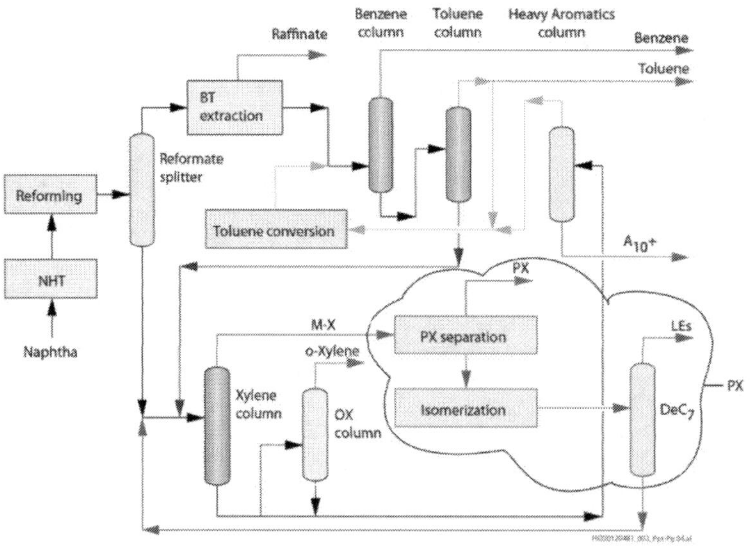

FLUOR

Feed Selection – Naphtha Cutpt; Pygas Treat

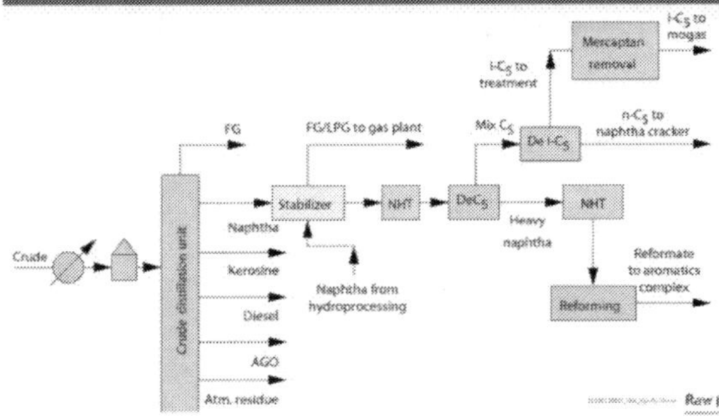

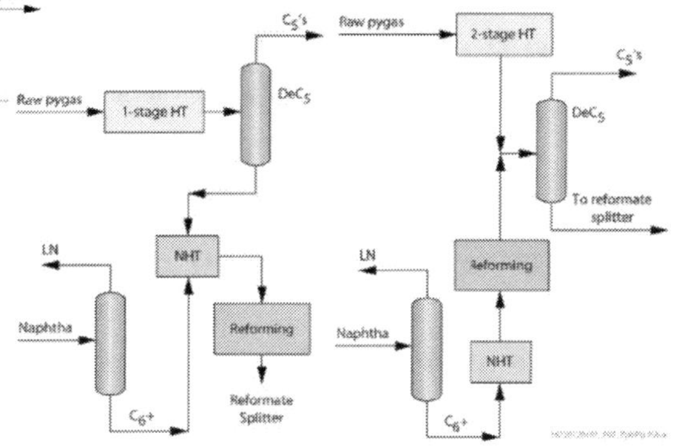

4

Case Study – 1
590 kta Benzene; Sell Excess Xylene; Return C9+ HA to Refinery

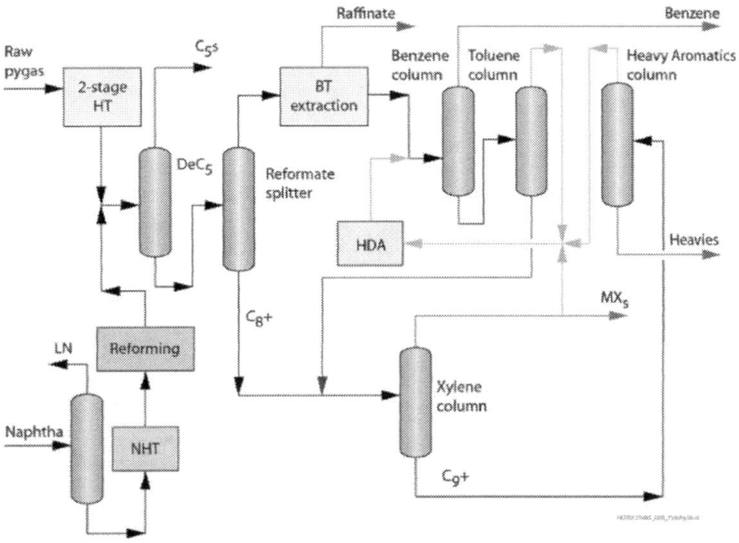

5

TABLE 1.1 Feed Stock Naphtha

Feedrate	KMTA	2,410	
	BPCD	65,860	
Composition	wt%	Properties	
Paraffins (P)	85.4	Specific gravity (15°C)	0.651
Naphthenes (N)	10.9	Total sulfur, ppmw	65
Aromatics (A)	3.7	Total nitrogen, ppmw	0
Total	100	Mercaptan sulfur, ppmw	120
		ASTM D86 distillation curve	
		Volume %	°C
		IBP	28
		30%	49
		50%	59
		70%	78
		FBP	141

BPCD: Barrels per calendar day ppmw: Parts per million in weight

TABLE 1.2 Feed Stock Raw Pygas

Feed Rate, KMTA	447
Mercaptan sulfur, ppmw	50
Composition	wt%
C_4s	0.1
C_5s*	67.1
C_6 NA (nonaromatics)	12.2
Benzene	13.1
C_7 NA	1.1
Toluene	4.0
C_8 NA	0.1
Styrene	1.1
C_8 aromatics	0.5
C_9 aromatics	0.7
Total	100

*Includes C_5 recycles.

TABLE 1.3 BTX Producton In Reforming

	HN to Reformer	Reformate
Total, KMTA	1,291	893
P, wt%	74	13
N, wt%	19	2
A, wt%	7	85
N+2A, wt%	32	–
RONC	–	99.7
Composition	wt%	
Benzene	0.6	16.1
Toluene	2.6	32.9
C_8 aromatics	2.4	24.0
C_9^+ aromatics	1.0	11.7

TABLE 1.4 Pygas C_{6+} to Reformate Splitter

Composition	wt%
C_6 NA	36.8
Benzene	40.4
C_7 NA	3.5
Toluene	12.3
C_8 NA	0.2
EB+MXs	4.5
C_9^+ aromatics	2.3
Total Flow, KMTA	100 143

TABLE 1.5 Aromatics Targets vs. Sources, KMTA

Aromatics	Targets	Sources		Demand	
		Reformate	Pygas	Excess[1]	Short
Benzene, 99.9 wt%	590	143	58		389
Toluene	0	294	18	312	
A_8 (MXs and EB)		214	6	(Note 1)	
A_9, A_{10}		104	3	(Note 1)	

Note (1) Benzene target to be achieved by converting all of the toluene and some heavier aromatics via HDA.

HDA Conversion. The following reactions take place in the thermal hydrodealkylation reactor to produce 389 KMTA ("Demand" in Table 1.5) of additional benzene:

Toluene + H_2 Benzene + Methane, and

(C_8+ Aromatics) + H_2 → Toluene + Methane -

TABLE 1.6 HDA conversions[1,2]

Units	$C_7H_8 + H_2$ --> $C_6H_6 + CH_4$		$C9H_{12} + 3x\ H_2$ --> $C_6H_6 + 3x\ CH_4$		$C_8H_{10} + 2x\ H_2$ --> $C_6H_6 + 2x\ C_{H4}$	
	Toluene	Benzene	A_9	Benzene	A_8	Benzene
kg-mole	1	1	1	1	1	1
kg / kg-mole	92	78	120	78	106	78
KMTA	–312	+265	–14	+9	–170	+125

[1] For simplification purpose, side reactions have been neglected.
Product yields approach stoichiometric for the thermal hydrodealkylation process.
[2] "–" means consumption and "+" production.

Case Study – 2
600 kta p-x; 330 kta (min) Benzene; 150 kta Toluene for Mogas Blending; Sell Excess Xylene; Return C9+ HA to Refinery

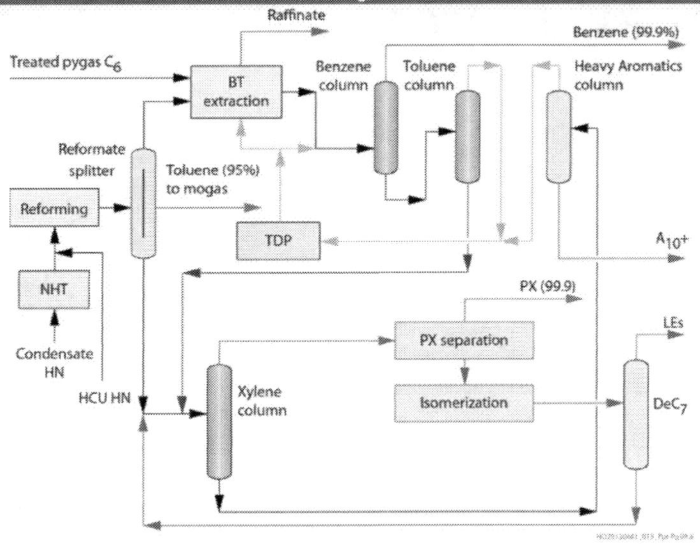

FLUOR

8

TABLE 2.1 Feed Stock				
		Crude Naphtha	Condensate Naphtha	HCU Naphtha
Feed Rate	BPCD	17,100	40,320	3,190
	KMTA	1,030	1,715	102
Specific gravity	15°C	0.723	0.731	0.741
Properties				
Total paraffins	wt%	66	61	49
Total naphthenes	wt%	23	25	40
Total aromatics	wt%	11	14	11
N+2A	wt%	45	53	62
Total sulfur	ppmw	400	2,000	< 0.5
Total nitrogen	ppmw	125	550	< 0.5
ASTM D86, vol%	°C			
IBP		71	65	100
10%		83	80	101
50%		99	99	110
FBP		155	151	128
Composition	wt%			
Benzene		1.2	3.7	1.8
Toluene		3.1	4.9	5.9
EB		0.9	0.6	0.4
MXs		3.2	4.1	2.6
C_9+ Aromatics		2.2	0.2	0

TABLE 2.2 Treated Pygas C_6 Cut		
Total sulfur	ppmw	1
Total nitrogen	ppmw	1
Components	KMTA	wt%
C_6 non-aromatics	29	17.2
Benzene	138	82.8
Total	167	100

TABLE 2.3 BTX Production in Reformer		
	HN to Reformer	Reformate
P, wt%	61	21
N, wt%	26	1
A, wt%	13	78
N + 2A, wt%	52	–
RONC	–	104
Total, KMTA	1,810	1,564
Composition	wt%	
Benzene	3.6	11.7
Toluene	5.0	29.6
EB	0.6	5.2
MXs	4.0	24.6
C_9+ aromatics	0.2	6.7

Naphtha Feed Selection:
Condensate HN is the highest priority feed
HCK HN a supplemental feed
Depentanized crude HN is lower priority feed

Naphtha cut point:
145 C (293 F) to recover xylene precursors

Pygas:
C6 to supplement reformate for Bz production
C7 retained in Pygas to aromatics
C7+ is routed to Mogas blending

FLUOR

9

TABLE 2.4 Aromatics Targets vs. Sources, KMTA

Aromatics	Targets		Sources		Demand	
	Chemical grade, min 99.9 wt%	Fuel grade, approximately 95 wt%	Reformate (Table 2.3)	Pygas C_6 (Table 2.2)	Excess	Short
Benzene	> 330	–	184	138		> 8
Toluene	0	150	463	0	313 (1)	
Xylenes	600 (PX)	–	385	0		233 (3)
EB	0	–	82	0	82 (2)	
A_9^+	–	–	105	0	105 (1)	

(1) Can be converted to benzene and xylenes in TDP unit
(2) Dealkylated to benzene in Isomerization unit
(3) PX recovery is assumed to be 97 wt%

TABLE 2.5 TDP-Toluene and A_9 Conversion (1,2)

Units	Transalkylation				Disproportionation				
	C_7H_8	$+C_9H_{12}$	-->	$2C_8H_{10}$	$2C_7H_8$	-->	C_8H_{10}	+	C_6H_6
	Toluene	+ A_9	-->	Xylenes	Toluene	-->	Xylenes		Benzene
kg-mole	1	1		2	2		1		1
Kg/kg-mole	92	120		106	92		106		78
KMTA	–30	–40		+70	–283		+163		+120

(1) For simplification purpose, side reactions are neglected
(2) "–" means consumption and "+" production.

TABLE 2.6 EBConversion (1, 2)

Units	EB Dealkylation	
	EB --> Benzene + C_2H_4	
	C_8H_{10}	C_6H_6
kg-mole	1	1
Kg/kg-mole	106	78
KMTA	–82	+60

(1) Side reactions are neglected
(2) "–" means consumption and "+" means production

FLUOR

10

Case Study – 3
780 kta p-x; 400 kta (min.) Benzene; Return C9+ HA to Refinery

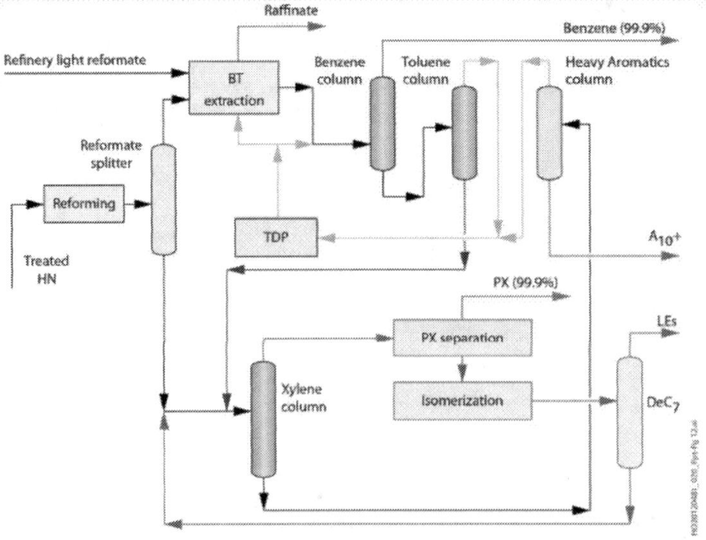

FLUOR

11

TABLE 3.1 Feed Stock - Treated Naphtha

Feed Rate	BPCD	37,500
	KMTA	1,564
Properties		
Total Sulfur	ppmw	< 0.5
Total Nitrogen	ppmw	< 0.5
Total Paraffins	wt%	44
Total Naphthenes	wt%	44
Total Aromatics	wt%	12
N+2A	wt%	68
Distillation Curve	°C	D86
IBP		45
10%		90
50%		113
FBP		175

TABLE 3.2 Refinery Light Reformate

Components	wt %
Benzene	39.0%
Toluene	2.2%
Non-aromatics	58.7%
Total Flow, KMTA	100.0% 316

TABLE 3.3 BTX Production in CCR

	HN to CCR	Reformate
P, wt%	44	15
N, wt%	44	0
A, wt%	12	85
N+2A, wt%	68	-
RONC	-	106
Total, KMTA	1,564	1,361
Composition	wt%	
Benzene	1.3	9.3
Toluene	4.5	24.8
EB	0.7	4.9
MXs	3.6	23.1
C_9 Aromatics	1.9	19.1
C_{10} Aromatics	0.2	3.9
C_{11} Aromatics	0.2	0.3

Naphtha Cut Point:

SRN is pre-fractionated to preserve the benzene precursors in the aromatics feed.

The end point of the naphtha is 175C (or 347F), to recover all of the xylene precursors and maximize C9 aromatics precursors in the aromatics feed.

FLUOR

TABLE 3.4 Aromatics Targets vs. Sources KMTA

Aromatics	Targets	Sources		Demand	
	min. 99.9 wt%	Reformate (Table 3.3)	Refinery Light Reformate (Table 3.2)	Excess	Short
Benzene	> 400	126	123	–	> 151
Toluene	–	337	7	344[1]	–
Xylenes	780 (PX)	314	0	–	490[3]
EB	0	67	0	67[2]	–
A_9	–	260	0	260[1]	–
A_{10}	–	53	0	53[1]	–
A_{11}	–	4	0	–	–

(1) Converted to benzene and xylenes in TDP unit
(2) EB isomerized to xylenes in isomerization unit
(3) PX recovery is assumed to be 97 wt%

TABLE 3.5 TDP - Toluene and A_9 Conversion[1,2]

	Transalkylation					Disproportionation				
Units	C_7H_8	+	C_9H_{12}	--->	$2C_8H_{10}$	$2C_7H_8$	--->	C_8H_{10}	+	C_6H_6
	Toluene	+	A_9	--->	Xylenes	Toluene	--->	Xylenes		Benzene
kg-mol	1		1		2	2		1		1
Kg/kg-mole	92		120		106	92		106		78
KMTA	–80		–104		+185	–414		+239		+176
					Dealkylation					
Units	$C_9H_{12} + H_2$		--->		$C_7H_8 + C_2H_6$	$C_{10}H_{14} + H_2$	--->	$C_7H_8 + C_3H_8$		
	A_9 --->				Toluene	A_{10}	--->	Toluene		
kg-mol	1				1	1		1		
Kg/kg-mole	120				92	134		92		
KMTA	–155				+119	–45		+31		

(1) For simplification purpose, side reactions are neglected
(2) "-" means consumption and "+" production

TABLE 3.6 EB Conversion[1, 2]

Units	EB isomerization	
	EB - - -➤ Xylenes	
	C_8H_{10}	C_8H_{10}
kg-mole	1	1
Kg/kg-mole	106	106
KMTA	−67	+67

(1) Side reactions are neglected
(2) "−" means consumption and "+" means production

FLUOR

Summary – Optimization Handles

- Feedstock selection and routing
- CCR operating severity
- Product Targeted Technologies:
  - Toluene Conversion (TDP, Transalkylation, HDA)
  - Xylene Isomerization with EB Dealkylation / Isomerization
  - PX Separation (Adsorption, Crystallization)

Typical CAPEX (for PX)

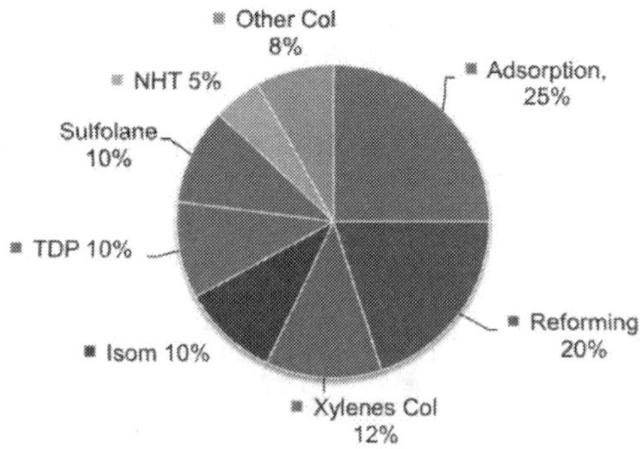

FLUOR

Membrane Applications for Commodity Scale Chemicals

Alexia Finotello, Scott Matteucci, Mark Brayden, John G. Pendergast
The Dow Chemical Company

October 2012
AIChE RPTC – League City, TX

The Dow Chemical Company

Diversified chemical company, harnessing the power of science and technology to improve living daily

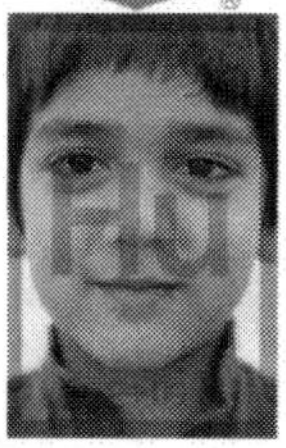

1897	2000	2008
Founded Midland, MI	Acquired Union Carbide	Acquired Rohm & Haas

Dow by the numbers

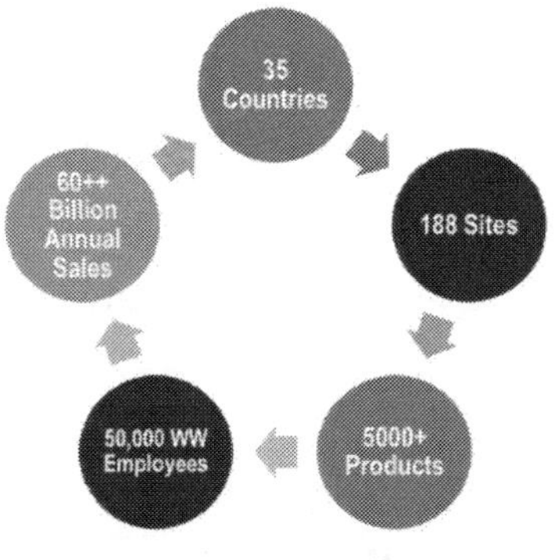

Unique perspective of commodity landscape and specialty chemicals

Large Scale
Ethylene and polyethylene
Chlorine and caustic
Ethylene oxide and ethylene glycol

Specialty Scale
Metal organics, tri-methyl gallium, tri-methyl indium
OLED materials
Chemical mechanical planarization products, photoresists…

Importance of Efficiency to a Global Dow

DOW

- World's largest (non-utility) industrial consumer of power and steam

3,700 MW of electricity required to operate Dow plants = **3,700 MW** of electricity Used by San Francisco, San Diego, & Oakland combined

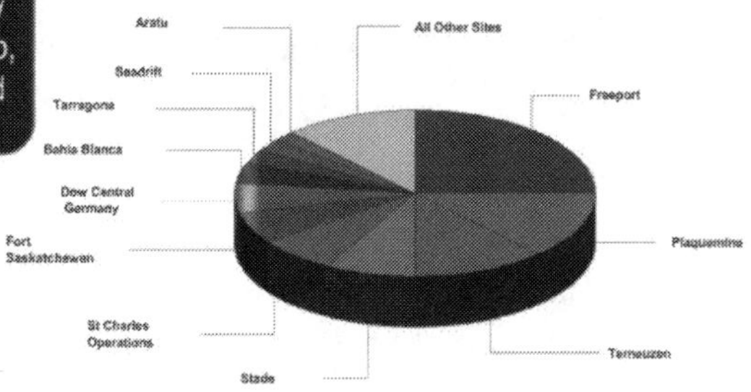

- World's largest producer of ethylene
- World's largest producer of chlorine
- Feedstock demand is **800,000 barrels/day**
- Leading innovator in cogeneration
  - Increased efficiency with reduced impact on the environment
  - Uses **20-40% less fuel**
- Self-generates **~75%** of all power & steam
- Operates over **$6.2 billion** in energy assets & supports **$2.5 billion** in JV assets

Page 3 2012: For External Release

Sustainable Chemistry

DOW

- Integrated look at chemicals, processes and products
  - *Maximize energy efficiency*
  - *Maximize the efficiency of raw material use*
  - *Minimize the use of scarce resources*
  - *Minimize the return of anthropogenic substances to nature*
  - *Improve safety of manufacturing processes*
  - *Not damage the environment*
  - *Significantly enhance benefits to society*
- Will change the way that some products are made
  - We are seeing "sustainability issues" raised
    - » Zero water discharge requests
    - » Solvent recovery, solvent recycle
- It's all about balance

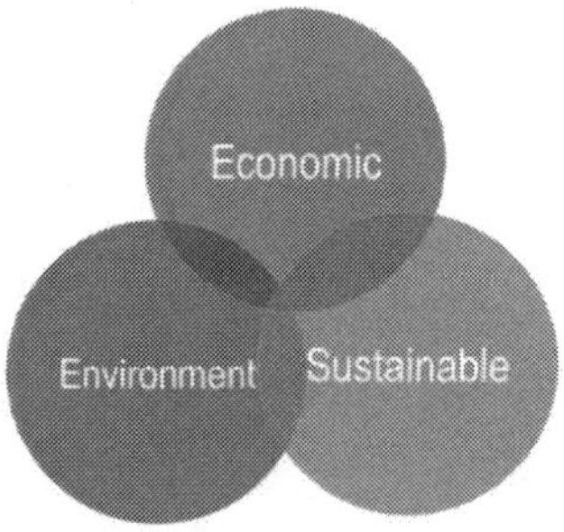

Page 4 2012: For External Release

Challenges for Separations

- Improve & Support Current Process
  - H_2 Removal
  - Olefin / Paraffin
  - Ar Recovery
  - Water Recovery

  (H₂ Removal and Olefin / Paraffin:) Standard is cryogenic distillation
  = energy & capital intensive
- Evolutionary Transformation
  - Next Generation Dehydrogenation Technologies
- Revolutionary Transformation
  - Carbon Neutral Processes and Products

Current Dow Feedstock Utilization

Starting Materials = Ethylene & Propylene

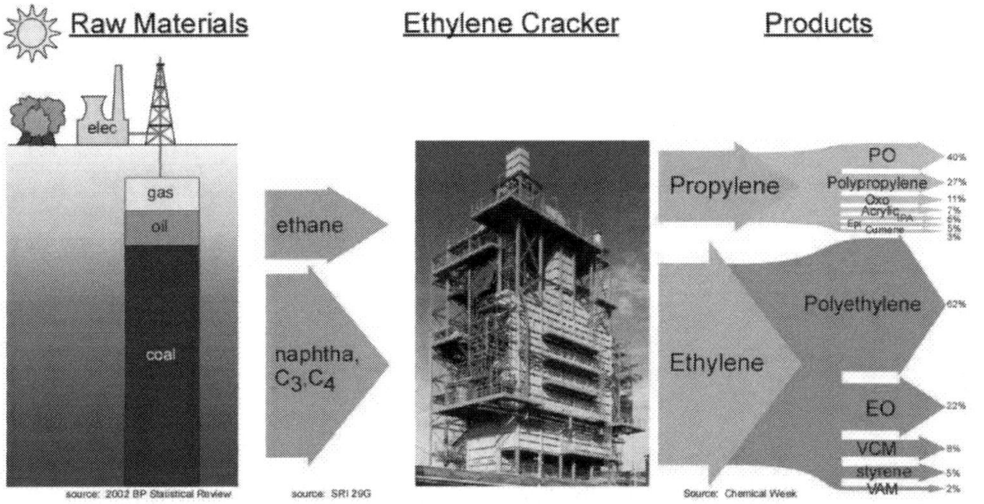

Scale is the issue for Dow

Dow is the top ethylene producer in the world

- 14 Dow crackers worldwide worth over $15 billion in capital
- Dow crackers globally convert over 5 million pounds of feedstock every hour
- Feedstock demand is **800,000 barrels/day** -- $15-20 billion/year
  - 40-50% of Dow's total annual operating costs and expenses

Olefin/Paraffin

Olefin / Paraffin

- Prime products are distributed as very pure products
  - First stage "concentrator" is of value
  - Hybrid separations with distillations are promising
  - Absorption and/or Membranes

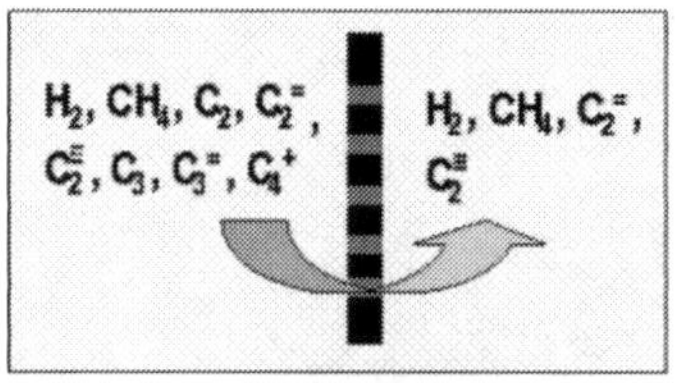

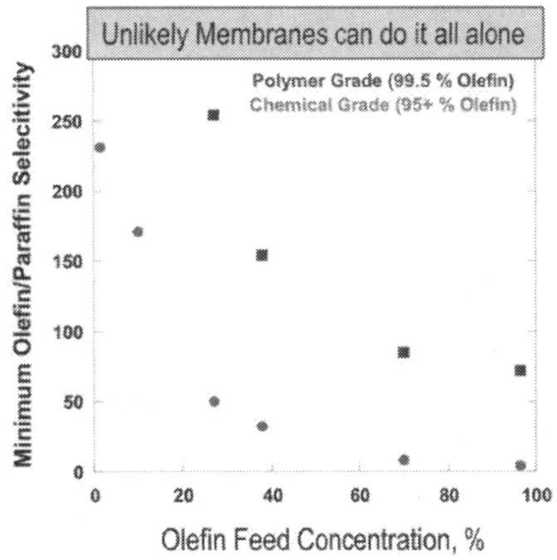

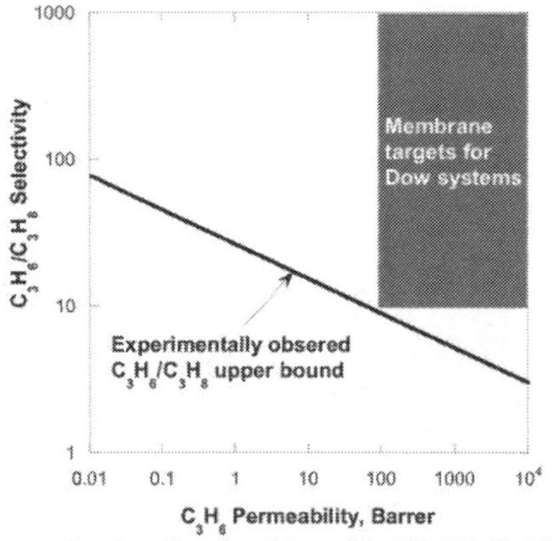

Upper bound from: Burns & Koros, JMS, 2003, 22(2), 299-309

A Problem with H_2

- High level in the thermal cracking process product stream
  - Approximately 30 mole % produced at low pressure (30-40 psi total pressure)
- Must be compressed and refrigerated to separate by distillation.
  - Pay thermodynamic penalty.
    - » Purity requirement: low ethylene losses.
  - Limited use for high pressure hydrogen in an ethylene plant
    - » Olefins plants can use 60-80 % purity hydrogen for its internal processes.
      - Methane as balance.
    - » Infrastructure is typically not in place to recover a high purity product or to distribute it.
  - Therefore, large fraction ends up as a fuel gas at low pressure.

Improve & Support Current Process

<u>**H_2 Removal**</u>

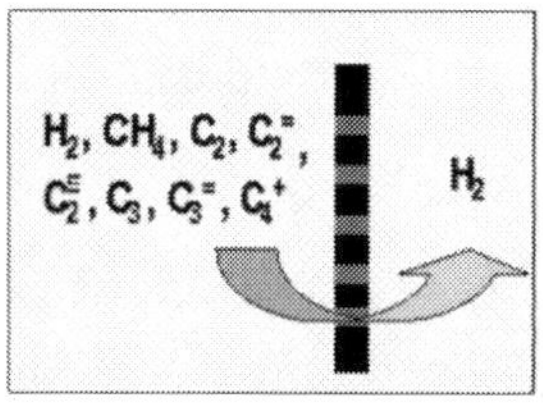

- H_2 production is a strong function of feedstock
- Efficient removal of H_2 reduces energy <u>and</u> capital
- H_2 removal becomes more important as feedstocks become lighter
- Incremental capacity and energy improvements are critical
  - Incremental improvement supports future research
- H_2 is a "waste gas" in many processes
- H_2 generally sells at a premium to energy
- What is desired:
  - High H_2 flux
  - Almost any selectivity will help

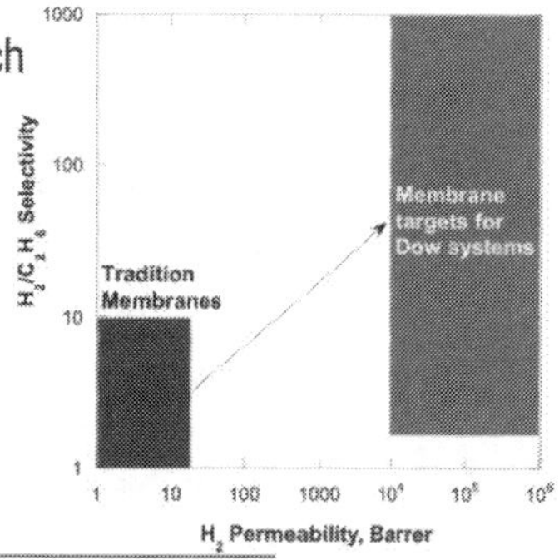

Argon Purge from Oxygen Consumers

- Consumers of oxygen often purge a stream that is rich in Argon
- Generally purged to recover fuel value
- Argon is usually the 0.5% in the 99.5% purity Oxygen
- Argon purge often concentrated in the purge by a factor of ~20/1 or greater than in air

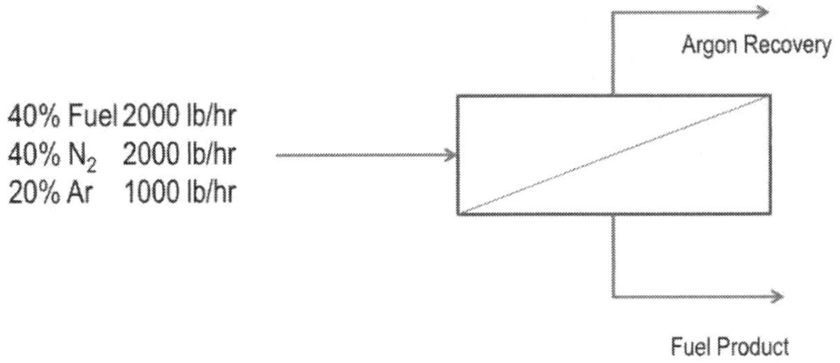

Ar from Hydrocarbons

- Recovery of fuels = Recovery of feedstocks
  - Hydrocarbons are worth more to TDCC unburned
- We produce enriched Ar
  - No Ar business in TDCC, but other may find use
- What TDCC needs from a membrane
  - Moderate to high flux
  - Prefer to permeate hydrocarbons if feedstock poor
  - Prefer to permeate Ar if hydrocarbons stream is feedstock rich

Water Recovery from Steam / Furnace Vents

- Plant Steam is valuable and is vented
  - Steam contains low grade heat (currently low value to recover)
  - High purity water in steam is quite valuable
  - Boiler and furnace vents contain water and latent heat
  - Challenging because going below the dew point causes corrosion
- What is desired is low cost separations system that recovers the water and the heat
- Classic problem: lots of energy, but low quality
- Partial savings 360 tera- Btu could be recovered from heat currently lost if the technology was implemented throughout the United States. UT-B ID 200902330
- Problem recognized by NETL, DOE, ... others

Water Recovery from Steam / Furnace Vents

- What is required:
  - No selectivity required, but condensation and recovery are needed.
  - High flux is most important attribute
  - Affordability also critical
- Working mode of porous membrane is critical for water vapor transportation.
- High permeate flux and high separation ratio could only be achieved in a capillary condensation mode.([1])

(1) Presented by GTI at NETL Meeting Pittsburg, Pa. October 2008

Gulf Needs Propylene

- Dow boosting USGC ethane cracking capability
  - *Dow Chemical says it will build a previously announced ethylene plant at its Freeport, TX complex to take advantage of low-cost feedstocks available from increasing U.S. supplies of shale gas.* (Chemical Week, April 19, 2012)
- But demand for propylene remains high
  - Less produced when cracking ethane versus other feedstocks like propane or naphtha.
- Propylene needs to be met with on-purpose production.
  - *In December 2011, Dow and UOP signed a technology licensing agreement to use UOP's C3 Oleflex process to produce propylene from propane. Dow also signed catalyst supply and performance guarantee agreements with UOP.* (Chemical Week, April 19, 2012)

Page 15 2012: For External Release

Evolutionary Transformation (Propane Dehydrogentation)

Next Generation Dehydrogenation Technologies

Dehydrogenations limited by thermodynamics

- Need to remove hydrogen or olefin to drive to higher conversion
  - 40% conversion at 550°C; 60% conversion at 625°C

Membranes to selectively remove hydrogen or olefin

- must function at high temperature (>500°C)
- non-reactive to olefins
- high flux to drive equilibrium
- Must be affordable

Membranes to selectively control oxygen addition

- Oxygen permeable membranes for Oxy-dehydrogenation
  - Temperature stable
  - Facilitate only desired reactions
  - Safety issue if membrane fails

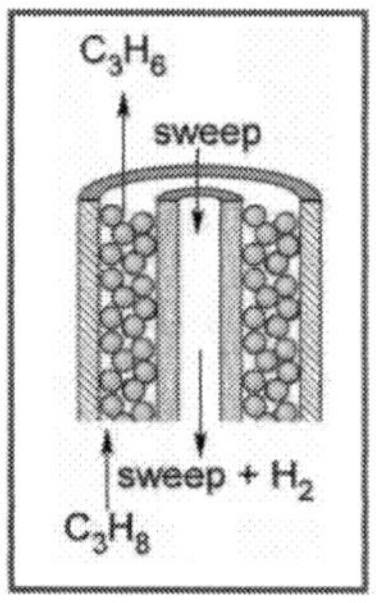

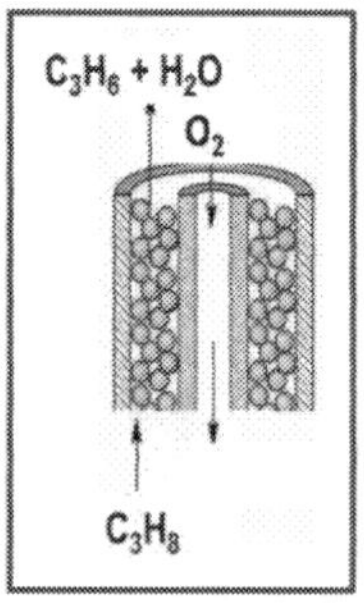

Page 16 2012: For External Release

Final Thoughts

- **Numerous Applications Exist in Dow Chemical that could benefit from membranes**
  - **High flux is almost always more important than high selectivity**
- **Breakthrough / step change catalysts and processes required for many Dow Chemical products**
  - ***Separations hold the key to many areas***
- **Need Fundamental technology development / understanding**
  - **Consistent, sustained focus**
- **Industry, Government, & Academic partnerships necessary to tackle long-term research issues**

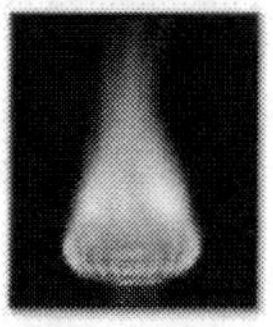

Thank You

Wadie Malaty – Senior Technologist - Ethylene

Imran Muhammad – Lead Process Engineer - Ethylene

AIChE 4th Regional Process Technology Conference, League City, Tx - October 4-5, 2012

Technip

take it further.

Feed Stock Flexibility – Maximizing Ethane in Existing Liquids Crackers

2

Technip

Maximizing Ethane Cracking-Factors to be Considered

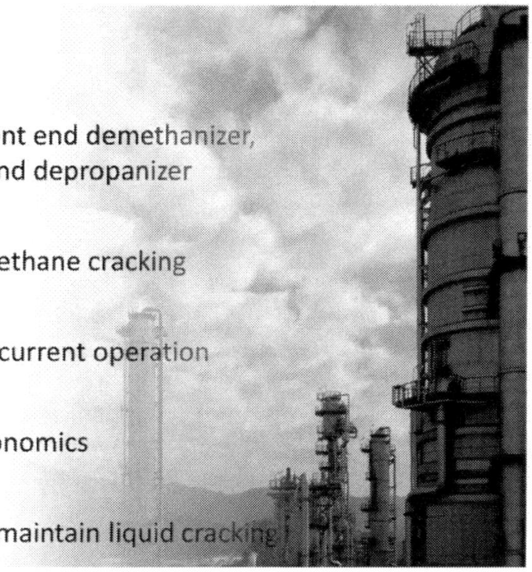

- Ethane availability and price
- Liquid cracker configuration, front end demethanizer, front end deethanizer or front end depropanizer
- Capacity increase in addition to ethane cracking
- Available equipment margins in current operation
- Products shift – local market economics
- Future feed flexibility (ability to maintain liquid cracking capability)

Technip

3

Products Shift – Ethylene and Propylene

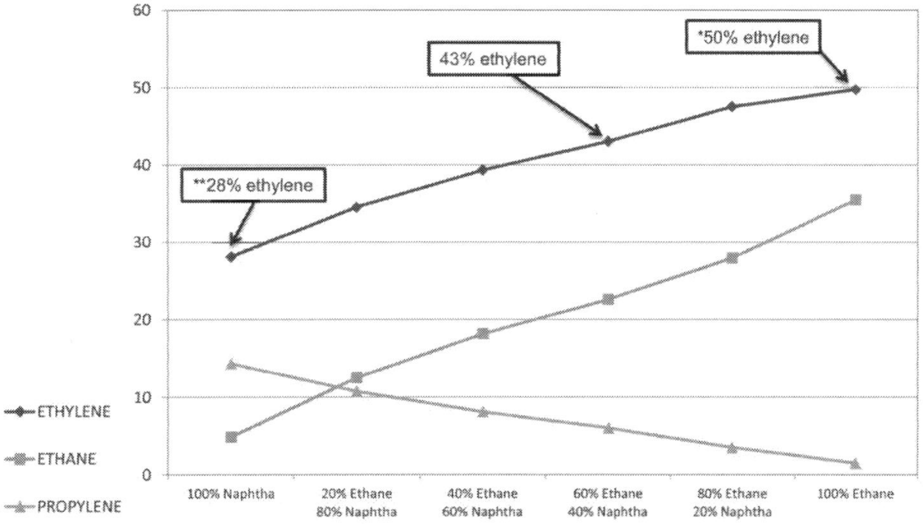

***Ethylene yield will higher depending on ethane % conversion**
****Ethylene yield will be higher depending on saturates % and cracking severity**

Technip

4

Light Products Shift

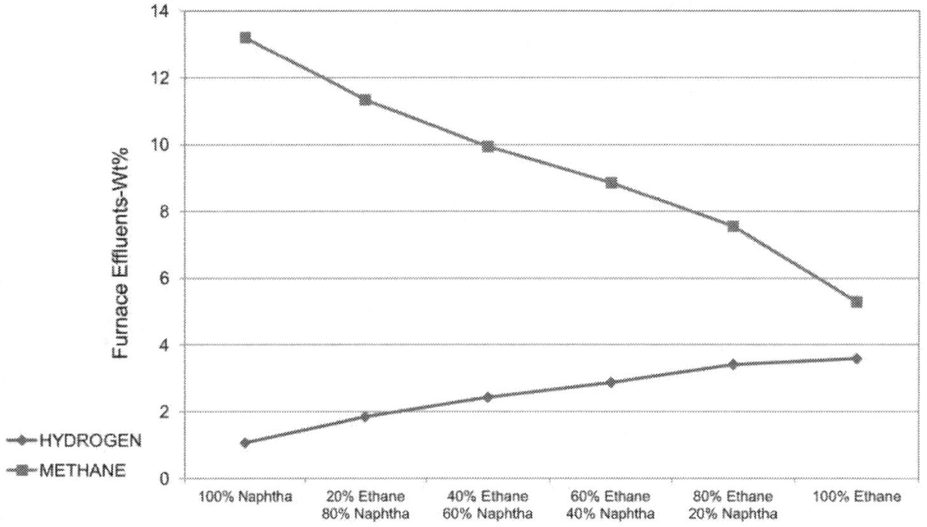

Technip

5

Heavy Products Shift

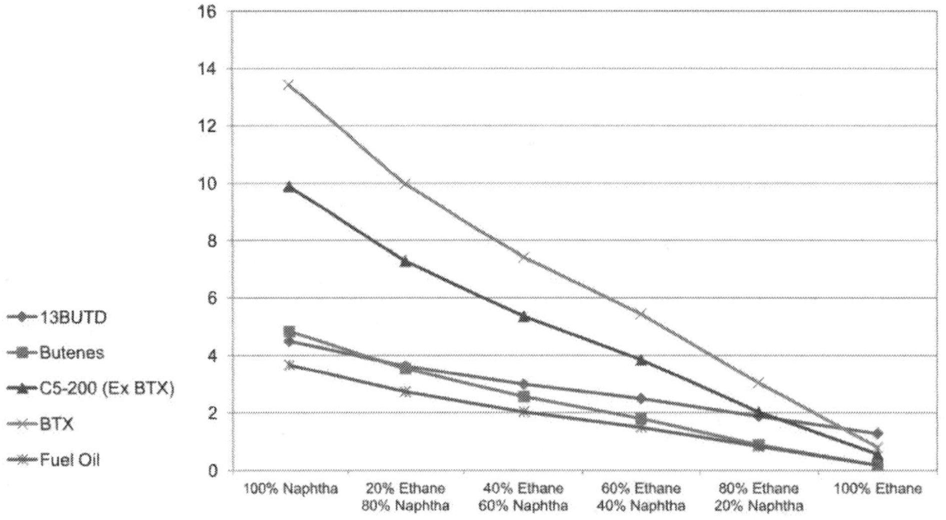

Technip

6

Primary Areas to be Considered

- Ethane feed conditioning (drying)
- Furnaces
- Quench oil tower operation
- Dilution steam
- Cracked gas compression
- Chilling train and demethanization
- C2 splitter (deethanizer &BE C2 hydrog. as needed)
- Refrigeration (C3R and C2R)

Technip

7

Impact of Going Lighter – Cracking Furnaces

- Reduced run length
- Capacity limitation
- Convection section limitation
- TLX outlet temperature - can be reduced further
- Lower feed throughput – due to better ethylene yield (venturis and CVs)

Technip

8

Possible Modifications – Cracking Furnaces

- Revamp liquid furnace to light feed furnace
- Evaluate recoiling vs. unmodified coils
- Addition of TLX-2 in BFW preheating or SHP production
- Convection section modification or replacement
- Combustion system upgrades
  - Evaluate need for new ID fan
  - Review burners
- Flexibility retained to crack liquid feed

9

Technip

Considerations – Quench Oil System

- Fuel oil production drops
- Lower amount of middle and heavy F.O. components
- High residence time
- High viscosity
- Agglomeration to PNA and tar
- Lower dilution steam production
- Higher import -MP steam
- System shutdown

SOLUTION

Flux oil injection

10

Technip

Going Lighter - CG Compressor Considerations

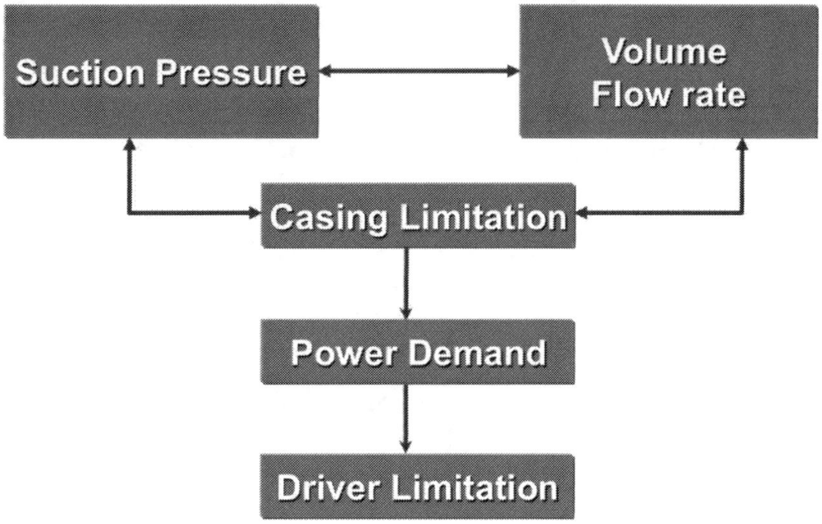

Technip

11

Going Lighter – CG Compressor

Advantages

- Better ethylene yield
- Lower mass flow of C.G

Disadvantages

- Low Mol. Wt. ~30 ~19
- High volumetric flow rate
- Higher required head
- Higher 1st stage suction pressure
- Lower selectivity (esp. for liquid cracking)

$$Head = \frac{1545}{MW} ZT_1 \left(\frac{n}{n-1}\right)\left[\left(\frac{P_2}{P_1}\right)^{\left(\frac{n-1}{n}\right)} - 1\right]$$

Technip

12

CG Compressor – Possible Modifications

Depend on

- Any capacity increase required?
- CGC performance curves
- % of ethane cracking
- 1st stage suction pressure

- **Addition of booster stage**
- **Addition of a wheel**
- **Steam turbine upgrade**

Technip

13

Considerations - Expander

- Methane mol% reduces from 25% to 12%
- Feed flow rates to demethanizer and expander drop
- Expander/compressor efficiency drops
- A lower refrigeration available to chilling train

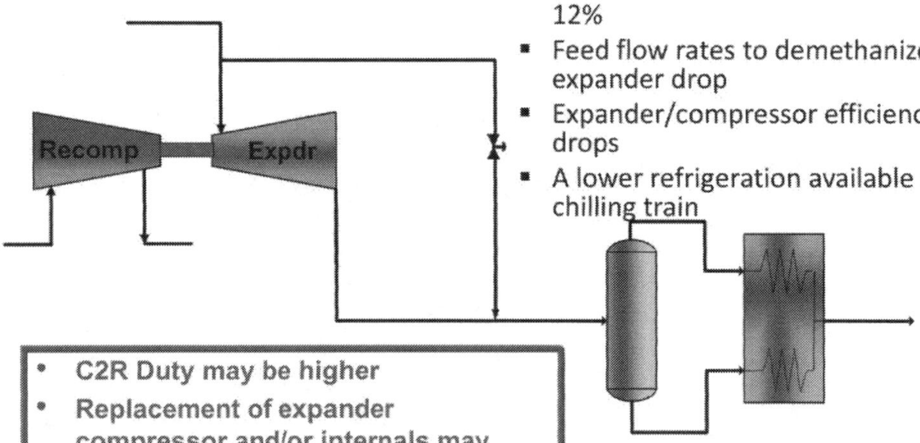

- **C2R Duty may be higher**
- **Replacement of expander compressor and/or internals may need to be considered**
- **Flexibility to revert back to heavy feed by re-installing original internals**

Technip

14

Considerations – Ethylene Tower

- Typically higher ethylene production
- Lower purity - ethane % higher in feed to tower
- A higher reflux ratio required
- Higher vapor liquid traffic in tower
- Extra capacity on trays – trays hydraulics

Potential replacement of existing trays with high capacity trays

Technip

15

Light Feed Impact – C2R & C3R Compressors

Credits

- Liquid ethane refrigeration credit (C3R)
- Recycle ethane refrigeration credit (C3R)
- Lower duty - DC1 condenser (C2R)
- Any margins on compressors?
- Any margins on approach T?

Additional Loads

- Higher duty - C2 tower condenser (C3R)
- Expander efficiency - lower
- C2R condenser duty - may be high

- **C2R compressor – may be limiting due to lower credit from expander**
- **C3R compressor - may be OK due to liquid ethane credit**

Technip

16

Further Actions – Going Lighter in Existing Liquid Cracker

- Furnace area study required
- Quench oil system to be checked for operation suitability
- Review of hydrogenation reactors (front-end or back-end) operation and catalyst volume due to ethane cracking
- Simulation studies required for recovery section areas including refrigeration compressors

17

Technip

Questions

18

Technip

Thank you

www.technip.com

Technip
take it further.

Developing Large Petrochemical Facilities

Sanjeev Kapur
Apex PetroConsultants

Best in Class Facilities

Well Planned

- Access to Feed
- Access to Resources
- Market Access
- Infrastructure
- Competitive Technologies
- World Scale Capacities
- Access to Financing
- Viable and Attractive Alternates

Well Implemented

- Alignment with Overall Business Objectives
- Reliable and Predictable Implementation
- Low Life Cycle Costs
- Successful Start-up for brining Facility to Commercial Production
- Maximizing Asset Utilization
  - Flexibility for changing feed and product slate

Success Factors

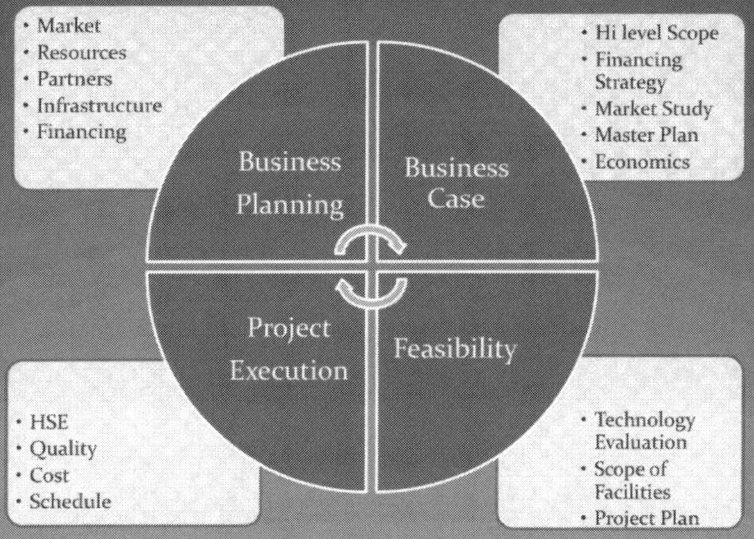

Discussion Topics

- Growth Drivers for US Petrochemical Industry
- Industry Success Factors
- Mega Project Challenges
- Summary

Growth Driven by Shale Developments

- Access to Cheaper Feed
  - Competitive Advantage relative to Low Cost Regions
- Growth in Global Demand
  - Export Market
- Developed Infrastructure

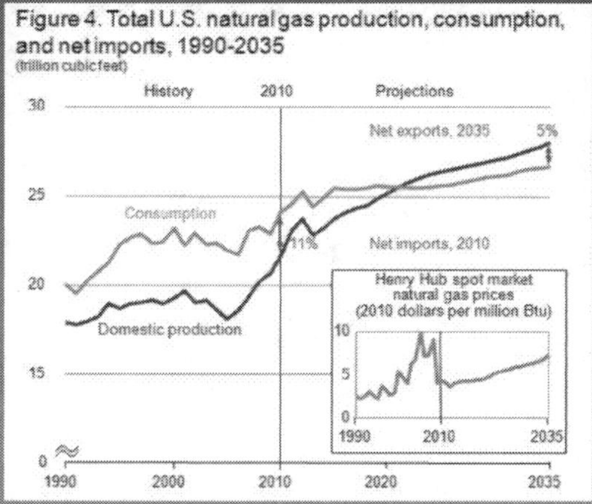

Source: EIA

Shale Provides Largest Growth Source in US

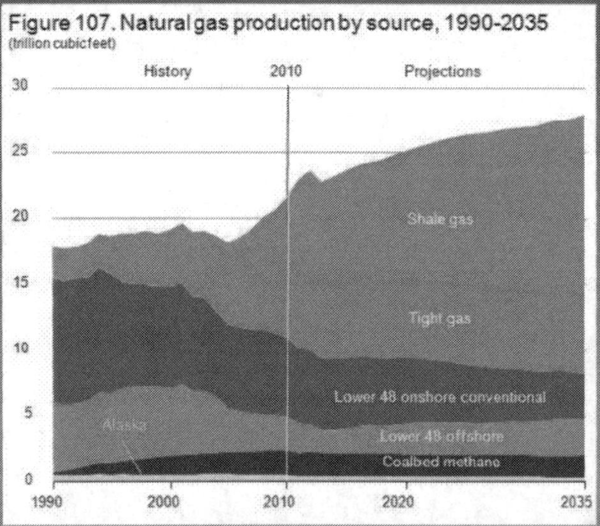

>40 % Share

Source: EIA

Petrochemical Feedstock Pricing Factors

- Strong Asian Demand
- ME imports are up
- MENA unrest
- Strong South American Imports
- Shale Gas Supply
- Export oriented growth in US/Canada
- Liquefaction capacity surges

·Gas prices in NA expected to stay below $30 per barrel of oil equivalent

·Global gas prices expected to lose value relative to oil

Driving Growth of US Petrochemical Industry

Petrochemical Industry Success

- Economies of scale
- Integrated Facilities
- Technology
  - readily available
- Competitive pressures/market conditions dictate decisions
  - Access to resources
  - Market access
  - Product diversification
  - Capital efficiency
  - Resource utilization efficiency
  - Infrastructure and logistics
- Effective Planning & Execution of Facilities

. Raw Material Advantage
. Customer Relations

It's all about Cost

Economic Incentives – Integrated Complexes

- **Leverage Feedstock**
  - Upgrade low value streams to high value products
  - Lower Transfer Pricing of Feedstock
- **Positioning**
  - Backward Integration → secures feedstock
  - Forward Integration → secures market share
- **Stability Over Value Chain**
  - Diversified Product Slate Dampens Cyclic & Non-cyclic Profitability Impact
  - Flexibility to Dynamic Market Demands ("Bigger Tool Box")
- **Capital, Operating Cost & Resource Optimization**
  - Shared Infrastructure, Storage & Utilities
  - Lower Logistics & Energy Cost
  - Minimize Overhead & Waste

It's a mega project

Top Challenges of Mega Project Execution

- Execution complexity – multiple EPC project in one
  - Interfaces increase exponentially
- Execution platform
  - Experienced resources
  - Effective project organization/decisions structure
  - Shared and achievable goals & expectations

Planning is Vital

Challenges & Risks

RECENT PAST IN USGC REGION

- Limited major investments of the large scale (> 1 Billion US$ in CAPEX)
- Limited number of Large Investments implemented at same time

Resource Constraints

Material and Equipment Supply	Skilled Construction Labor	Experienced and Qualified Professionals

Success Can't be Taken for Granted

- Early Planning
- Secure Resources
- Incentive Programs for achieving desired results
- Promote Effective and Timely Decision Process
- Shared and Achievable Goals

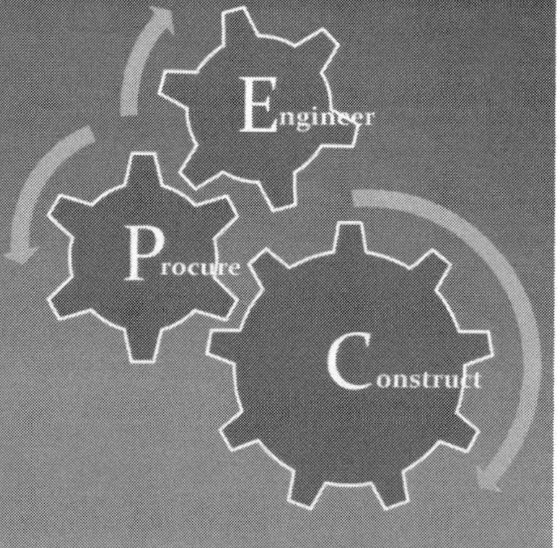

The Shale Hydrocarbon Surge
Implications for Refining, Petrochemicals

Praveen Gunaseelan, Matt Thundyil

VANTAGE POINT
ENERGY CONSULTING

Transcend Solutions

AIChE 4th Regional Process Technology Conference
League City, TX | October 5, 2012.

Cautionary Statement

- This document has been prepared by Vantage Point Advisors and Transcend Solutions (hereafter, the "Authors") for presentation at the AIChE Regional Process Technology Conference in League City, TX on October 4-5, 2012. The findings and recommendations in this document should be treated as preliminary and should be reviewed, validated, and updated by means of more detailed market assessments. Possession of the presentation does not carry with it the right of publication.
- This document contains various forward-looking statements and content that are based on information currently available to and assumptions made by the Authors. Although the Authors believe that the expectations reflected in such forward-looking statements are reasonable, they cannot give assurances that such expectations will prove to be correct. Such statements are subject to a variety of risks, uncertainties and assumptions. If any of these risks or uncertainties materialize, or if underlying assumptions prove incorrect, the market's actual potential may vary materially from the outlook that the Authors estimated or anticipated. NO IMPLIED WARRANTY OF MERCHANTABILITY OR FITNESS FOR A PARTICULAR PURPOSE SHALL APPLY.
- Among the key risk factors that may have a bearing on the outlook conveyed are:
  - Economic weakness globally and/or in regions that are important to the markets discussed
  - Fluctuations in energy commodity prices and production volumes due to various factors
  - The failure of projects to materialize or progress according to the developers' intended schedules
  - Changes in the regulatory environment that have a bearing on the markets discussed

Outline

- Shale Hydrocarbons - An Introduction
- Implications for Refining and Petrochemicals
- Implications for Process Industries
- Recap

3

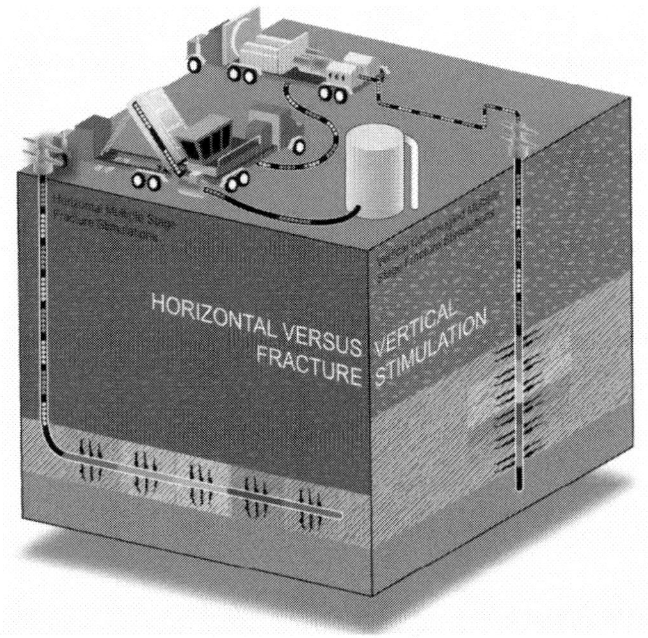

Shale Hydrocarbon Production

Horizontal Drilling & Hydraulic Fracturing are Key Enabling Technologies

Source: National Energy Board, Canada.

4

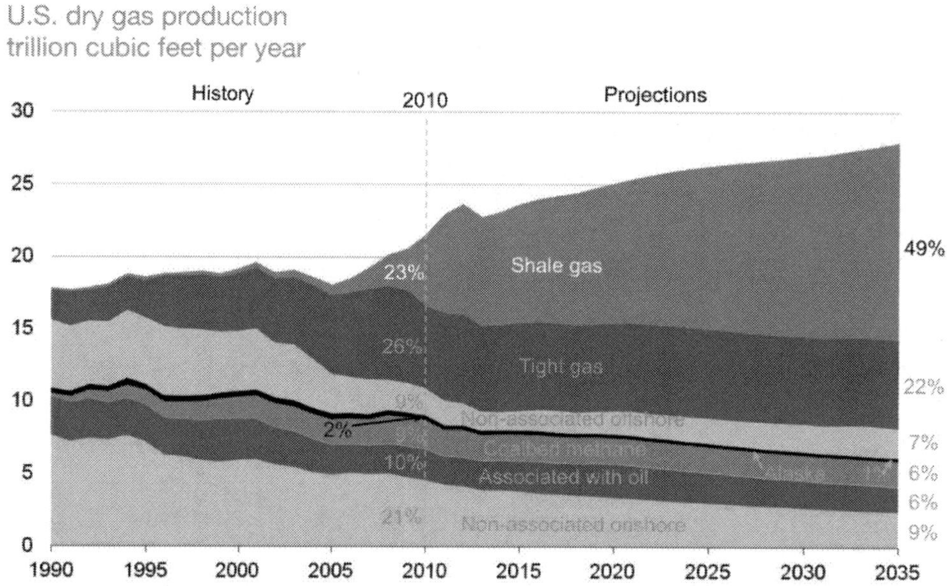

US Shale Gas Growth

Shale gas currently ~25% of US gas

Source: US EIA 2012 Annual Energy Outlook

5

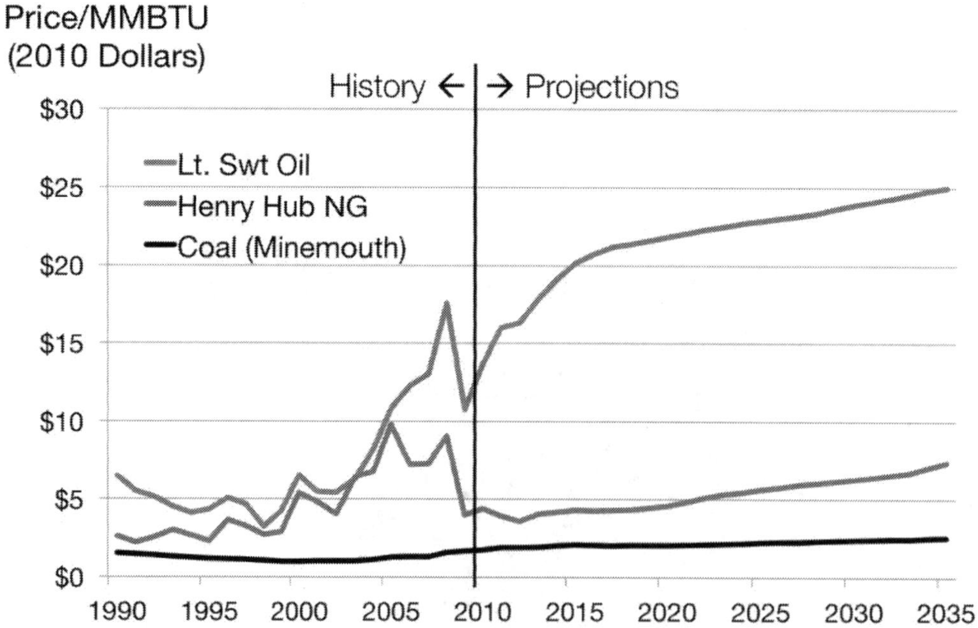

Oil to Trade at Premium to NG

Monetizing cheap NG will drive investment

Source: US EIA 2012 Annual Energy Outlook

6

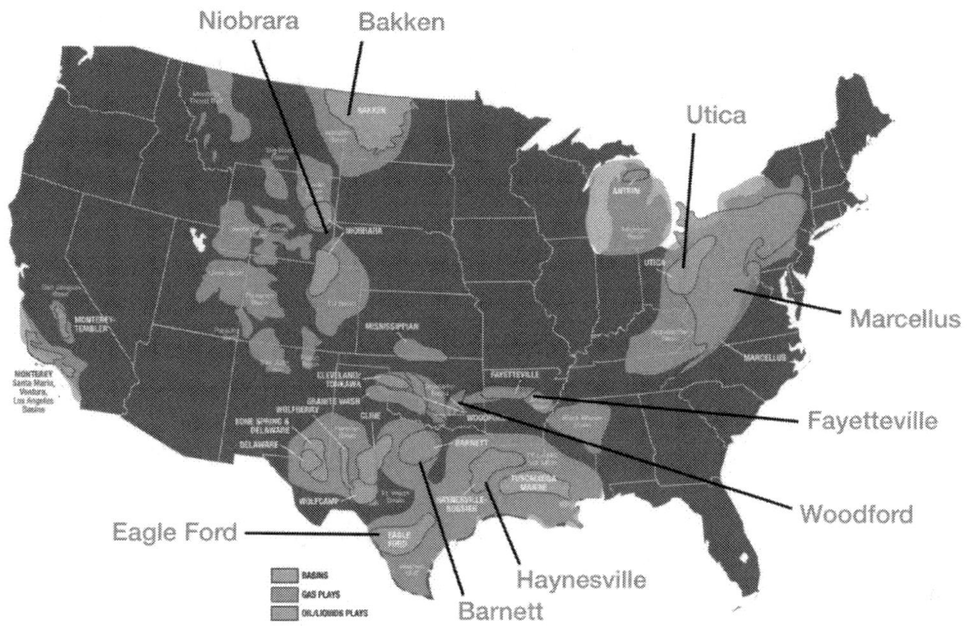

Major US Shale Plays

Producers moving to liquids plays

Source: PACWest Consulting Partners (Emphasis added)

7

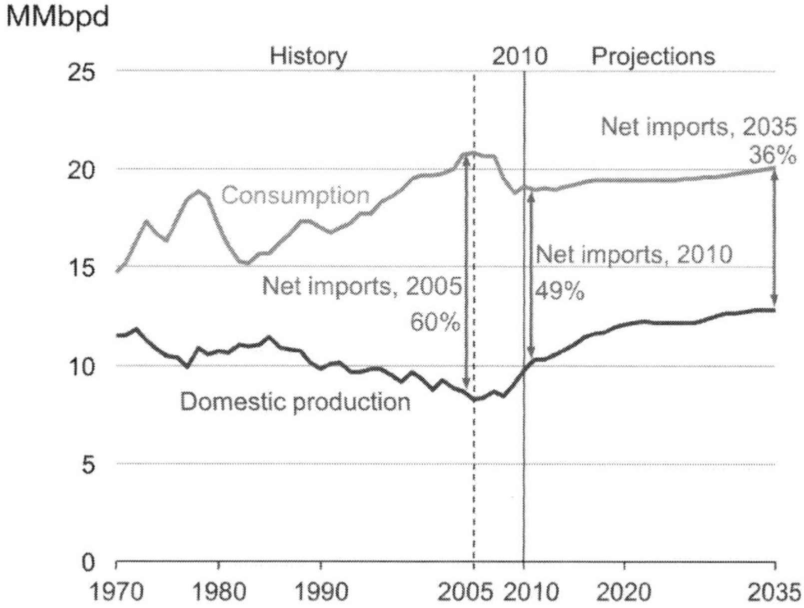

Shrinking US Oil Imports

Surging domestic supply, softer demand

Source: US EIA 2012 Annual Energy Outlook

8

Shale Oil is NOT Oil Shale

Mature oil in low permeability reservoir	Immature oil (kerogen) trapped in shale
Produced via horizontal drilling, hydraulic fracturing	Produced by retorting the shale and cooling the vapors
Economical at current energy prices	Highly capital- and energy-intensive
Bakken, Eagle Ford, Utica, Niobrara, etc.	Green River Formation (CO, UT, WY)
Increasingly referred to as Tight Oil	*Increasingly referred to as shale oil?*

9

Implications for Refiners

Market Trends

Refining	Shale HC sector
Oil price high, margins tight	Natural gas price low
Gasoline growth weak, diesel 2%	Producers moving to liquids plays
US product exports on the rise	Ethane growing as ethylene feed

Increasing interplay between both markets

Widening oil-natural gas price differential

Both sectors in transition

11

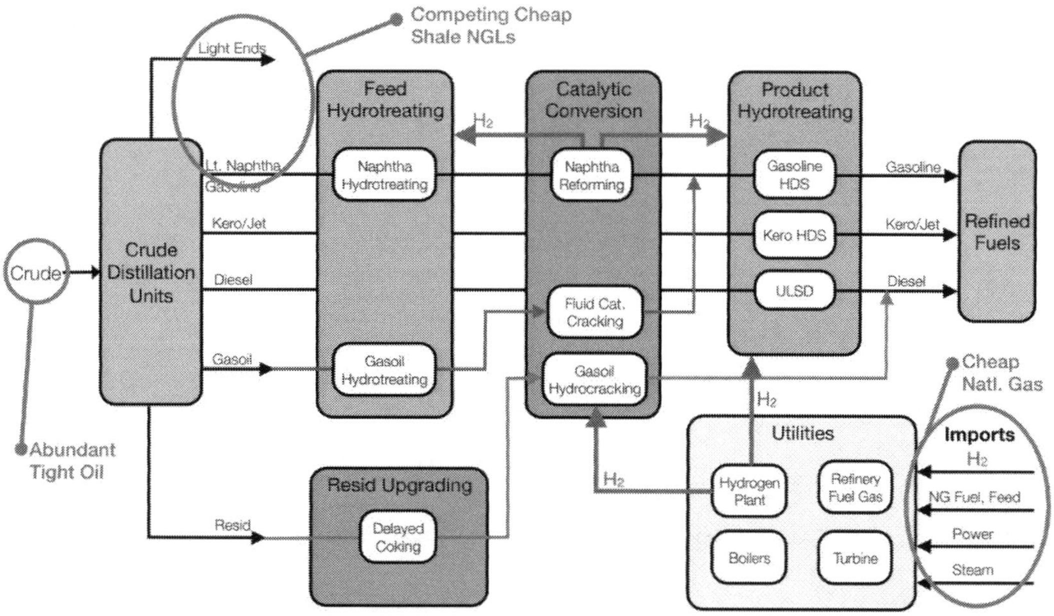

Shale Hydrocarbons: Implications for Refiners

Cheap gas and plentiful light oil: Drivers for US refining

12

Market Trend	Implications for Refiners	Opportunities
Growth in Domestic Light Tight Oil	• Access to the oil not a given • Tight oil displacing light imports	• Study optimal access route(s) • Evaluate processing tight oil
Cheap NGLs from Shale Gas	• Ethylene plants favoring ethane • Capacity loss for C3,C4 olefins • Capacity loss for aromatics • LPG surplus to exacerbate	• Alt. outlets for light naphtha • FCC propylene (e.g., ZSM-5) • Boost aromatics production • LPG disposition alternatives
Cheap Natural Gas	• Cheaper fuel NG • Lower-cost H_2	• Switch fuel to NG; Co-generation • Use H_2 for vol. gain, upgrading

13

Implications for Petrochemicals

Impact of Steam Crackers Switching to Ethane

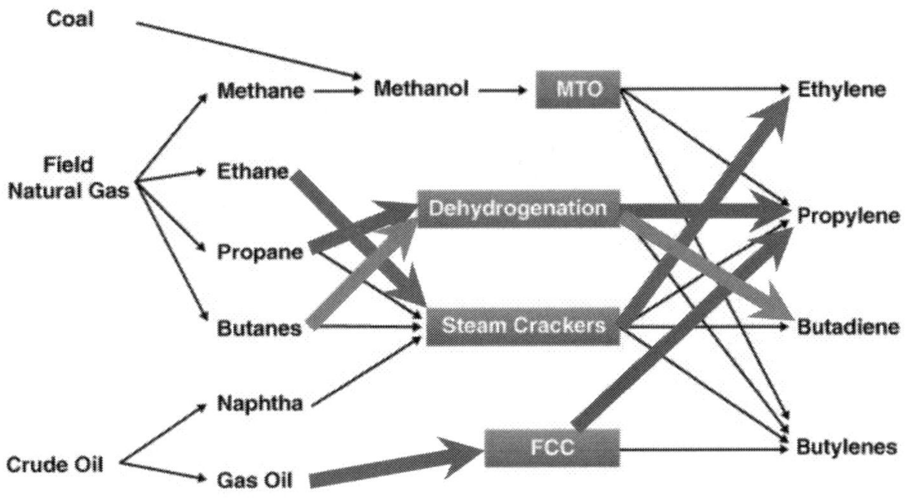

Figure Source: UOP website (Emphasis added.) 15

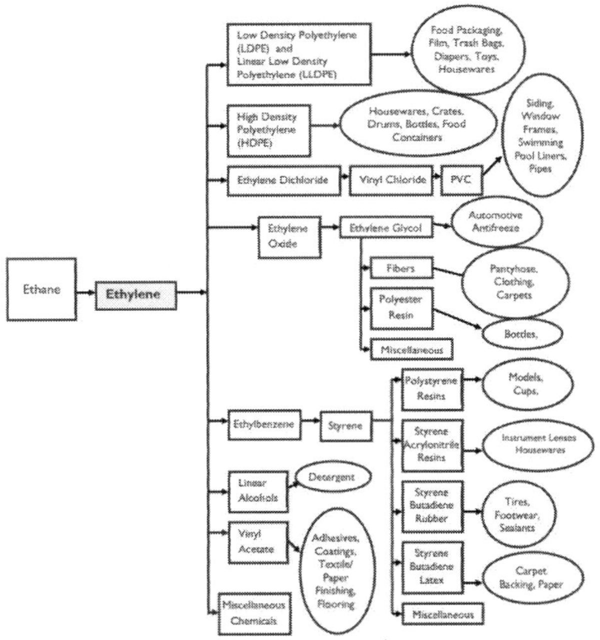

Ethylene Value Chain

Downstream benefits of switch to cheap ethane.

Source: Shale Gas Report, American Chemistry Council, March 2011. 16

Market Trend	Implications for Petchem Manufacturers	Opportunities
Plentiful cheap ethane from shale gas	Potential to lower ethylene costs	Switch cracker feed to ethane
Loss of supply capacity for C3,C4 olefins and aromatics	Supply-demand imbalances and resulting price swings	• On-purpose C3/C4 dehydro • Backup sources (e.g. refinery)
Cheap Natural Gas	Cheaper fuel NG	Switch fuel to NG, Co-generation

17

Shale Gas Implications for Process Industries

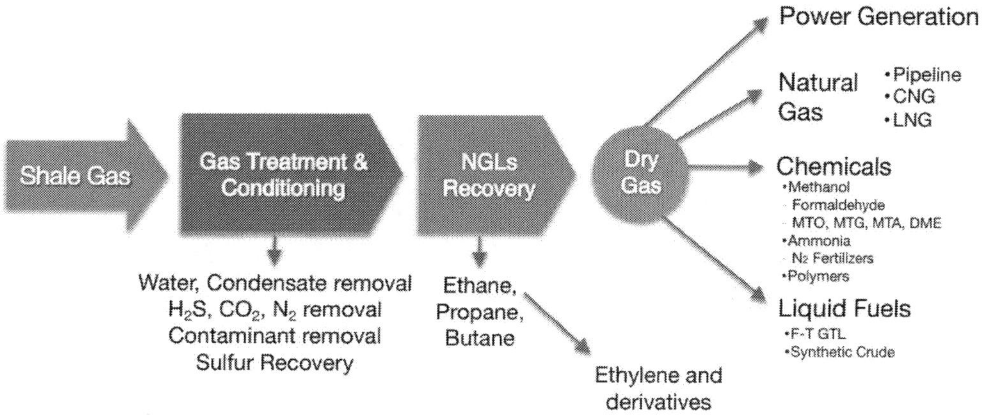

Shale Gas Implications for Process Industries

Source: Gunaseelan paper on associated gas monetization, GPA 2009

19

Recap

Highlights

- Shale hydrocarbons are resurrecting US oil & gas production, reserves
- Widening oil/NG price gap driving shale liquids and interest in LNG, GTL
- US Refiners benefiting from cheap NG, tight oil, and cheap H_2
- Cheap ethane driving a US Petrochemicals resurgence
- Cheap NGLs creating a surplus of LPG, light naphtha
- Lost supply of C3, C4 olefins, aromatics creating market opportunities
- **The Shale Hydrocarbon Surge: A Driver for Process Industries**

21

Questions?

Praveen Gunaseelan
+1-832-429-7650
praveen@vantagepoint-energy.com

Matt Thundyil
+1-936-697-9100
matt.thundyil@gmail.com

VANTAGE POINT
ENERGY CONSULTING

Transcend Solutions

Refinery-Wide Topics for Profitability, Reliability and Safety

Mike Siravo

FLUOR®

1

OBJECTIVE

- **Raise your awareness of the problems lurking among the many activities performed in a complicated facility like an oil refinery**
- **Activities like:**
  - ↗ Turnarounds
  - ↗ Unit Monitoring
  - ↗ Capital Projects

FLUOR®

2

METHODOLOGY

- Review several case studies to identify learning's
- Discuss what went wrong and why

FLUOR®

3

Common Themes in Incidents

- Poor communication
- Lack of attention to detail
- Violation of procedures

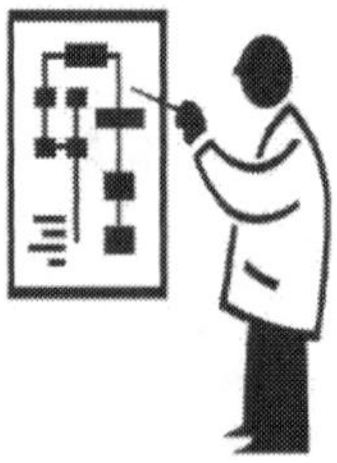

FLUOR®

4

Case Study 1

An Operational Mistake

(Know & Follow Procedures)

FLUOR®

5

Background

- Refinery Crude Unit was nearing the end of its four week shutdown/turnaround
- Lack of crude processing over the turnaround period had significantly reduced the rates of those refinery units that stayed on-line; units like the FCCU, Alky and Hydrocracker
- These units and others were able to operate due to inventories of feedstock that were generated before the Crude Unit was shutdown
- These inventories were getting low

FLUOR®

6

Background

- Gasoline blending was also reliant on inventories of certain blend components
- These inventories were also getting low
- So it was **imperative** that the Crude Unit start up on time.

FLUOR

7

Background

- **Any delay in S/U was unacceptable**
- **There was a very strong sense of urgency for this startup to begin and be completed**
- **This was the refinery personnel's most important task and everyone was focused on meeting schedule**

FLUOR

8

The Activity Description

- One of the activities during the turnaround was to perform a light oil wash of some of the heavy oil equipment
- Disposal of this wash oil was necessary to proceed with commissioning of the unit for startup
- In the evening, it was discovered that the original plan for disposing of the contaminated wash oil was no longer an option. The trucks scheduled to haul away the oil were not coming.
- A scramble to identify an alternate disposal method began

FLUOR®

9

The "Alternate Solution"

- Some ingenious soul discovered an old, idle line that could be used to connect the Crude Unit vessel holding the oil to a heavy oil tank in the Offsites area of the refinery
- Sometime in the past, the need for flexibility had added these connections for a reason no one could now remember
- This <u>appeared</u> to be great news

FLUOR®

10

"Alternate Disposal Method"

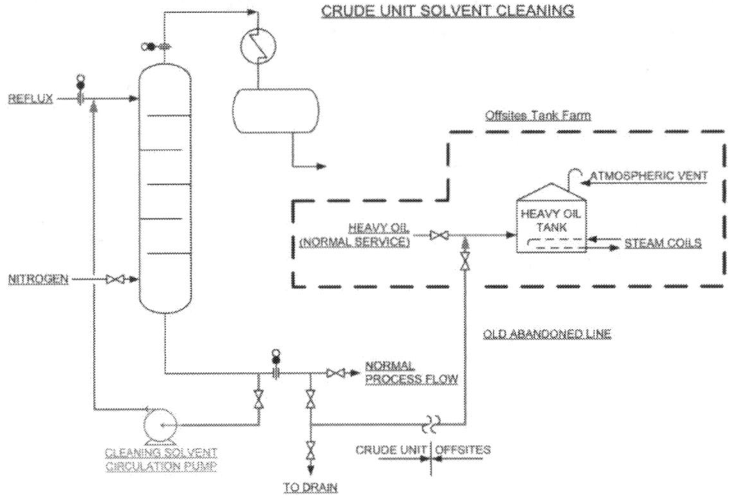

FLUOR.

WHAT HAPPENED NEXT

So with their hastily prepared plan in mind the:

- Crude Unit contacted the Tank Farm area
- They learned that the tank was empty
- But, the Tank Farm foreman would not take the oil transfer because:
  - He did not know the oil properties
  - His procedure placed limits on flash point
  - This was not a normal operation and he was nervous
  - His first instinct was that this was not safe

FLUOR.

WHAT HAPPENED NEXT

- A standoff occurred
  - They believed there was **no time** to get a sample analyzed
- But …. after an hour or so the Tank Farm foreman relented
- Why did he give in?
  - Because he did not want to be the person that delayed the Crude Unit start up. He did not know if he could defend that action.
  - Because he was trained that "Offsites" existed to support the operating units and should never hold them back

FLUOR.

13

WHAT HAPPENED NEXT

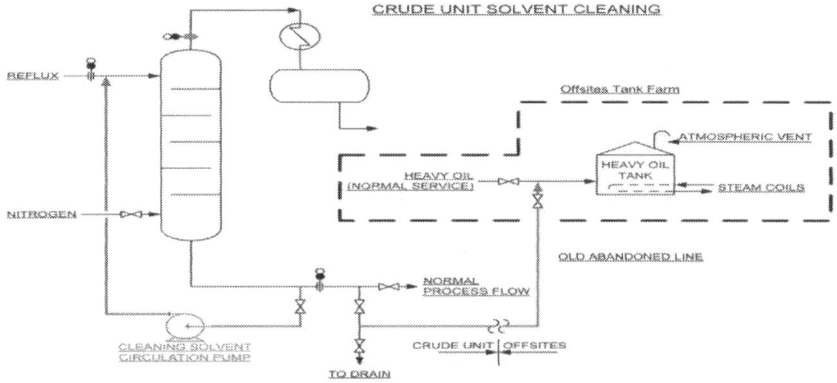

- The light oil transfer was made from Crude Unit to the atmospherically vented, steam heated Offsites heavy oil tank

FLUOR.

14

DID THE RISK PAY OFF? (Not So Much)

- Within 6 hours the tank exploded
- Roof was launched some distance damaging another tank
- The heat from the fire damaged the adjacent tanks
- The light oil was generating a vapor in the tank
- The ignition source was determined to be either pyrophoric iron in the tank or a thunderstorm that passed overhead around that time.

FLUOR®

15

Learnings:

- Management has to create the proper safety culture: Anyone has the right to stop work that is unsafe
- Always take the time to think it through and do it right
- Do not take risks
- Follow procedures (this incident would have been avoided)
- Involve next level of management in non-normal situations

FLUOR®

16

Case Study 2

Poor Unit Monitoring

and

Poor Change Management

FLUOR®

Background

In this example:

- A diesel hydrotreater reactor was loaded with the latest generation of high activity catalyst
- The predicted catalyst life was 2 years
- In first 3 months, rapid deactivation was observed by the unit foreman (the reactor temperature requirement was increasing too quickly over the early weeks of the run)

FLUOR®

Hydrotreater System Sketch

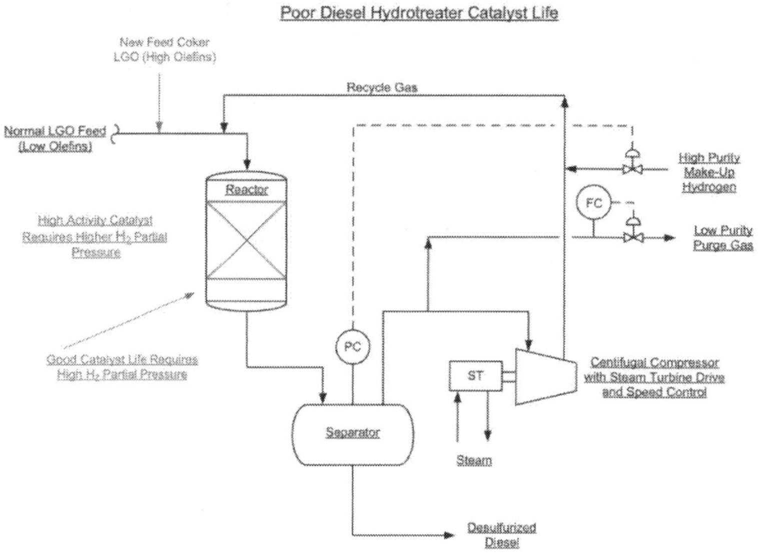

FLUOR

19

Study Team was Formed

Study Team Findings:

- Deactivation rate was actually <u>three times higher</u> than expected, and increasing rapidly
- Catalyst was now predicted to reach EOR after only 6 months, 18 months earlier than planned
- Primary finding: Low Hydrogen partial pressure in reactor outlet causing gum and coke formation

FLUOR

20

Why was hydrogen PP low?

FLUOR®

21

What had changed

- **Feed now contained more cracked stock**
  - This increased hydrogen consumption
- **Higher activity catalyst:**
  - Increased the hydrogen consumption as aromatics saturation increased somewhat over past runs
- **Recycle Gas Rate was lower than design due to:**
  - Very low molecular weight gas in the centrifugal compressor
  - Compressor speed was below target

FLUOR®

22

More Findings

- The hydrotreater actually had marginal but barely acceptable hydrogen PP in previous run
- Changes made for this run reduced hydrogen partial pressure in the reactor outlet by 25% - 35%
- This was enough to reach a point of rapid deactivation

FLUOR®

23

What Happened Next

As a result:

- Catalyst lasted only <u>8 months, total</u>
  - -At reduced feed rate
  - -With improved compressor performance

- Fresh catalyst needed to be procured on very short notice (at premium price, $$$)

FLUOR®

24

What Happened

Additionally:

- An unplanned shut down for catalyst change was required
- This resulted in a disruption to the refinery operating plan just prior to entering the profitable gasoline season ($$$)
- The total cost of the incident was <u>several million dollars</u>

FLUOR®

25

Learnings

- **This was a failure to manage change**
- **There was a need to thoroughly study impact of changes**
  - Feed stock quality change
  - Higher activity catalyst
- **Hydrotreater was not being properly monitored by Process group**
  - Why was the unit foreman the one who discovered the rapid deactivation
  - Why was compressor performance allowed to deteriorate unnoticed

FLUOR®

26

Case Study 3

Equipment Installation Failure

FLUOR®

27

Case Study 3

- FCCU debutanizer was being revamped to:
  - Increase capacity and to allow for increased vapor and liquid traffic for improved separation
- Improved separation was needed to:
  - Recover additional C4's from the column bottoms gasoline, to reduce its blending vapor pressure
  - And reduce the C5 content of the overhead product to improve Alkylation Unit performance

FLUOR®

28

Debottleneck Study Findings

The revamp study found:

- Reboiler, condenser and reflux pumps all had the necessary spare capacity
- Bottom/Stripping section trays had the required capacity
- Top section trays were the only bottleneck to increased capacity and improved separation

FLUOR®

29

Debottleneck with Packing

The revamp plan:

- During the FCCU turnaround, replace top section trays with structured packing beds with liquid distributor trays above each bed

FLUOR®

30

Debutanizer System Sketch

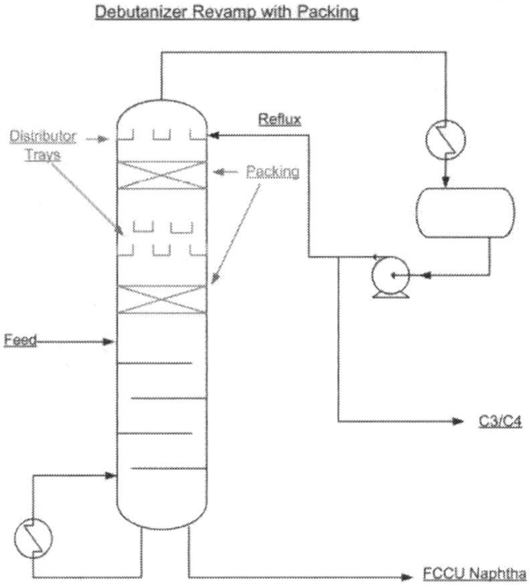

FLUOR®

31

How did the Revamp Work

So how did the revamp work:

- Post-Start Up: Initially the Debutanizer performance was good
- But … a problem developed when feed rate approached 60% of design
- An alarm occurred on the top packed bed for high differential pressure
- Overhead and bottoms product sampling confirmed a problem as the quality of separation was very poor.

FLUOR®

32

What Happened

- Investigation began while FCCU ran at only 60% rate
- The throughput of the entire refinery was being held back
- The only viable theories were plugging of the packed beds or design error
- An FCCU shut down for internal inspection was unimaginable

FLUOR®

33

What Happened

- Regardless, after 5 days of study with no answers the FCCU was S/D to open and inspect the debutanizer
- Upon looking in the manways, the problem was obvious

FLUOR®

34

Collector-Redistributor Tray as Intended to be Installed

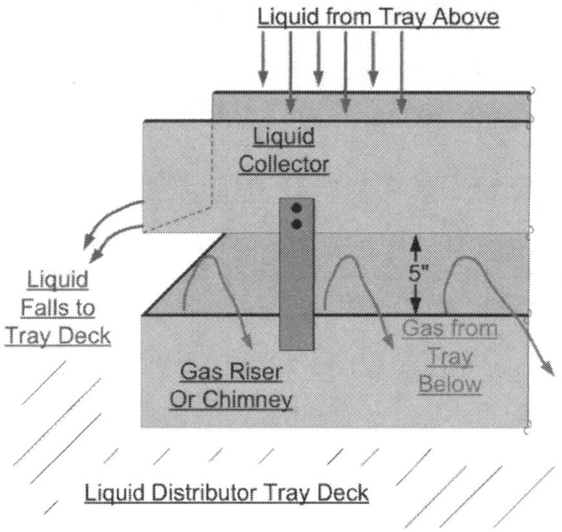

FLUOR®

35

Tray as Actually Installed

Liquid Collector is Open Downward (Upside-Down)

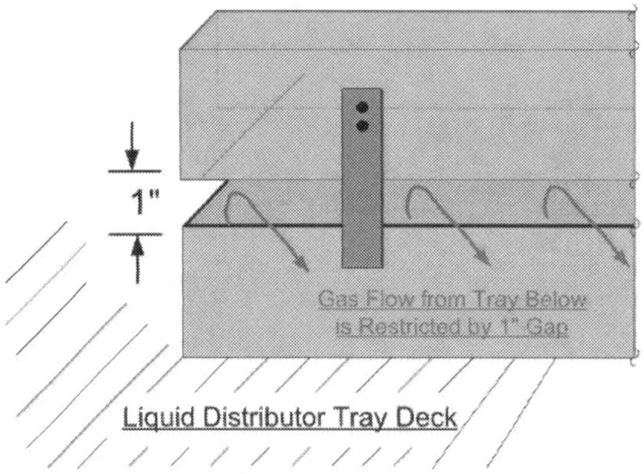

FLUOR®

36

Why Did This Matter

- The reduced area for gas flow increased tray dP
- The increase in tray pressure drop caused liquid level to rise
- Eventually tray liquid level exceeded height of the chimneys
- This restricted gas flow more which further increased pressure drop and liquid level
- Ultimately level rises and floods the entire top packed bed

FLUOR®

37

What was the Cost

- **The tray was fixed at little expense**
- **Cost of lost production was $2-3MM**

FLUOR®

38

Learnings:

- The process design engineer was reassigned and uninvolved in the equipment installation
- The unit engineer was new and mostly unaware of the project
- No process person inspected the newly installed equipment
- Supervision did not manage the transition from one engineer to the other so a fundamental inspection never happened

FLUOR®

39

Learnings:

- Poor communication among engineering staff
- Poor attention to detail

FLUOR®

40

QUESTIONS? ? ?

Mike Siravo

Fluor, Sugar Land

281-263-8460

FLUOR®

41

Global Biofuels Outlook:
Policy, Market and Technology Trends

October 5, 2012

Global Biofuels Center
GBC

HARTENERGY

Who We Are

★ HARTENERGY
Office Locations
www.hartenergy.com

- Established in 1973
- Privately Owned
- 10 Global Offices
- 180 Employees
- Headquartered in Houston

HARTENERGY

Breadth of Downstream Coverage

Comprehensive Resources & Analyses on All Fuel Related Matters

Hart Energy knows the Global Fuels Market!

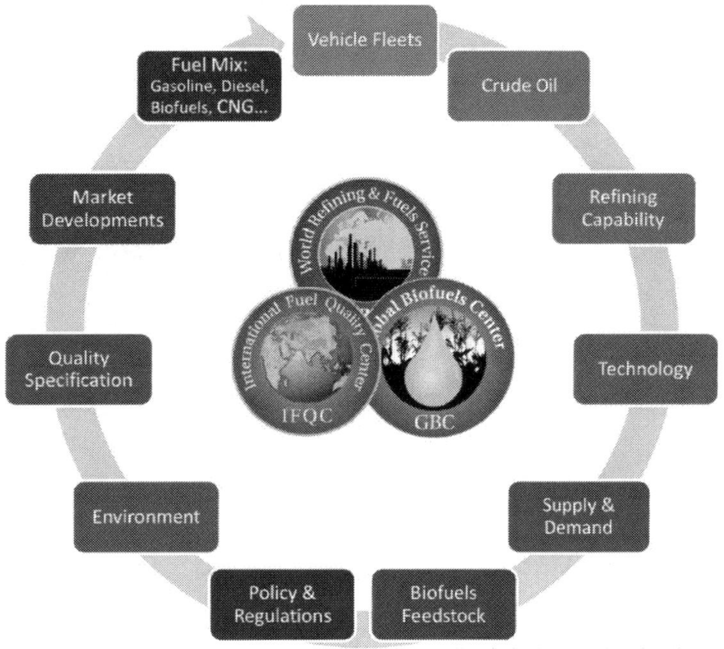

HARTENERGY

Presentation Overview

Petroleum Product Overview
- Regional Product Demand Growth & Share
- Refined Product Market Share

Global Biofuels Overview
- Global Trends
- Global Mandates
- Next Gen Biofuels

Ethanol
- Supply and Demand
- U.S. Ethanol & RFS2
- Brazilian Ethanol Supply & Demand

Biodiesel
- Supply & Demand
- Biodiesel in the EU Countries
- Biodiesel in the Major Producing Countries

Conclusions

HARTENERGY

Petroleum Product Overview

HARTENERGY

Regional Petroleum Product Demand Growth

Or...Why Do Biofuels Still Matter?

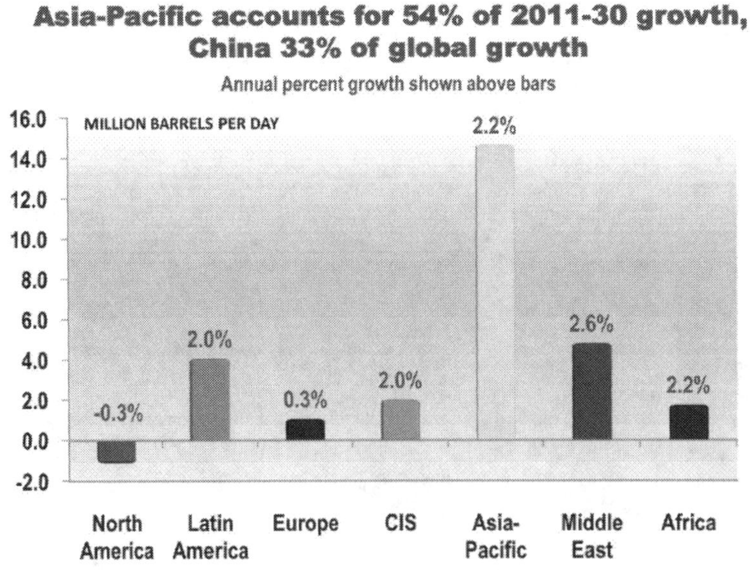

Source: Hart Energy's World Refining & Fuels Service, 2011

HARTENERGY

Refined Product Market Share

Distillate dominates refined product market, accounting for nearly half of refined product growth

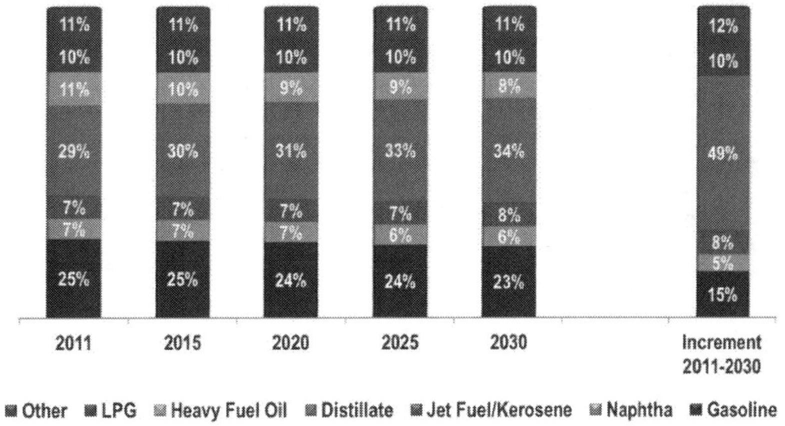

Source: Hart Energy's World Refining & Fuels Service, 2011

HARTENERGY

Global Biofuels Overview

HARTENERGY

Global Biofuels Trends and Outlook

Biofuels Largely Rely on Food-Based Feedstocks
- Biodiesel and renewable diesel prices are forecasted to be uncompetitive with petroleum products and depend on troubled mandates

Difficulties in Implementing Far-Reaching Biofuels Programs in the U.S., EU
- ILUC is a headache for EU policymakers

Brazil Considered the "Linchpin" but Internal Market Uncertainties
- Internal market demand and constraints make export volumes cyclical and uncertain

Next generation/advanced biofuels largely not available
- Technologies still unproven and scale-up has been very slow

Despite these constraints: Biofuels consumption will continue to grow!
- Ethanol: Strongest growth from Asia Pacific but the U.S. and Brazil remain largest consumers
- Biodiesel: Strongest growth from Asia pacific but the EU remains the largest consumer

HARTENERGY

Global Biofuel Mandates in 2012

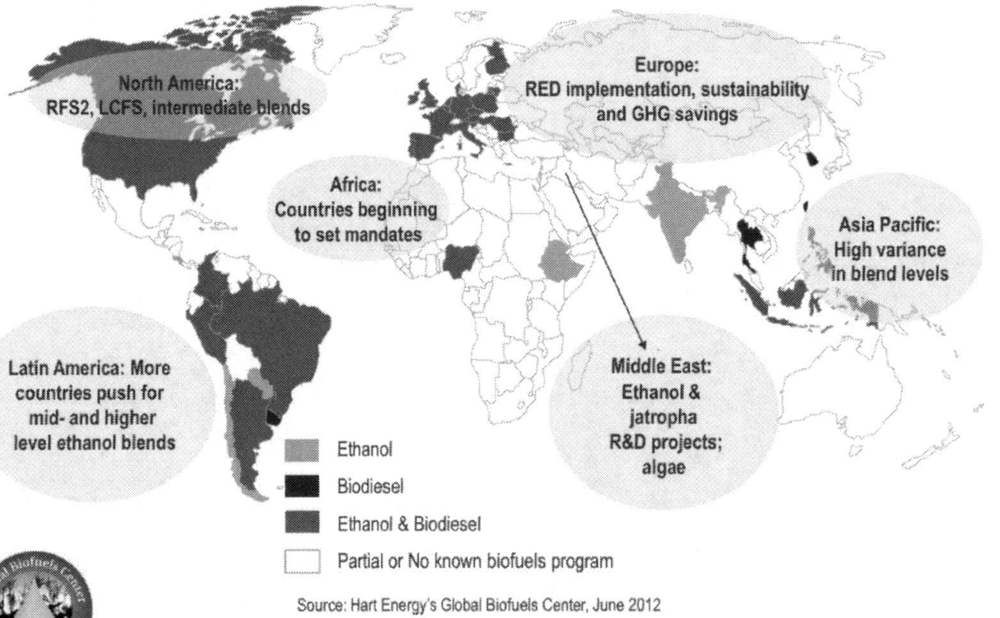

Source: Hart Energy's Global Biofuels Center, June 2012

HARTENERGY

Next Generation Biofuels Outlook

Hart Energy projects less than 800 million gallons commercial-scale production by 2020 in the U.S.

25 operating next generation biofuels pilot/demo plants in the U.S.

- 18 cellulosic ethanol plants with capacities ranging from 1,600 gallons to 1.5 million gallons per year
- 5 FT or HVO renewable diesel plants: undisclosed to 0.07 gallons to 5 million gallons per year
- 2 for biogasoline or biojet

2 operating commercial plants in the U.S.

- Gevo: 18 million gallons per year of butanol
- Dynamic Fuels: 18 million gallons per year of HVO renewable diesel

HARTENERGY

Evolution of Next Gen Potential Capacity

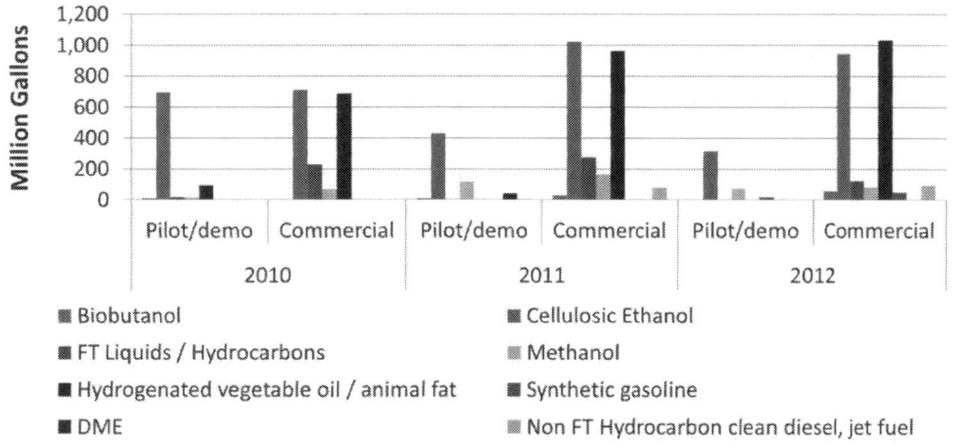

Note: Includes all capacity (operating, under construction and proposed)

Source: Hart Energy's Global Biofuels Center, June 2012

HARTENERGY

Ethanol

HARTENERGY

Global Supply & Demand for Ethanol

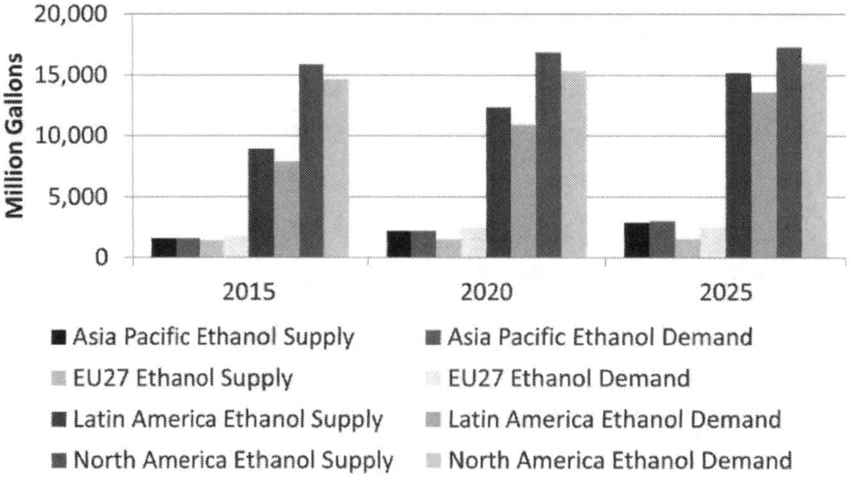

Source: Hart Energy's Global Biofuels Center, September 2012

HARTENERGY

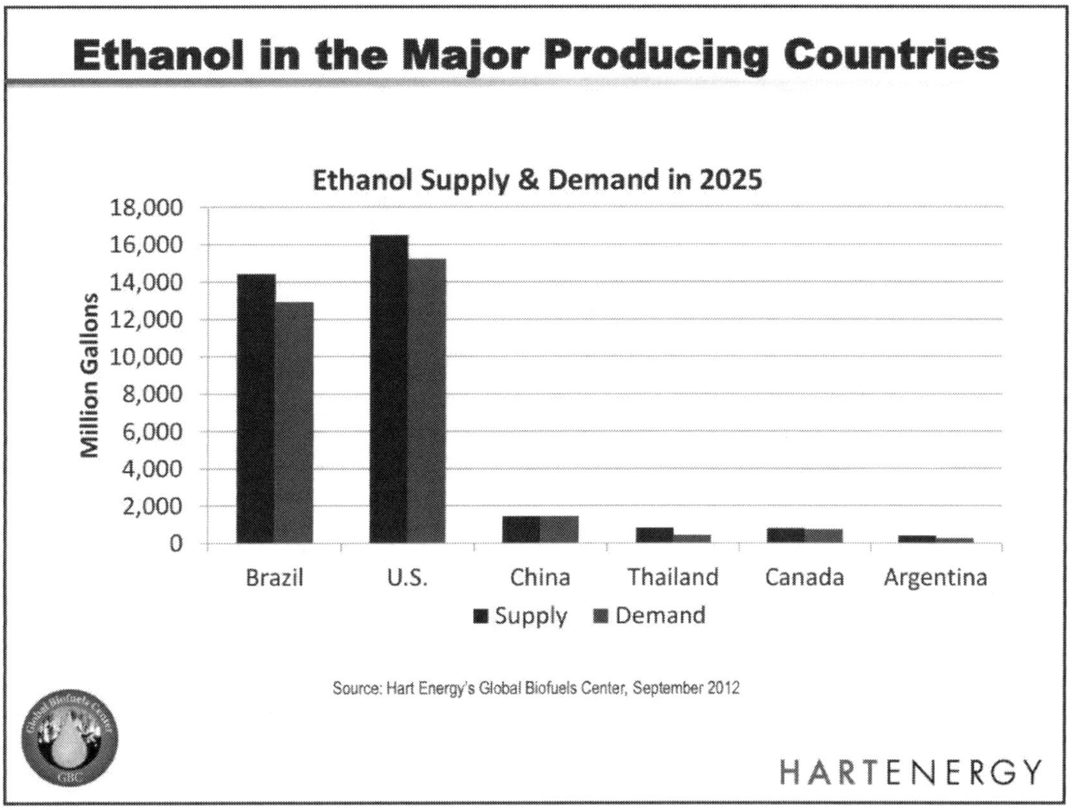
Ethanol in the Major Producing Countries
Ethanol Supply & Demand in 2025
18,000
16,000
14,000
12,000
10,000
8,000
6,000
4,000
2,000
0
Million Gallons
Brazil
U.S.
China
Thailand
Canada
Argentina
Supply
Demand
Source: Hart Energy's Global Biofuels Center, September 2012
HARTENERGY

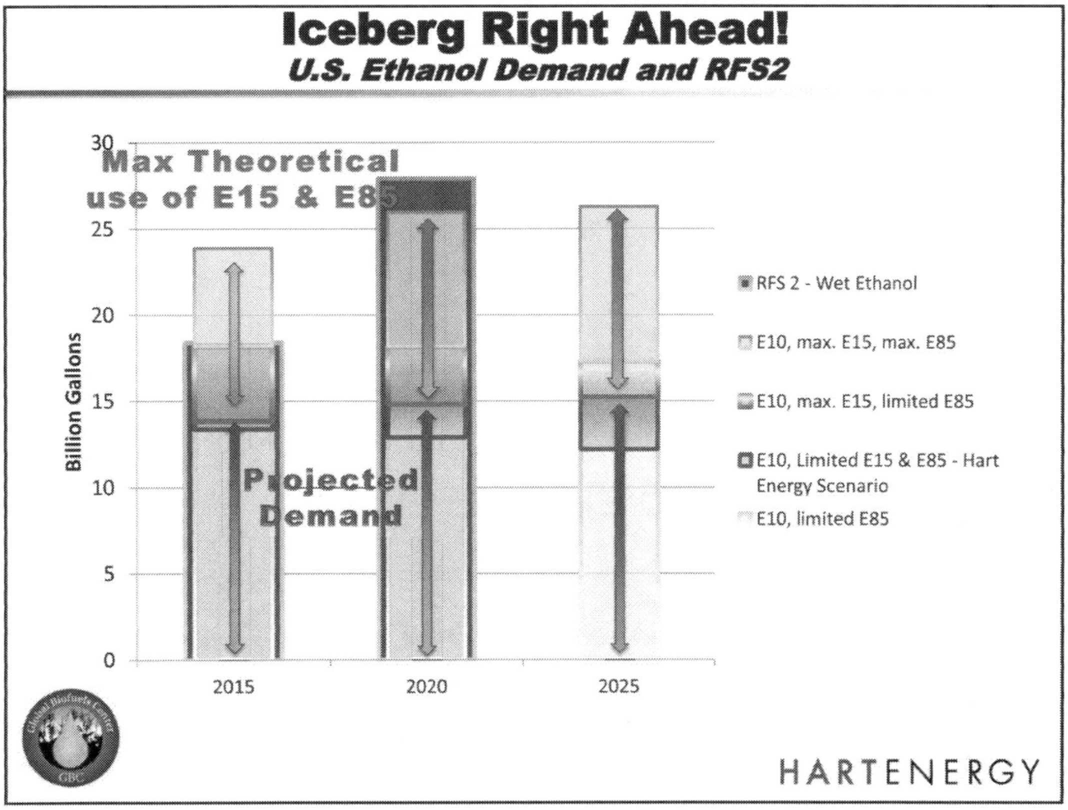
Iceberg Right Ahead!
U.S. Ethanol Demand and RFS2
30
25
20
15
10
5
0
Billion Gallons
Max Theoretical use of E15 & E85
Projected Demand
2015
2020
2025
RFS 2 - Wet Ethanol
E10, max. E15, max. E85
E10, max. E15, limited E85
E10, Limited E15 & E85 - Hart Energy Scenario
E10, limited E85
HARTENERGY

Ethanol Demand/Supply Balance

Upsurges in domestic demand – driven by softer prices – will periodically drive down net export availability. Even during peak years of availability, net exports will fall below most forecasts (e.g., Ministry of Agriculture, ICONE, UNICA, etc.)

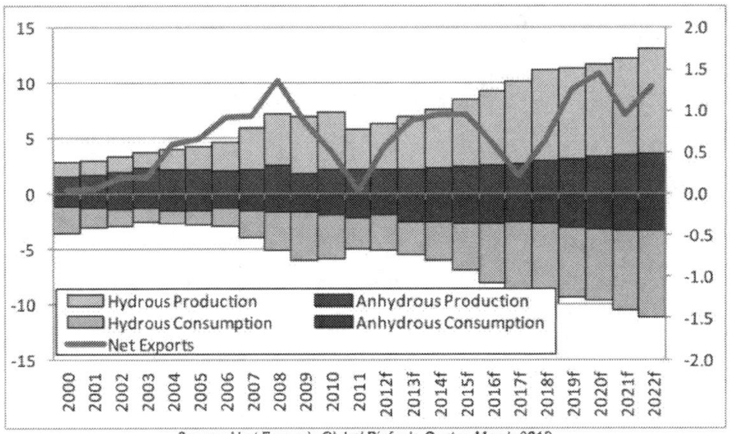

Source: Hart Energy's Global Biofuels Center, March 2012

HARTENERGY

Biodiesel

HARTENERGY

Global Supply & Demand for Biodiesel

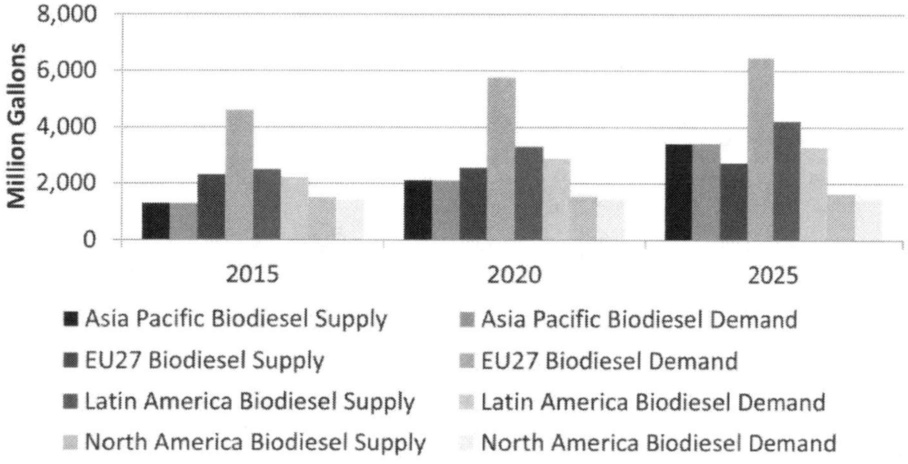

Source: Hart Energy's Global Biofuels Center, September 2012

HARTENERGY

Biodiesel in the Major EU Countries

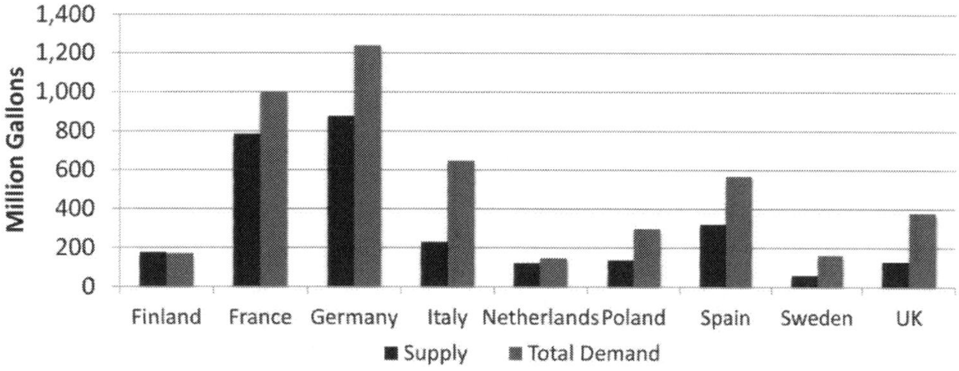

Source: Hart Energy's Global Biofuels Center, September 2012

HARTENERGY

Biodiesel in the Major Producing Countries

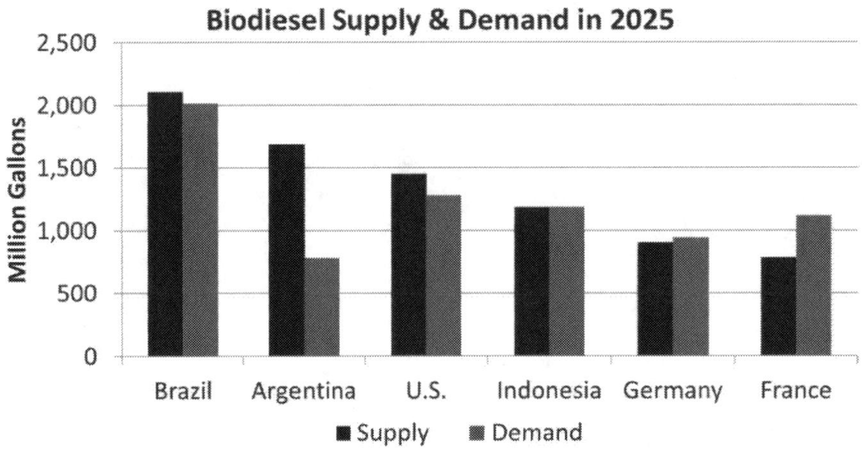

Source: Hart Energy's Global Biofuels Center, September 2012

HARTENERGY

Conclusions

HARTENERGY

Conclusions

Ethanol: No Shortage Until 2025

- The main ethanol consumer, the U.S., has hit a blend wall and cannot consume more ethanol until E15 penetrates the market. Vehicle fuel economy improvements will drive gasoline consumption down thus reducing even more the ethanol blending potential.
- The Brazilian ethanol market suffers from competition with the sugar industry
- The EU ethanol market is also limited by blending levels and declining gasoline consumption
- These factors have led to a reduced ethanol demand and potentially over supply until 2025

Biodiesel: Large Deficit Expected Until 2025

- The use of biodiesel in off-road applications creates a very large demand in the EU27
- Great uncertainties with regard to the sustainability of available supplies
- Argentina seen as major potential supplier to the global market, but hurdles seen in the domestic (price setting, taxes) and export markets (export taxes, sustainability)

Biofuels: Global Demand Equivalent to 5.4% by Energy of Gasoline and Diesel Pool by 2025

- 60% of the demand is to be met by ethanol

HARTENERGY

Thank You!

Paul Niznik
Biofuels Manager
+1.713.260.1073
pniznik@hartenergy.com

HARTENERGY

Addressing Tier 3 Regulations in a Declining Gasoline Market

Delphine Largeteau
Technology Manager
Axens
Houston, TX - USA

Many thanks to:
Annick Pucci

Agenda

- **Tier 3 overview**
- **FCC gasoline post-treatment solutions**
- **FCC feed pre-treatment solution**
- **Other options downstream of FCC**
- **Conclusion**

Ultra Low Sulfur Gasoline

EPA Proposal Expected in 2013

- **Average gasoline pool S: 10 ppm**
- **Per gallon S cap: 30 - 80 ppm?**

Expected Implementation 2015 – 2018
Incentives for early compliance

Current Situation in North America

- **Steady / declining gasoline demand forecast**
- **North Am. refiners typically octane long**
  - **Ethanol blending**
- **Gasoline margins < middle distillates**

  ➡ **Diesel projected to grow**

Addressing Tier-3 at minimum cost with incentive to adjust product slate

Tier 2 Approach in the U.S.

- 50-100 ppm S required from FCC naphtha
  - Depending on FCCN contribution to pool
- ~80% of refineries installed post-treaters

- Octane barrels high value early-mid 2000s

Tier 3 Approach

- 10-25 ppm S FCC naphtha required
- Other naphtha streams with S can no longer blend directly to gasoline pool (e.g. LSR)
- Many factors affect optimum solution
  - Adequacy of existing CFHT / post-treater
  - Future sour crude slate?
  - Octane value
  - H_2 availability
  - Solution in sync with FCC cycle
  - Targeted product slate

Agenda

- Tier 3 overview
- FCC gasoline post-treatment solutions
- FCC feed pre-treatment solution
- Other options downstream of FCC
- Conclusion

Prime-G+ Commercial Status

- **206 Prime-G+ licenses**

140 units already operating & 100% successful S/U

➡ **Most widely used FCC gasoline HDS technology**

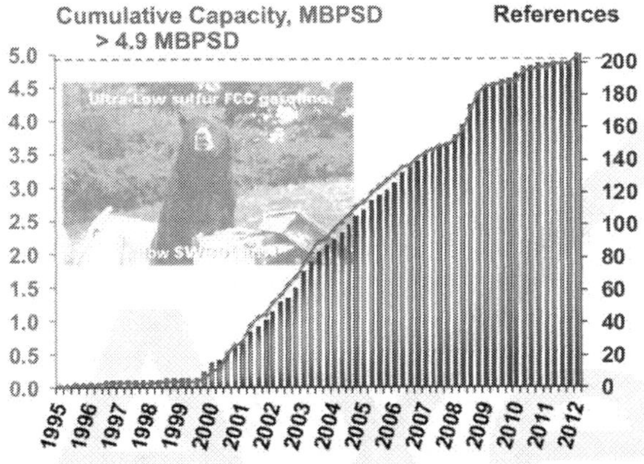

In 2003
Kirkpatrick Chemical Engineering Achievement Honor Award

"Chemical Engineering" magazine

FCC Gasoline Composition

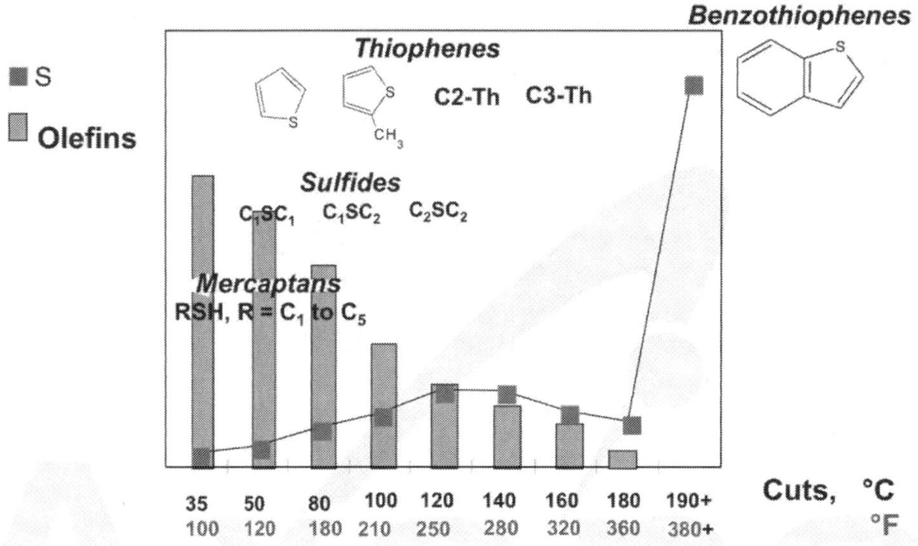

Irving Oil April 2012 – Proprietary & Confidential Information

Prime-G+ Typical Scheme 10-50 ppm S Specification

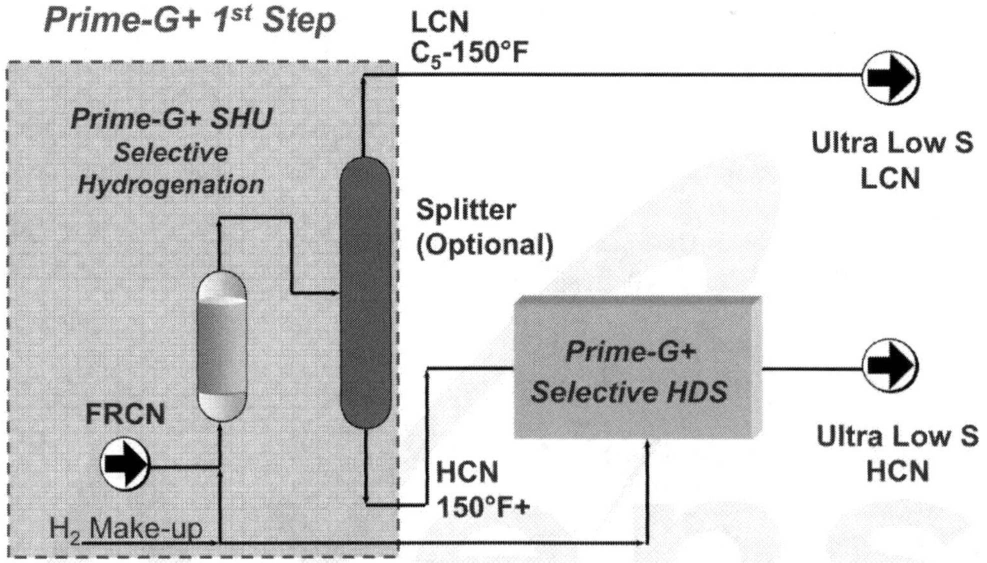

AIChE 2012 Regional Meeting – League City, TX

Prime-G+ SHU Principles

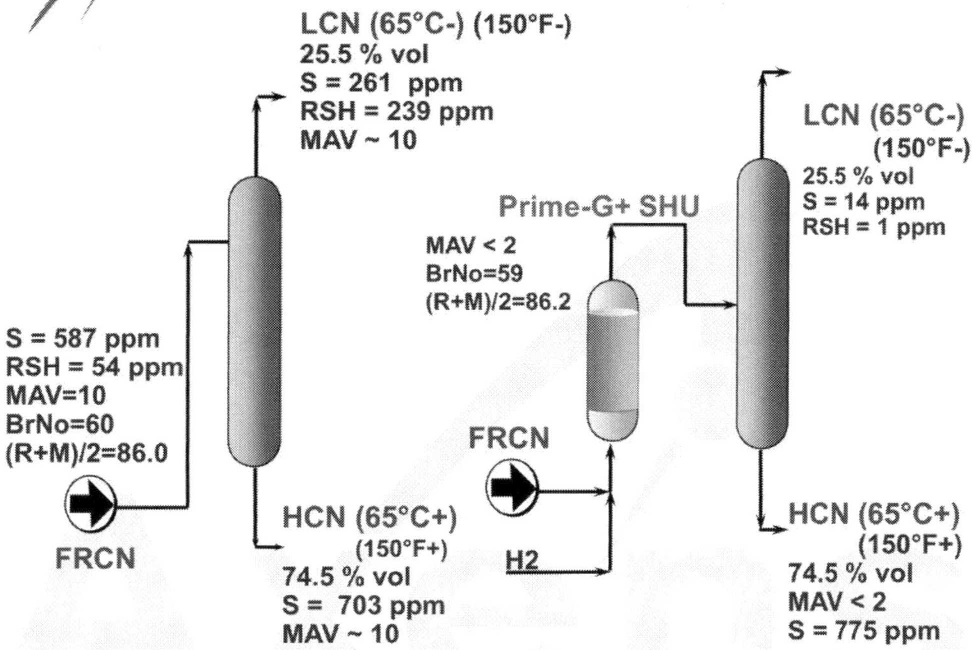

Design for 10 ppm Gasoline S

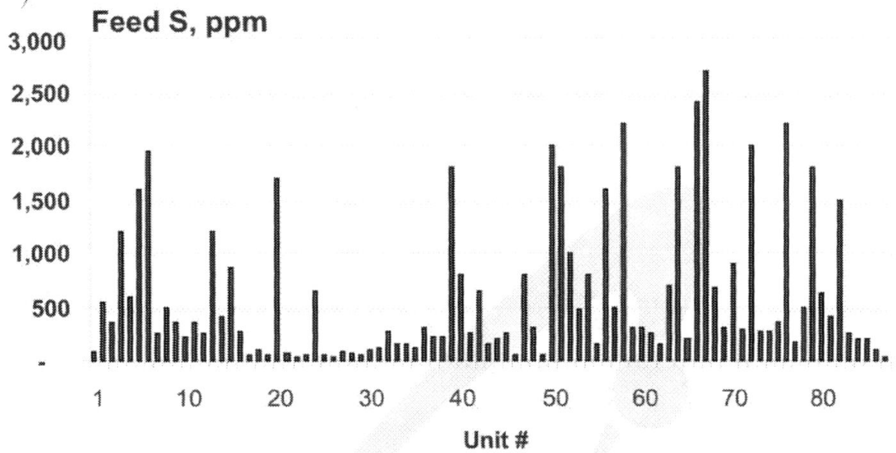

~ 90 Units designed for 10 ppm gasoline S

Median Feed S: 300 ppm

Feed S: 30 – 2,670 ppm

PG+ 1st Step Allows ULS LCN

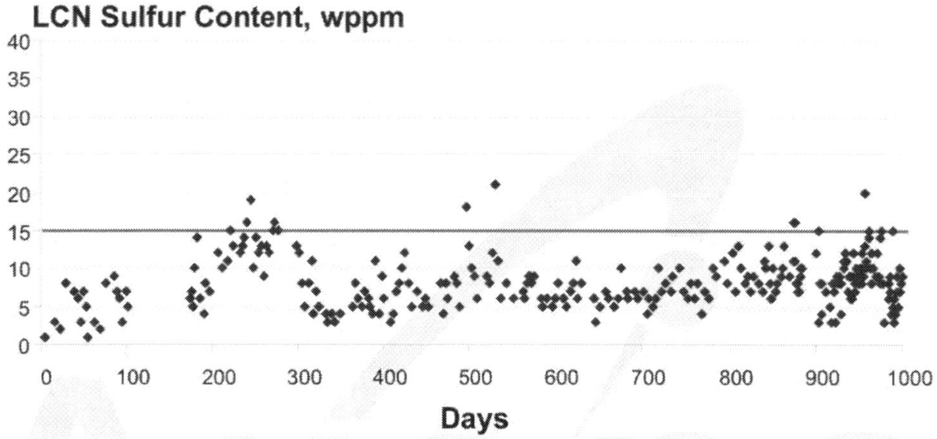

13 AIChE 2012 Regional Meeting – League City, TX

ULSG Spec Despite Varying Feed S

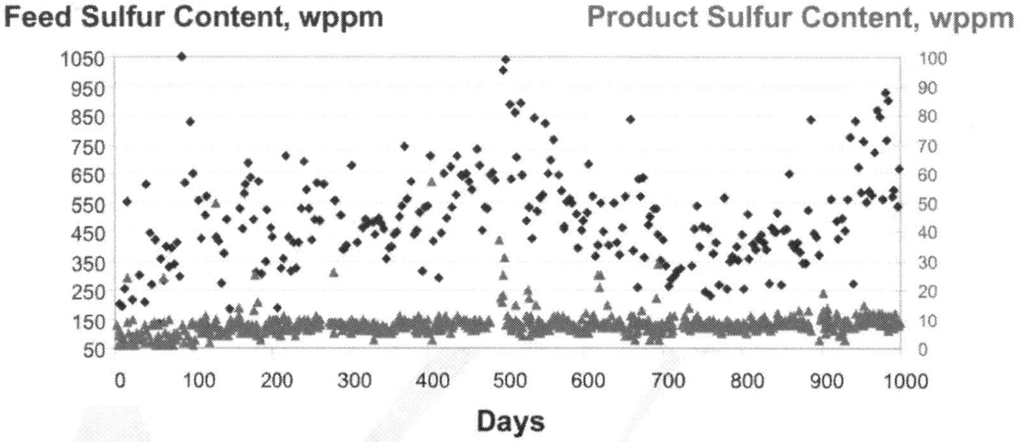

10 ppm operation achieved at 98-99% HDS

14 AIChE 2012 Regional Meeting – League City, TX

Post-treat Basic Strategy

- Basic strategy for existing post-treaters
  - **If splitter, modify LCN cut point to reduce LCN S**
  - **Increase severity in HDS reactor(s)**
- "Do Nothing" Consequences
  - **Reduced cycle length**
  - **Increased octane loss**
  - **Increased Hydrogen consumption**
- Multiple solutions exist
  - **Feed tailoring**
  - **Improved catalysts**
  - **Co-processing of orphan streams**
  - **Process scheme modification**

Prime-G+ Catalysts

PG+ Selective HDS ***PG+ Polishing***

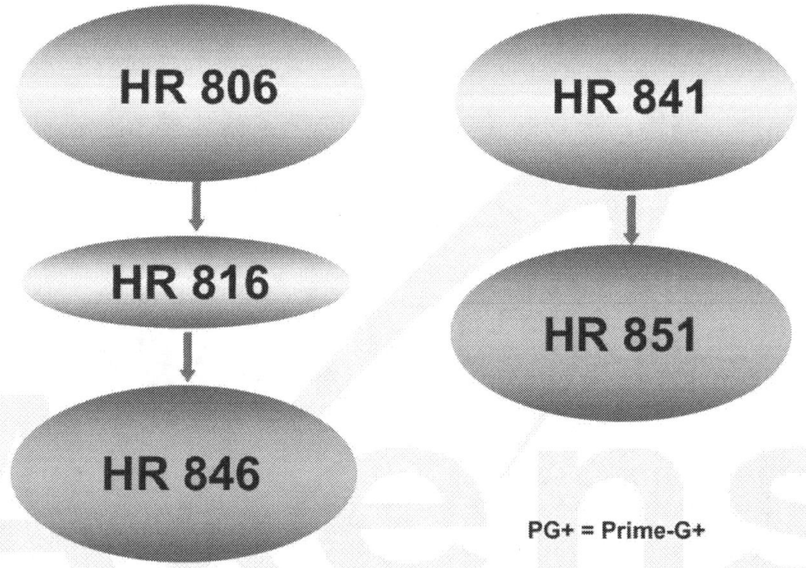

PG+ = Prime-G+

10 ppm S Using Existing Assets

- Co-Process Other S Rich Streams in PG+
  - **Raw Light Straight Run**
  - **Light Coker Naphtha**
  - **Pyrolysis Gasoline**
- Potential Impact
  - **Hydraulics limitation**
  - **Octane loss**
  - **Cycle length reduction**

Post-treater Revamp Options

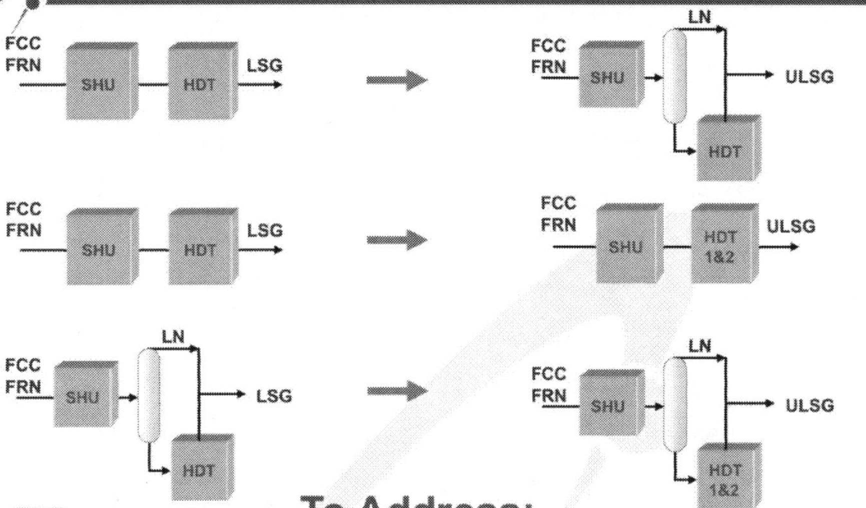

To Address:

- **Hydraulic bottleneck**
- **Product octane**
- **Cycle length**
- **Blending flexibility**

Innovative Option : OATS

Olefins Alkylation of Thiophenic Sulfur

- Sulfur Shift
  - **Addition of Thiophenes onto Olefins**
  - **Formation of Heavy Alkylated Thiophenes**

➡ ***Desulfurization of Light Fraction***

- Olefin Shift
  - **Dimerization of Light Olefins**

  Olefins Reduction in Light Fraction

➡ ***RVP Reduction***

Prime-G+ & OATS

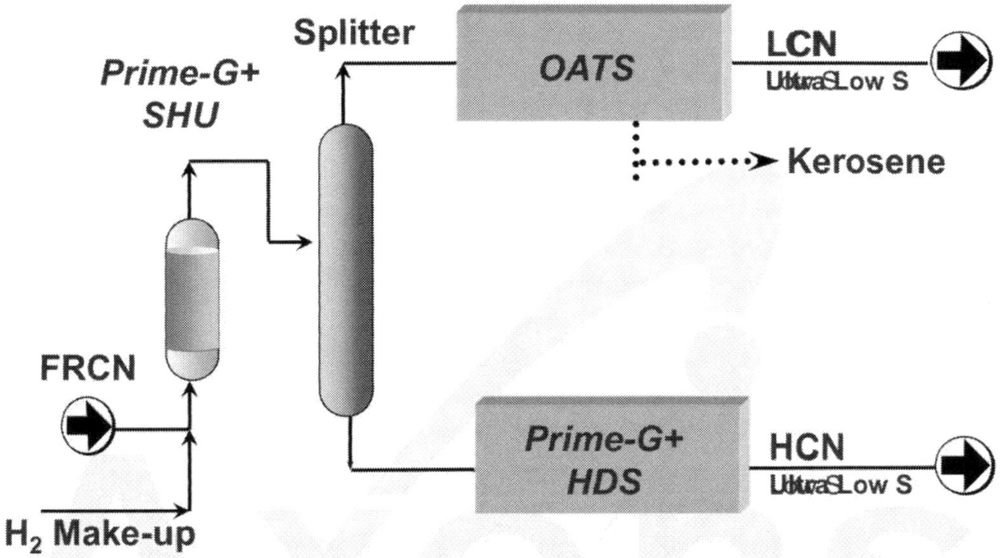

OATS Main Features

- **Fixed Bed Solid Acid Catalyst**
- **Mild operating conditions**
- **Low CAPEX / OPEX**
- **No Hydrogen**
- **Octane neutral**
- **Commercial experience**
  - ➡ **3 units S/U**

OATS Industrial Data

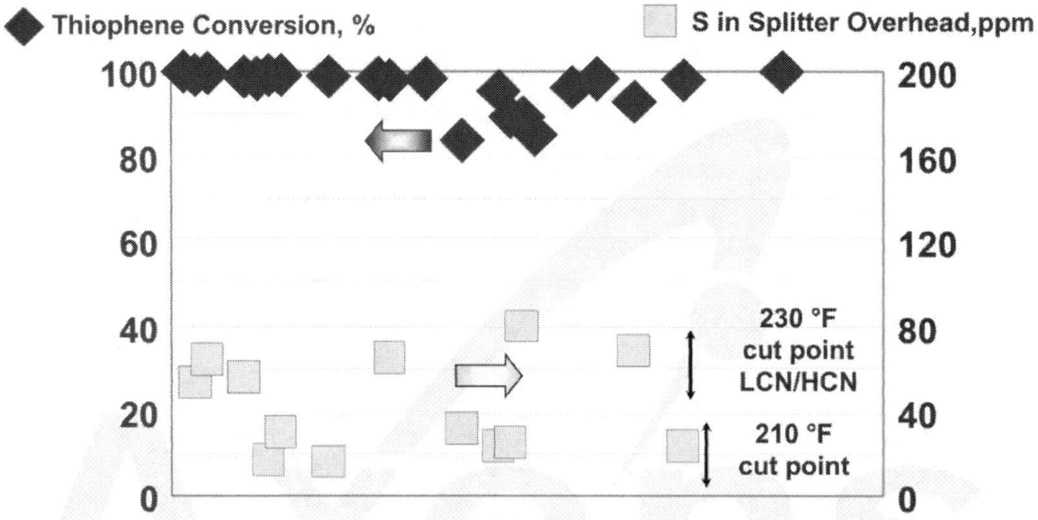

Agenda

- Tier 3 overview
- FCC gasoline post-treatment solutions
- FCC feed pre-treatment solution
- Other options downstream of FCC
- Conclusion

Refineries with CFHT

- Maintain Pre-treat Alone
  - FCC Feed S < 300 ppm to avoid post-treat
  - 4 to 5 year cycle length is challenging
  - Revamp can be very expensive
  - No margin for error
    - *Cycle length, operation, fractionation*
- Install or Revamp Post-treater
  - Often lowest cost option
  - Co-processing of other streams (LSR, LKN)
  - Pre-treat / Post-treat

    ➡ *Max flexibility*

Mild Hydrocracking

Revamp CFHT to MHC

- **MHC expands options for heavier crudes**
- **Diesel margins justify revamp into MHC**
- **Revamp strategy:**
  - ***Optimal solution is direct production of ULSD***
  - ***Minimize changes in reaction section***
  - ***Plan for increased H_2 cons. and quench***
  - ***Proper catalyst selection for new service***
  - ***Limit revamp to product fractionation***

FCC Pretreat & MHC Catalysts

HR 526	HR 568 HT 438 HR 504 HR 544	HR 548 HR 538 HR 508 HRK 658	HDK 766 HDK 776 HDK 786
CoMo	CoMoNi	NiMo	NiW
High HDS activity at very low pressure	Best HDS activity at low to high pressure	Best HDN activity at all pressures	Best Cracking activity for Mid Distillates
⇨ Lower H_2 consumption	⇨ Highest HDS activity	⇨ Highest HDN activity	⇨ Cracking + Selectivity

Pretreatment → Mild Hydrocracking

Agenda

- **Tier 3 overview**
- **FCC gasoline post-treatment solutions**
- **FCC feed pre-treatment solution**
- **Other options downstream of FCC**
- **Conclusion**

Case Study
Max Gasoline Configuration

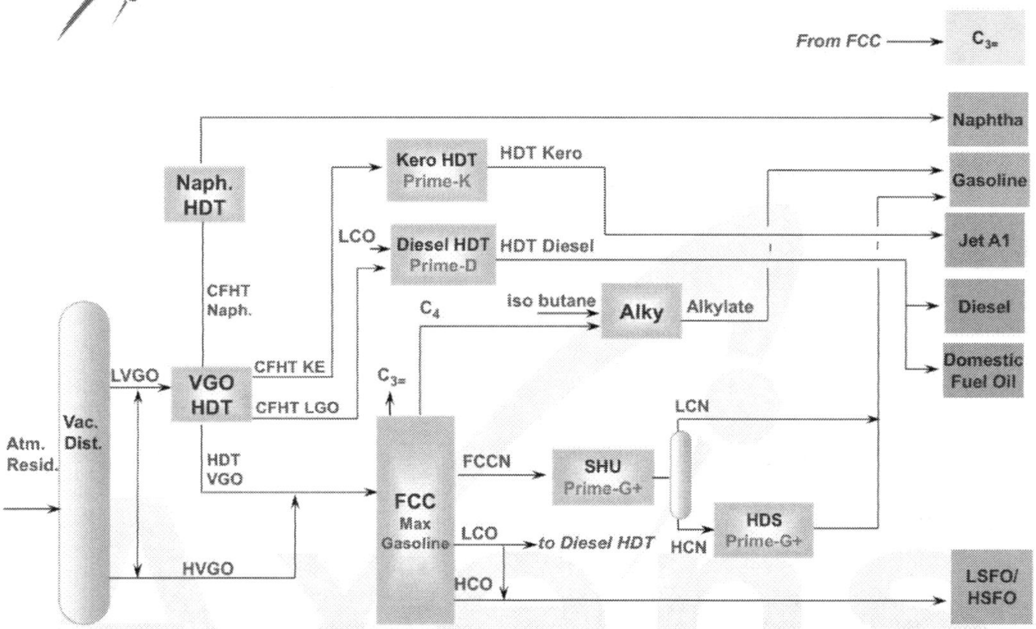

Liquid Products Yields Comparison (vol %)

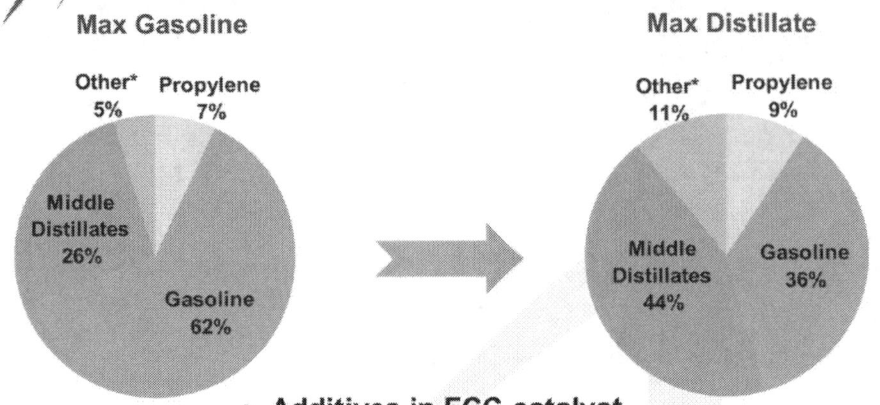

- Additives in FCC catalyst
- FCC in LCO mode
- C_5/C_6 PolyFuel
- FlexEne : Polygasoline recycle for C_3=
- Revamp CFHT to MHC

*Other : LPG, naphtha, slurry

PolyFuel™ Process

Conversion of Light Olefins to Distillates

- **Extension of C_4= oligomerization Polynaphtha™ experience**
  - *Commercially proven – 19 references*
- **Ideal for integration with PG+ 1st Step in Tier 3**
- **Direct adjustment of G/D Ratio**
- **Removal of C_5 olefins from gasoline pool**
  - *~10% of FCC gasoline converted to distillates*
- **High quality Jet after HDT**
  - *Very high smoke pt, good freeze pt*
- **Feed / Product flexibility**

Conclusion

- The primary investment for 10 wppm S gasoline will be post-treatment
  - Evaluation of existing assets capability critical to minimize investment
- Ongoing R&D and catalyst development
- Innovative technologies around the FCC maximize production of valuable products

Q & A

Thank you for your attention!

Axens
IFP Group Technologies

Visit our new website: www.axens.net

GHG BACT for Fired Heaters

2012 Regional Process Technology Conf – Dennis Ducote, Ian M. Glasgow, Enrique Jimenez, Ray Shipman, Ed Palmer

*The authors are employed by Mustang Engineering, L.P.

Wood Group Mustang

Agenda

- GHG Regulations
- BACT Top-Down Analysis
- Fired Heater Case Study
- Results and Conclusions

Wood Group Mustang

Introduction

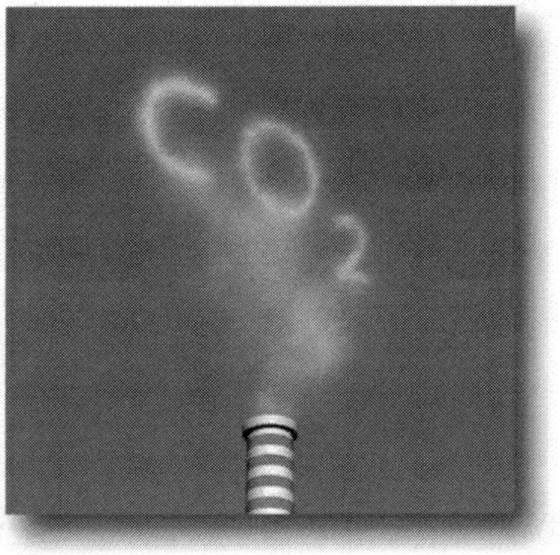

- GHG Emissions
  - CO_2
  - N_2O
  - CH_4
  - SF_6
  - HFC
  - PFC
- **Remain in atmosphere over long time horizons**
- **Pollutants not concentrated in area of emission source**
- **Global issue requires global solution**

Wood Group Mustang

Introduction

Energy-related carbon dioxide emissions, 1990-2011- US Emissions

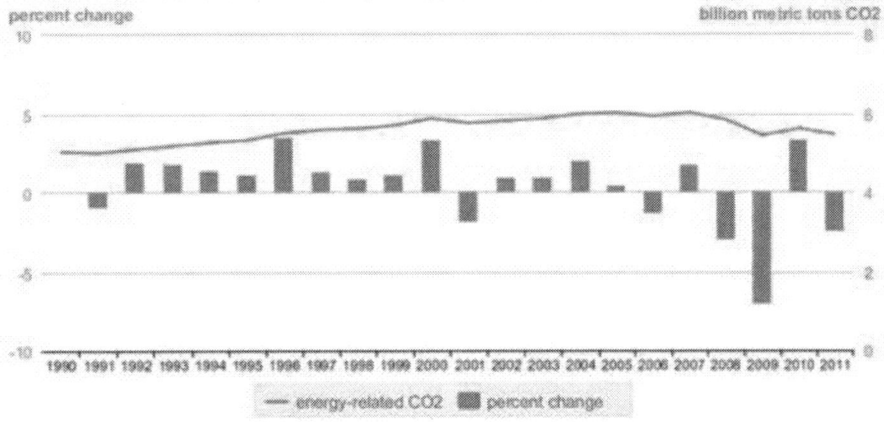

From EIA webpage; www.eia.doe.gov

Wood Group Mustang

GHG Regulations Worldwide

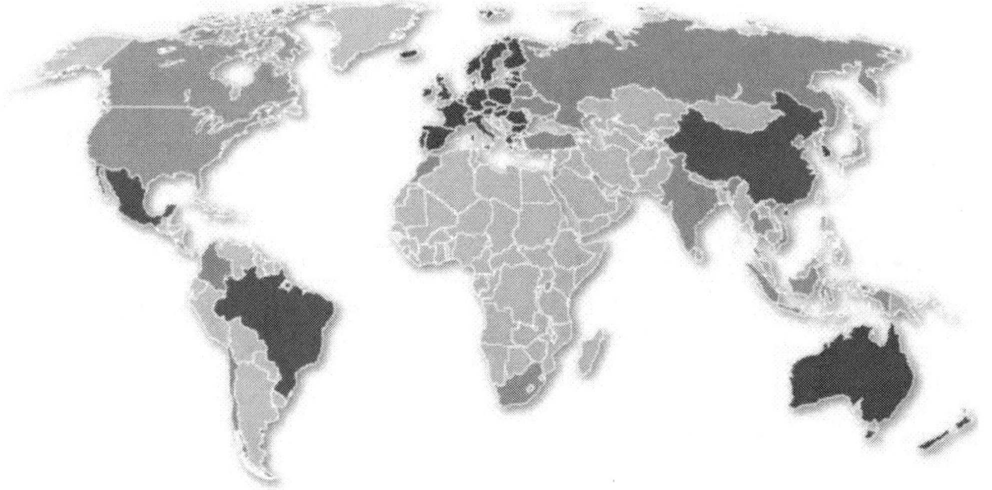

Source: Enhesa and Baker & McKenzie

Existing Legislation – EPA

- Covered facilities required to report GHG emissions
- Utilization of the Clean Air Act
  - 2007 Supreme Court Massachusetts vs. EPA decision
  - Endangerment exists
  - Mobile Sources

	2012	2014	2016	2018	2020	2022	2025
Passgr Vehicle (mpg)	33.8	36.0	39.5	41.6	44.8	49.0	56.2
Combined CAFE(mpg)	30.1	32.2	35.5	36.5	38.9	43.0	49.7
GHG (g CO_2/mile)	295	276	250	232	213	190	163

from EPA/NTHSA

    - Various credits available to reduce calculated GHG emissions
  - Stationary Sources
    - "Tailoring Rule" - redefines Clean Air Act emission limits for GHG emissions
      - 100,000 MT/yr (215 mmBtu/hr fired duty) for new facility
      - 75,000 MT/yr (160 mmBtu/hr fired duty) for revamp
    - Title V Operating Permits must include GHG
    - PSD Permits require BACT for GHG

Wood Group Mustang

BACT Top Down Analysis

- Step 1 – Identify all available control technologies
- Step 2 – Eliminate technically infeasible options
  - Carbon Capture **IS** considered technically feasible.
- Step 3 – Rank remaining control technologies
- Step 4 – Evaluate most effective controls and document results.
- Step 5 – Select the BACT.

from EPA BACT Guidance

Wood Group Mustang

Case Study Basis

PROCESS

- 150 MBPSD Notional Refinery
  - Hydrogen from on-site plant and CCR
  - FCC with Cat Feed HDS
  - Resid Upgrading with Delayed Coker
  - Flare Gas Recovery Installed
- Arab Medium Crude Feed
- MP Steam is 175 psig
- Incremental fuel is Natural Gas
- Fuel Gas – 975 Btu/scf LHV

ECONOMICS

- NG price : \$3.50/mmBTU
- Steam price: \$3.90/Mlb
- Power price: \$0.07/kWh
- Water Cost: \$0.10/Mgal
- Targeted 20% IRR
- 8,400 hours of operation pear year

Wood Group Mustang

Introduction – Refinery GHG Emissions

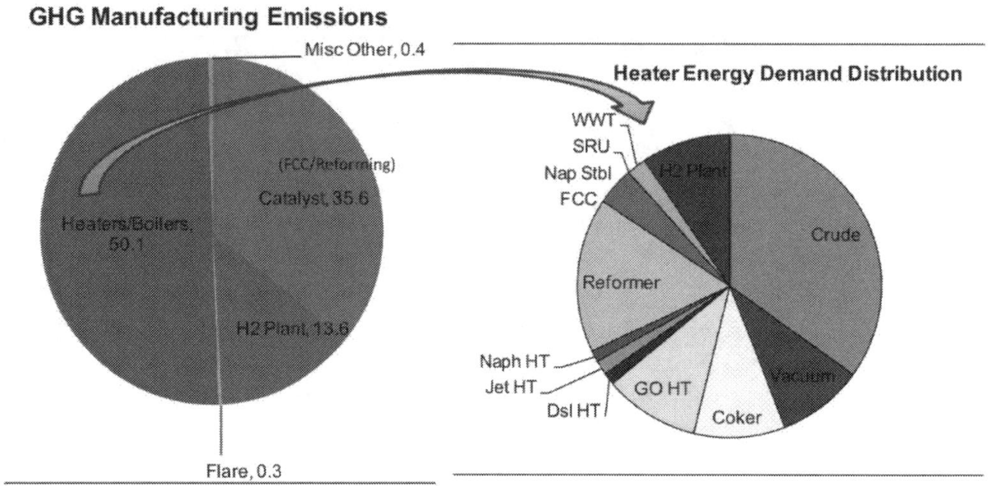

Wood Group Mustang

Crude Charge Fired Heater

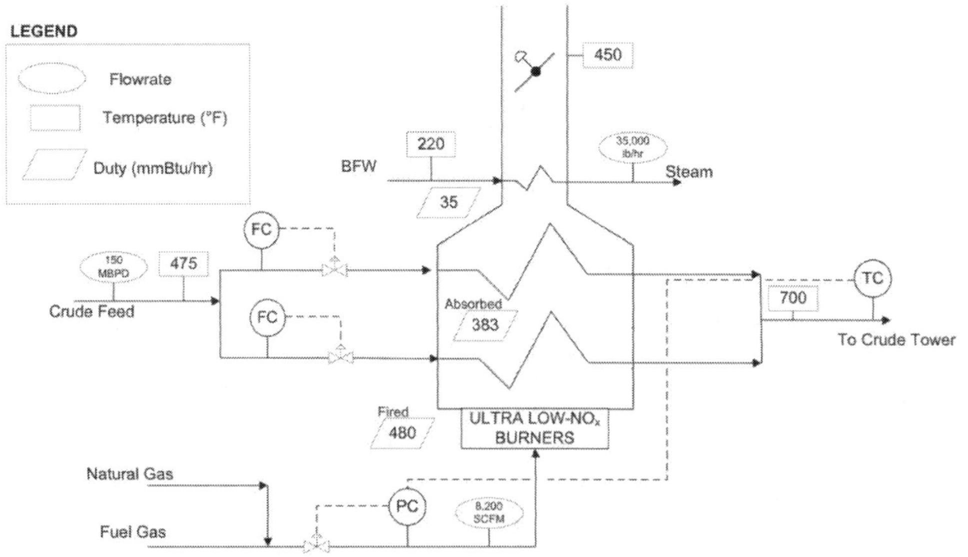

Wood Group Mustang

Paths to Reduced Emissions

- Fuel Switching
- Utilize Good Operating Practices
- Energy Efficiency
- Carbon Capture and Sequestration (CCS)

WoodGroup Mustang

Fuel Switching

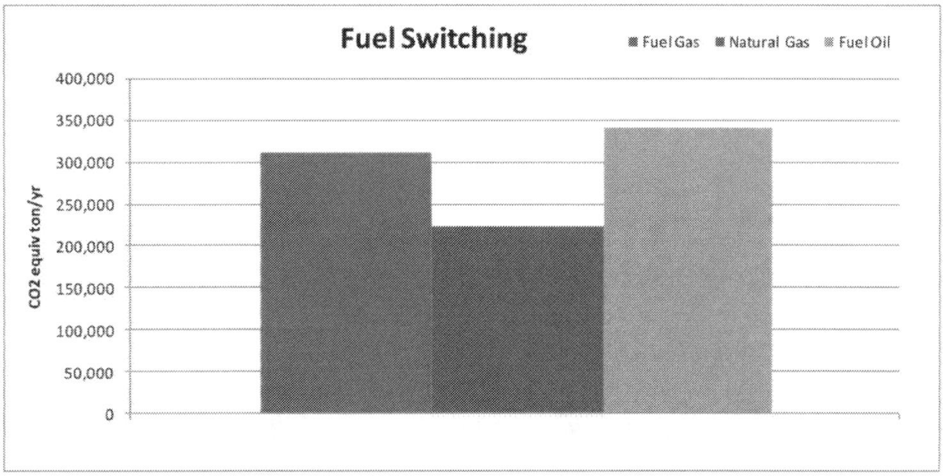

- Fuel Oil not used in practice in US (not available)
- If fuel gas is not burned in heaters, it is burned at Flare

WoodGroup Mustang

Paths to Reduced Emissions

- Fuel Switching
- Utilize Good Operating Practices
- Energy Efficiency
- Carbon Capture and Sequestration (CCS)

Good Operating Practices

- Tuning and Optimization
  - Control for stand-by operation
  - 85% reduction of standby losses*
- Maintenance Plans
  - Burners and Condensate Return
  - Fouling
  - 1 to 10% improvement possible*
- Fuel Supply
  - Consistency in fuel composition
  - Fouling precursors

** - from EPA White Paper on Refinery GHG Emissions*

Paths to Reduced Emissions

- Fuel Switching
- Utilize Good Operating Practices
- Energy Efficiency
- Carbon Capture and Sequestration (CCS)

Energy Efficiency

- Burner Design
- Combustion Air Preheat
- Waste Heat Recovery
- Process Heat Integration
- Excess Combustion Air Monitoring and Control

Wood Group Mustang

Burner Design

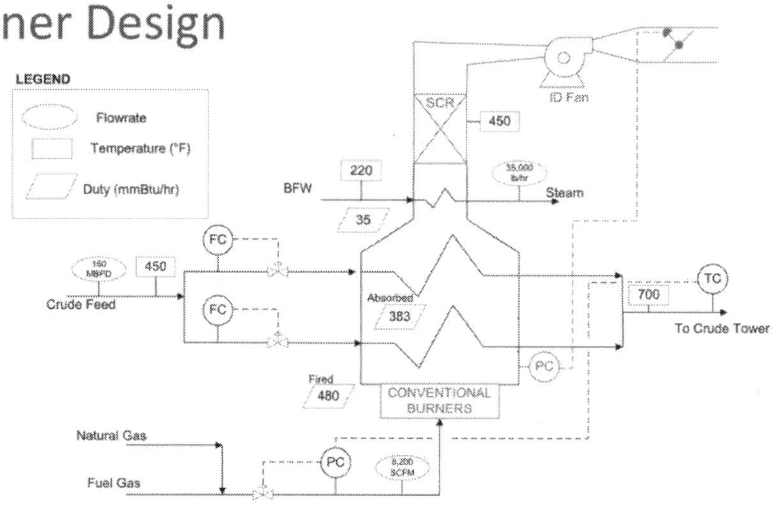

- Increased Radiant section heat transfer efficiency with Conventional
- For grassroot case no change in the fired duty due to Firebox design
- No change in GHG emissions with Burner selection
- Revamp scenario with fixed Firebox geometry and Ultra-Low NO_x
  - Increased Convection section duty available
  - Accept loss of Radiant heat duty

Wood Group Mustang

Energy Efficiency

- Burner Design
- Combustion Air Preheat
- Waste Heat Recovery
- Process Heat Integration
- Excess Combustion Air Monitoring and Control

Wood Group Mustang

Combustion Air Preheat

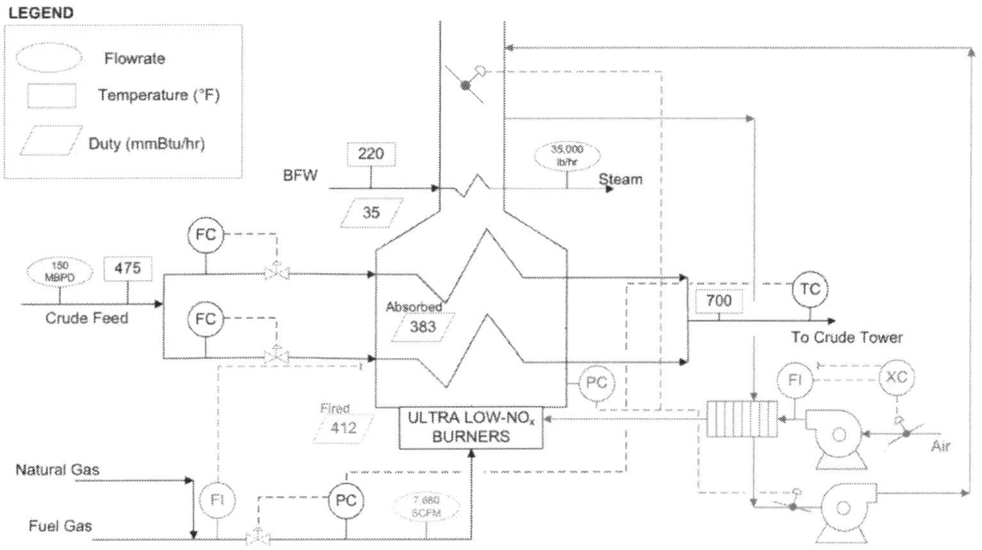

Wood Group Mustang

Combustion Air Preheat

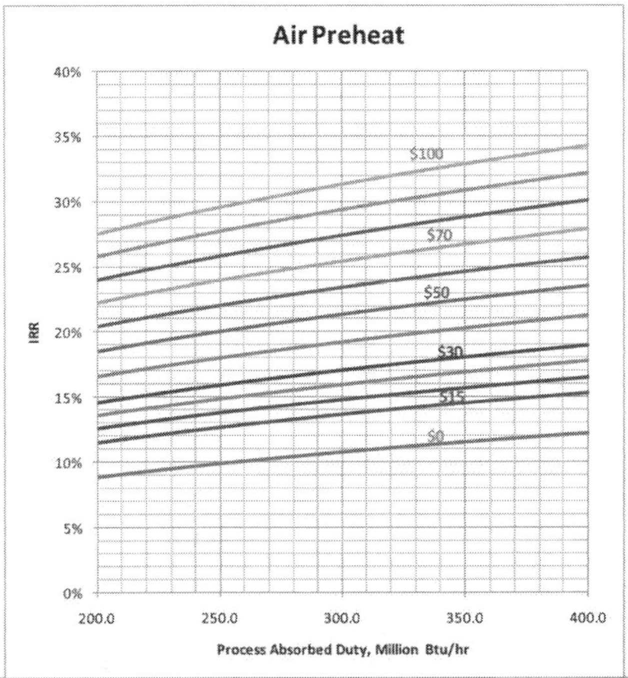

- Parameter is cost of CO_2 ($/metric ton)
- IRR is over 20% for $5.50/mmBtu Natural Gas

Wood Group Mustang

Energy Efficiency

- Burner Design
- Combustion Air Preheat
- Waste Heat Recovery
- Process Heat Integration
- Excess Combustion Air Monitoring and Control

Wood Group Mustang

Waste Heat Recovery

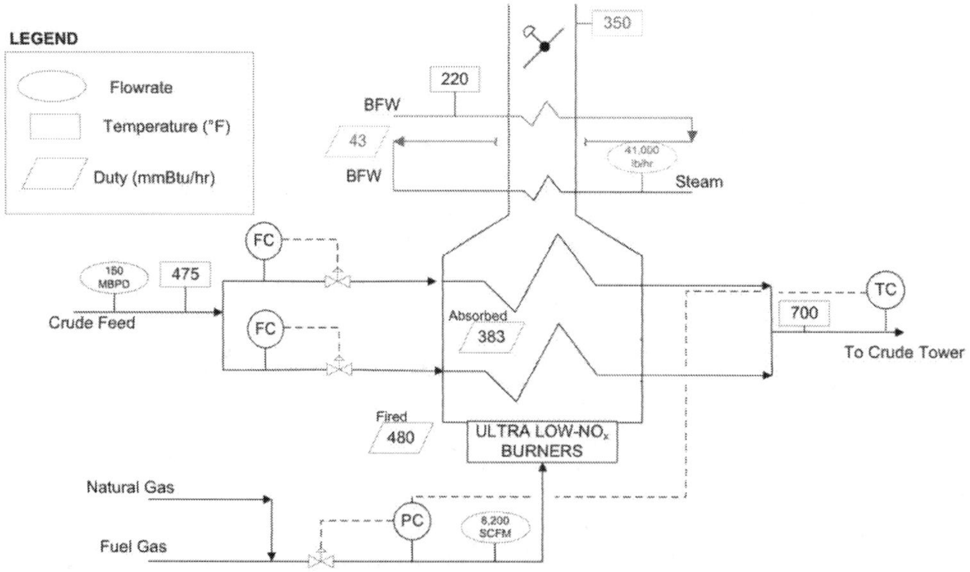

Wood Group Mustang

Waste Heat Recovery

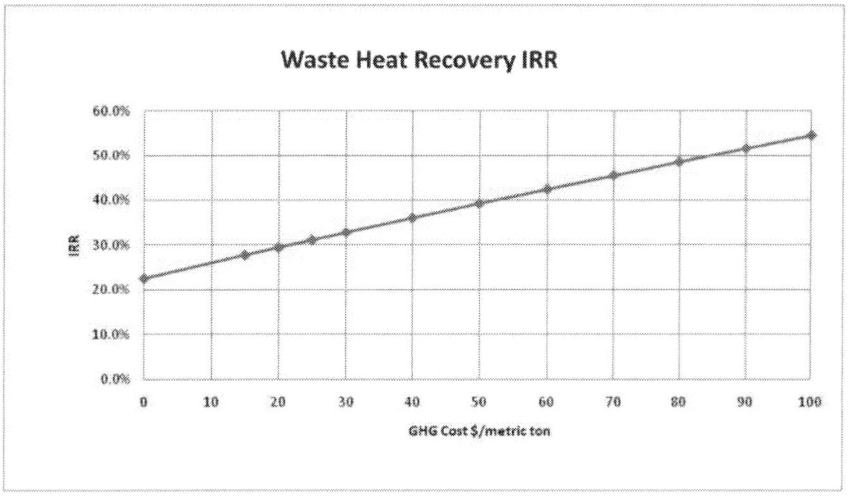

Energy Efficiency

- Burner Design
- Combustion Air Preheat
- Waste Heat Recovery
- Process Heat Integration
- Excess Combustion Air Monitoring and Control

Process Heat Integration

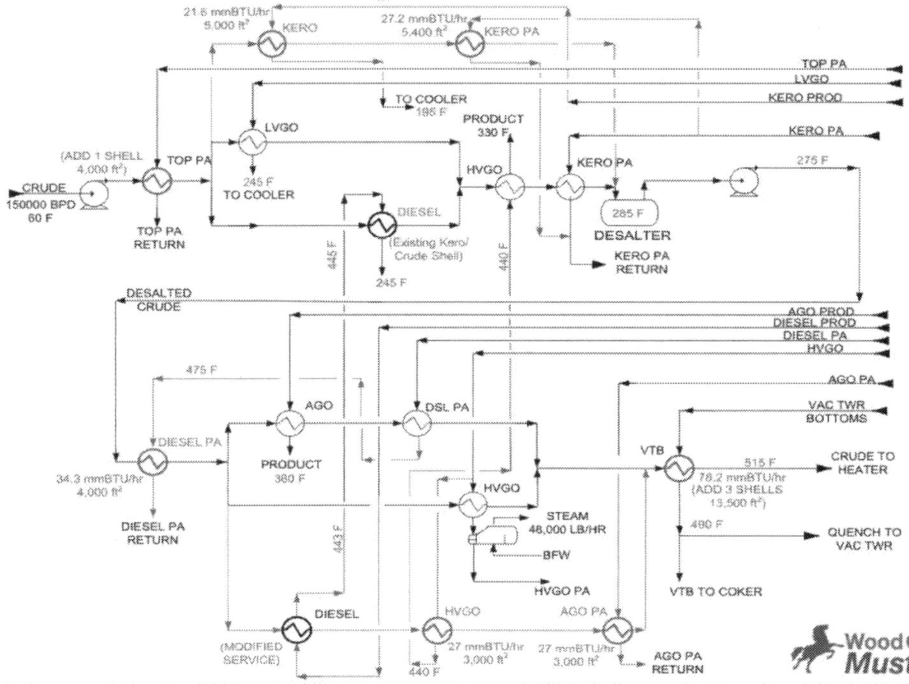

Process Heat Integration

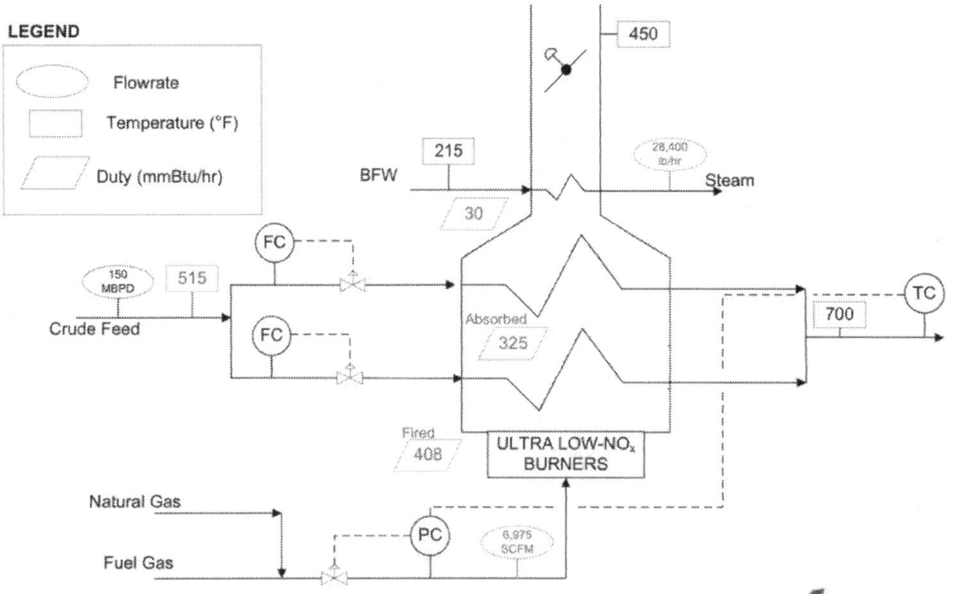

Process Heat Integration

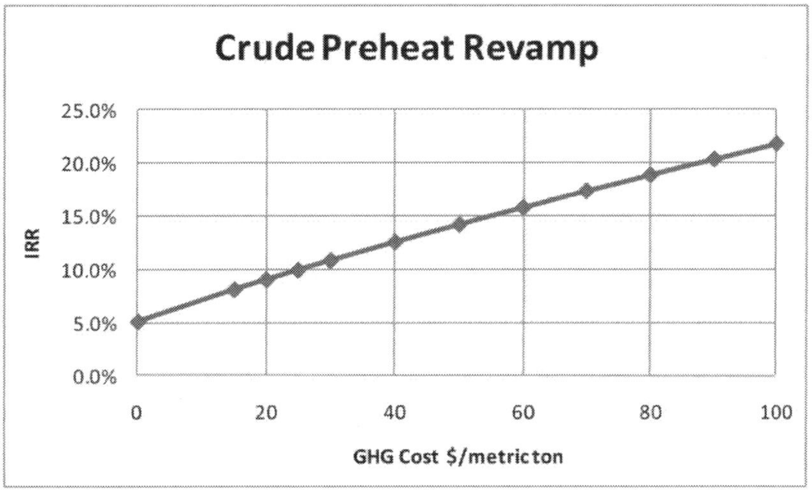

- Reduced steam generation also results from lower VTB rundown temperature

Wood Group Mustang

Energy Efficiency

- Burner Design
- Combustion Air Preheat
- Waste Heat Recovery
- Process Heat Integration
- Excess Combustion Air Monitoring and Control

Wood Group Mustang

Combustion Controls

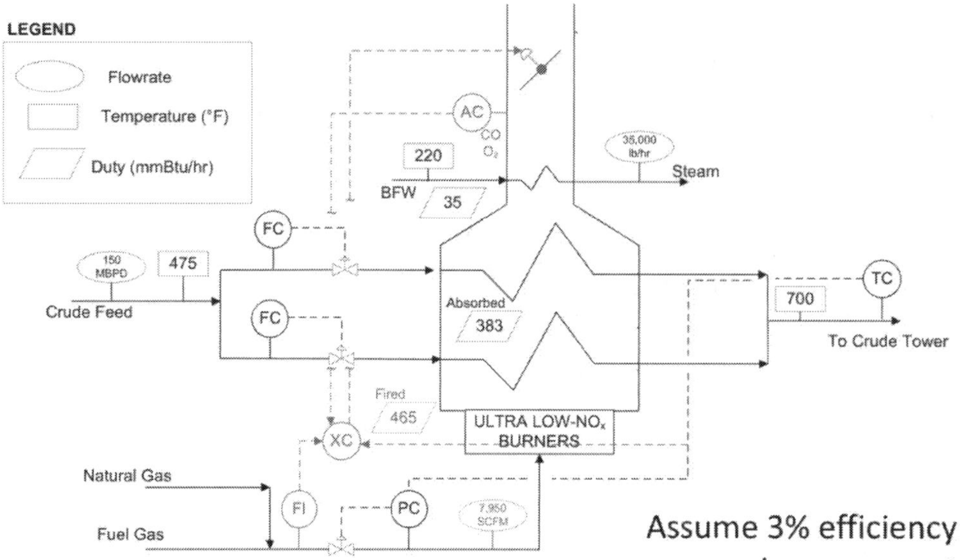

Assume 3% efficiency improvement

Efficiency Improvement ranges from 1% to 6%
from www.energystar.gov and EPA White Paper on Refinery GHG Emissions

Wood Group Mustang

Combustion Control

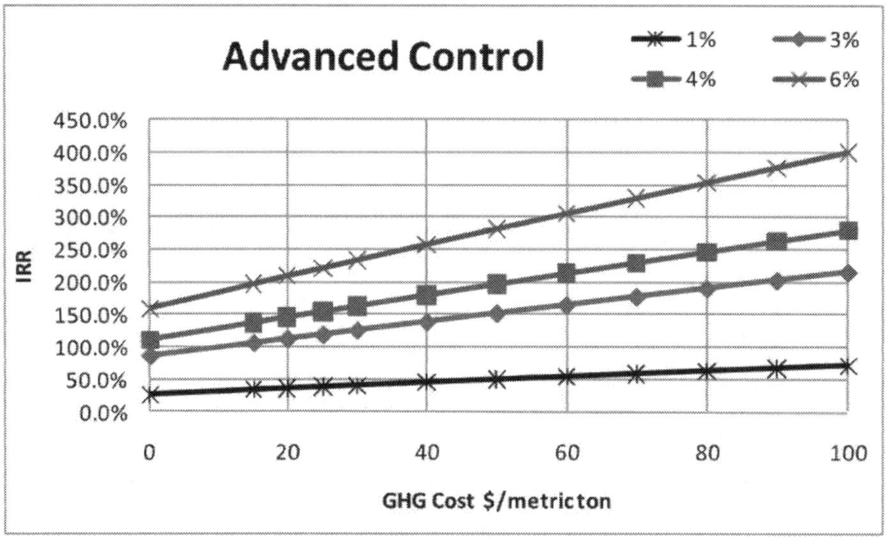

Wood Group Mustang

Paths to Reduced Emissions

- Fuel Switching
- Utilize Good Operating Practices
- Energy Efficiency
- Carbon Capture and Sequestration (CCS)

Wood Group Mustang

Carbon Capture and Sequestration (CCS)

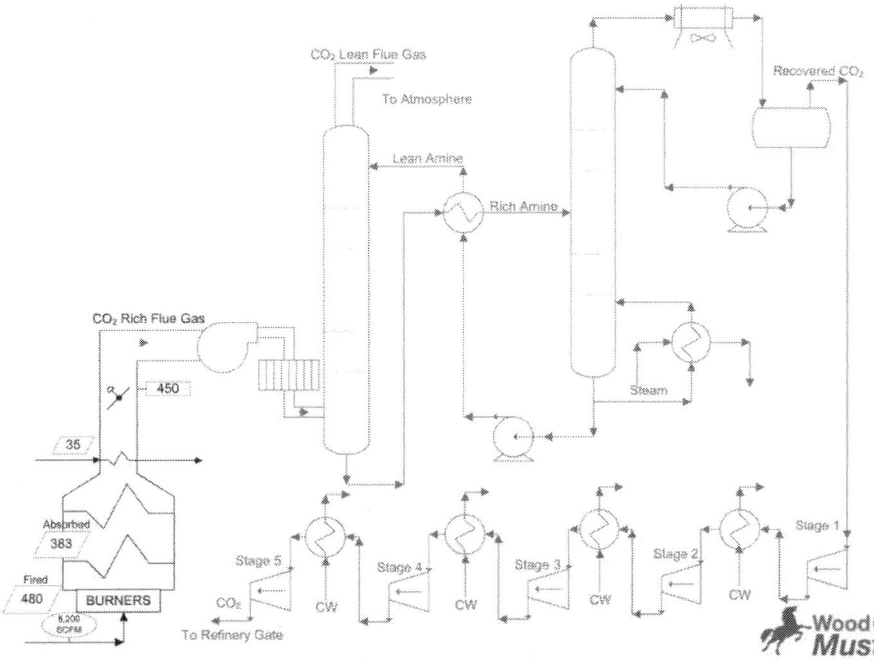

Wood Group Mustang

Carbon Capture and Sequestration (CCS)

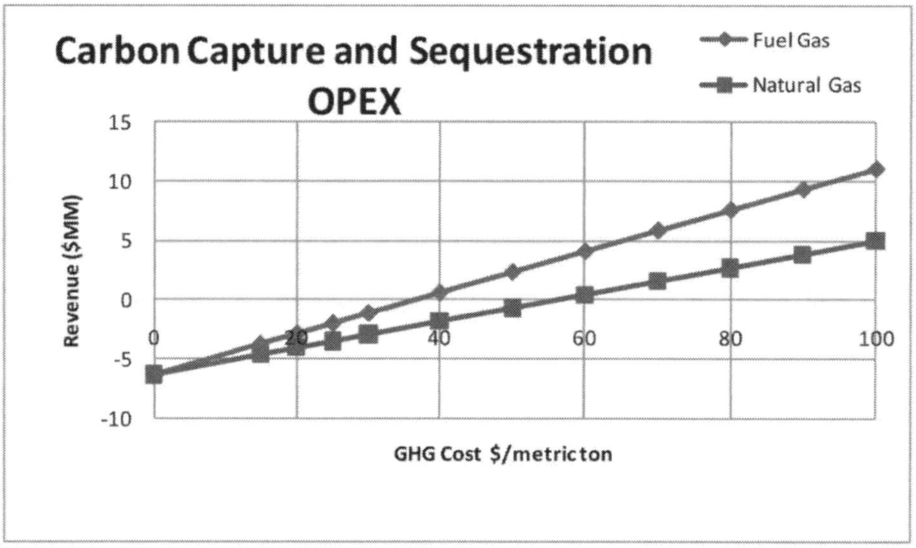

Wood Group Mustang

Carbon Capture and Sequestration (CCS)

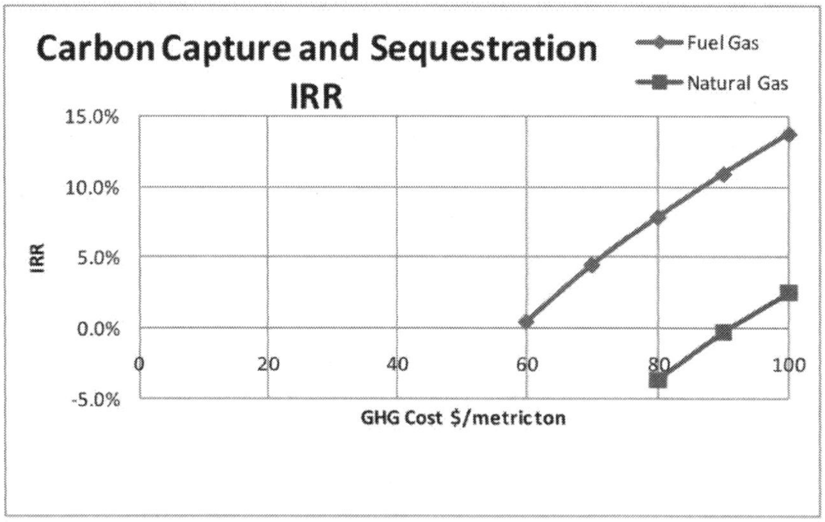

- Transportation excluded from analysis

Wood Group Mustang

Results

	GHG Emission Reduced (metric ton/yr)	20% ROI CO_2 Price $/metric ton
Burner Design	0	N/A
Air Preheat	13,950	$35
Waste Heat Recovery	7,500	$0
Process Heat Integration	29,800	$90
Combustion Control	7,950	$0
CCS	113,200	> $100

Wood Group Mustang

Results

	Is Option Available?	Cost	Used for BACT	Reason
Fuel Switching	Unlikely	Low to none	Unlikely	Fuel gas must be burned
Operating Practices	Unlikely to Limited	Low to none	Limited	Most done in practice
Energy Efficiency	Likely	High to Low	Likely	Most low cost options
CCS	Likely	High	Unlikely	Economics

Wood Group Mustang

Conclusions

- Energy efficiency will likely provide what (if any) required changes to reduce GHG emissions to meet BACT requirements
- Good operating practices are the low hanging fruit (if available)
- CCS is cost prohibitive

Wood Group Mustang

Acknowledgements

THI TULSA HEATERS INC.

ZEECO

OnQuest

CALLIDUS TECHNOLOGIES by Honeywell

TRAINING REVAMPING TROUBLESHOOTING FIS

WILLBROS Downstream

Wood Group Mustang

Does "Thermodynamic Intuition" have a Role in Applied Thermodynamics?

Paul Mathias
Fluor Corporation

4th AIChE Regional Process Technology Conference

League City, TX, 4-5 October 2012

FLUOR®

Motivation and Overview

FLUOR

- Powerful process modeling and analysis tools are readily available to today's engineers and scientists → big increase in productivity.
- An unintended consequence is that these powerful tools have caused a decrease in scientific and engineering expertise.
- Applied thermodynamics success requires "thermo intuition":
  - Ability to guess answers;
  - know what calculations will be most effective;
  - choose and evaluate data and property method;
  - ability to judge qualitative and semi-quantitative correctness of the computer's answers;
  - creative use of visualization to capture the essence of complex calculations.
- Thermo intuition and rigorous analysis are both needed.

1

Productivity Improvements Have been Wonderful – But the Unintended Consequence has been Diminishing Expertise

Explicit and Tacit Knowledge in Applied Thermodynamics

FLUOR

- Process modeling has become increasingly automated:
  - Big increase in productivity.
  - Increase in quality as computers have strong capacity for computation and minute detail.
- But there are downsides:
  - Decrease in expertise.
  - Computers miss subtle errors.
- *"A man cannot eat his cake and haue it stil."*• Or can he?
  - Develop and use our tacit (intuitive) knowledge.
  - Mathematics and computers should be our faithful servants.
  - Exploit the complementarity between humans and computers.

• John Davies, 1611

3

Tacit and Explicit Knowledge

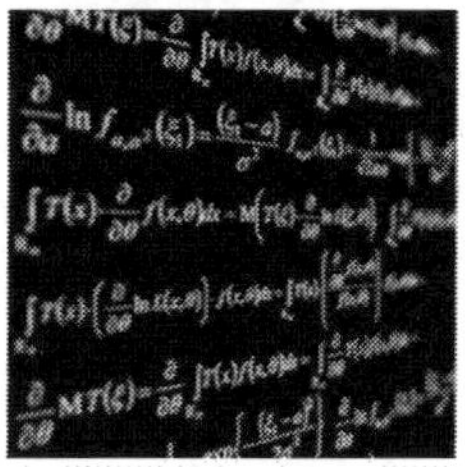

Explicit knowledge is knowledge that has been or can be articulated – typically emphasized in schools and universities.

Tacit knowledge in the form of informed guesses, hunches and imaginings comes naturally to human beings through our evolved brains, and our experience and immersion in subjects of expertise.

4

What is Tacit Knowledge?

- "*The Tacit Dimension*" Polanyi (1966)
  - Knowledge difficult (perhaps impossible) to transmit orally or verbally
  - "You know more than you can tell"
- "*Tacit and Explicit Knowledge*" Collins (2010)
  - Develops common language on tacit knowledge
  - Relational, Somatic and Collective
  - "Embedding" is necessary to knowledge acquisition
- "*Blink*" Gladwell (2005)
  - The power of thinking without thinking
  - "Thin Slicing"
  - Intuitive judgment is developed by experience, training, and knowledge
- "*Gut Feelings*" Gigerenzer (2007)
  - Usually scorned as fickle and unreliable
  - Takes advantage of our evolved brains, and are based upon rules of thumb - enables us to act fast, and often with astounding accuracy

5

Educators Glorify Explicit Knowledge

Neal Amundsen (1916-2011)

"Neal Amundson played a pivotal role in transforming the field of chemical engineering from glorified plumbing and phenomenological chemical recipes to a rigorous discipline that **translated scientific phenomena into engineering practice through sophisticated mathematical modeling**." Frank Bates (PNAS, **2011**, 180(18), 7285)

"It is the writing of the model that often has the most value," Jim Downs, Eastman Chemical Company.

6

Computation with Complicated Equations

The osmotic coefficient

(P4)

To practically apply the above equations one needs to derive the following coefficients and auxiliary functions.

Function F sums up terms dependent on the Debye–Hückel ionic strength

(P5)

Values of the Debye–Hückel coefficient, A, for the temperature range from 0 to 350 °C can be derived from the paper of Zemaitis et al. [29] or according to the Pitzer proposal [6] from Eq. (P6).

Easy for computers, but hard for us humans!

7

Executing Algorithms

FLUOR

Euclid's Algorithm

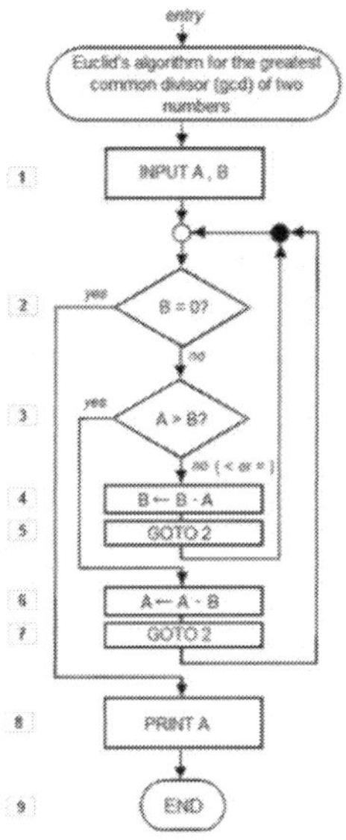

- Used for calculation, data processing and automatic deduction.
- Effective methods are expressed as a finite list for performing a well-defined task.
- Easy for computers, but impossible for us humans.

8

Repairing Broken Text – We're very good at recognizing patterns and using context

FLUOR

I cdnuolt blveiee that I cluod aulaclty uesdnatnrd what I was rdanieg. The phaonmneal pweor of the hmaunn mnid, aoccdrnig to a rscheearch at Cmabrigde Uinervtisy, is taht it dsenot mtaetr in waht oerdr the ltteres in a wrod are, the olny iproamtnt tignh is taht the frsit and lsat ltteer be in the rghit pclae. The rset can be a taotl mses and you can sitll raed it whotuit a pboerlm. Tihs is bcuseae the huamn mnid deos not raed ervey lteter by istlef, but the wrod as a wlohe. Azanmig huh? Yaeh, and I awlyas tghuhot slpeling was ipmorantt!

From "*Tacit and Explicit Knowledge*" Collins (2010)

Easy for us, but very hard for computers!

Good Intuitions are Superior to Logic

FLUOR

Miami is in Florida *and* San Francisco is in California.
San Francisco is in California *and* Miami is in Florida.

John and Peggy married *and* Peggy became pregnant.
Peggy became pregnant *and* John and Peggy married.

Mark got angry *and* Mary left.
Mary left *and* Mark got angry.

We invited friends *and* colleagues.

"Gut Feelings. The intelligence of the Unconscious,"
by Gerd Gigerenzer

10

Tacit Knowledge Should Lead and Direct Explicit Knowledge

FLUOR

I cdnuolt blveiee that I cluod aulaclty uesdnatnrd what I was rdanieg. The phaonmneal pweor of the hm… tisy, is taht it dsenot mtaetr i… h is taht the frsit and lsat ltteer b… an sitll raed it whotuit a pboerlm. Tihs is bcuseae the huamn mnid deos not raed ervey lteter by istlef, but the wrod as a wlohe. Azanmig huh? Yaeh, and I awlyas tghuhot slpeling was ipmorantt!

Human Expertise and Intuition

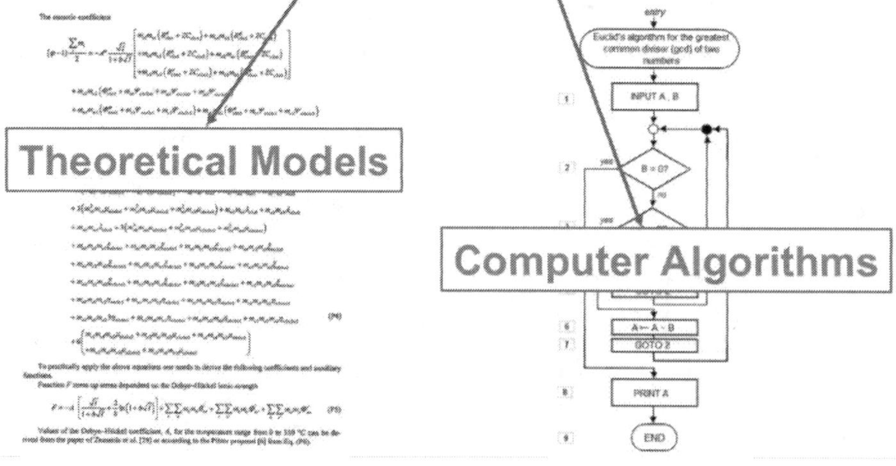

11

Modern Simulation Software promises the Easy Integration of Many Disciplines

FLUOR

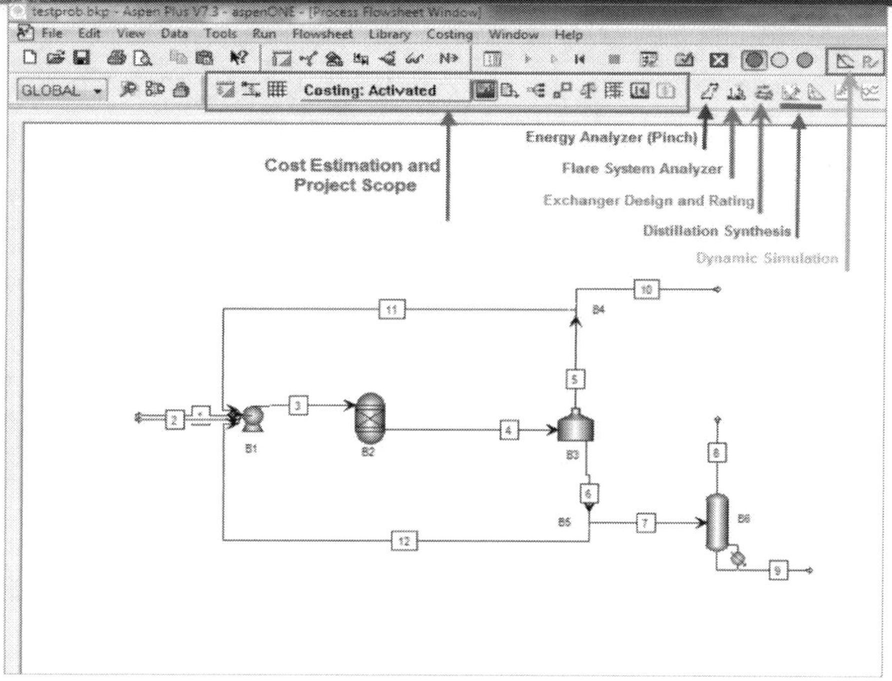

12

Examples

FLUOR

- Physical Properties
- The McCabe-Thiele "circle"
- CO_2 Capture
- Solubility of Water as hydrate in Compressed CO_2
- Water Dew Point of Thermal Cracker Product

13

Cox (1923) Chart – n-Alkanes

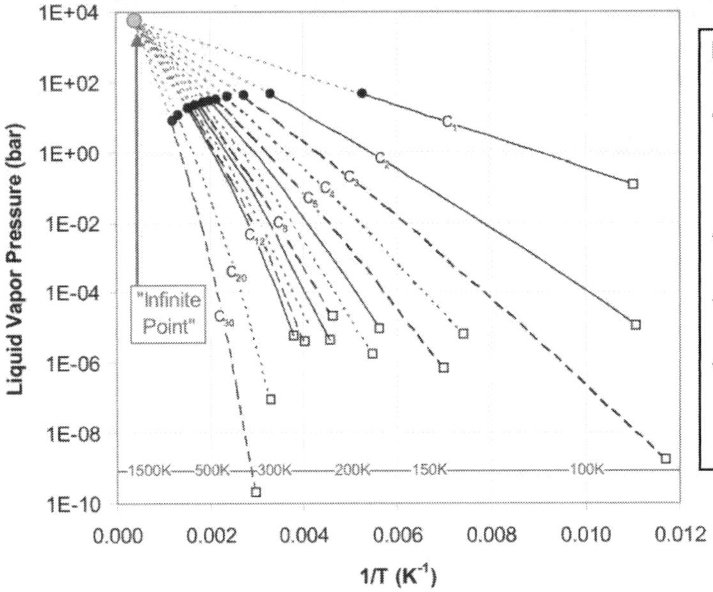

Benefits:

- **Patterns are regular and "expected" – increased confidence in data reliability and accuracy.**
- **Vapor pressures within a family rarely cross.**
- **Clear pattern for T_c and P_c, but not for T_t and P_t**
- **"Infinite Point" valuable for pattern recognition and for property estimation.**

14

Othmer (1940) Ratio Plot – Alcohols Relative to Water

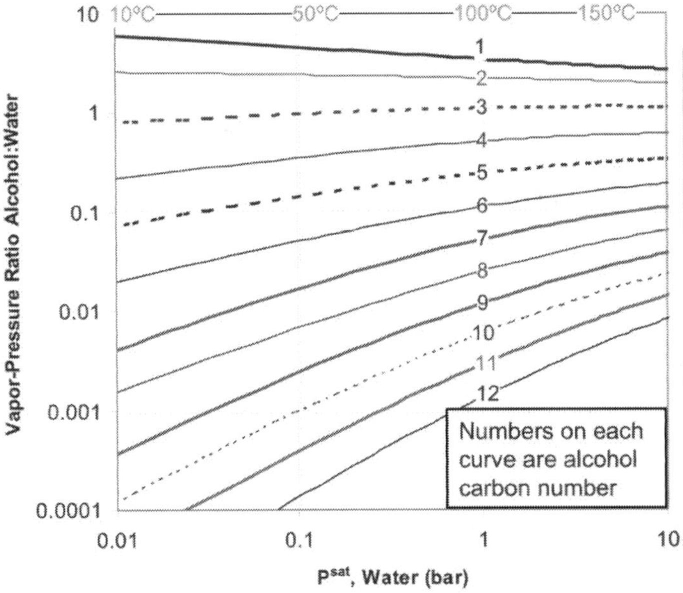

Benefits:

- **Patterns are regular and "expected" – increased confidence in data reliability and accuracy.**
- **Visual insight into relative vapor pressures and enthalpies of vaporization.**

15

Infinite-dilution activity coefficients in water (at 75ºC)

FLUOR

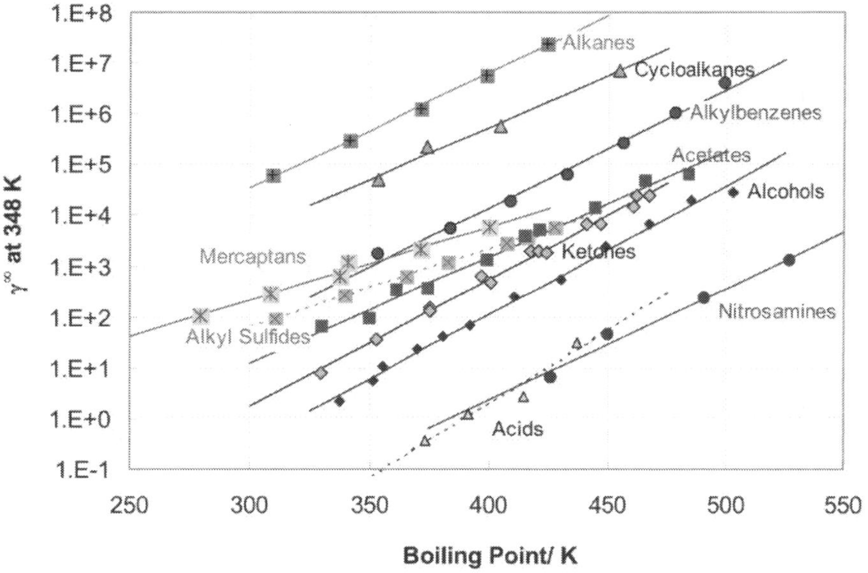

16

Relative Volatility with Respect to Water (at 75˚C and Low Solute Concentration)

FLUOR

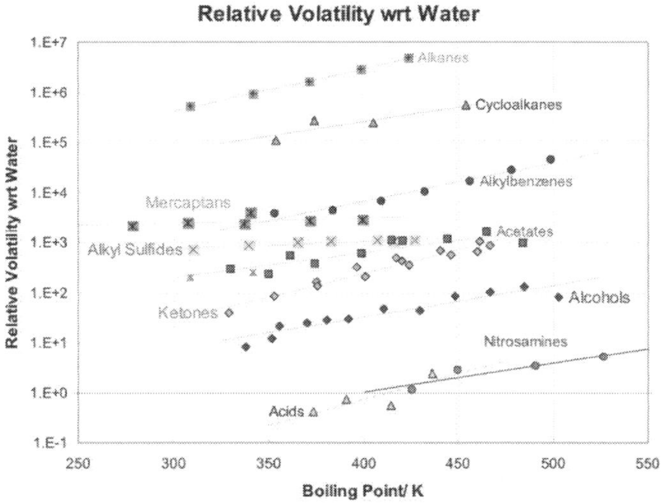

Benefits:

- **Patterns help build our intuition.**
- **Visual insight into distillation for water systems.**
- **Applicable to gasification and biofuels separation trains.**

"The eye can see at a **glance** what may not be easy to deduce from columns of numbers and a few statistical parameters." (Douglas Ambrose, University College London, 1985)

17

Examples

- Physical Properties
- The McCabe-Thiele "circle"
- CO_2 Capture
- Solubility of Water as hydrate in Compressed CO_2
- Water Dew Point of Thermal Cracker Product

18

Brief History of Distillation Modeling Technology

1550	Alchemists – before the dawn of exact science – had practical understanding of distillation principles (e.g., several products, semi-continuous feeds, reboiling, partial condensation).
1889	Sorel developed theoretical equations (theoretical stage, mass balance, energy balance and equilibrium equations).
1922	Lewis, McCabe-Thiele introduced CMO (constant molal overflow) approximation and graphical solution – needed in pre-computer age. Intuitive and reasonably accurate, became dominant design and analysis tool for several decades.
1951-Today	Advances in computers and solution algorithms enabled rigorous simulations. Today's capability (multicomponent systems, heat effects, chemical reactions, mass-transfer) → rigorous designs.
2003	Kister observed that tower malfunctions not decreasing, and one reason is that rigorous simulations are used as a "black box."
Today	Renewed recognition that visualization – McCabe-Thiele diagrams, residue curves – necessary for superior designs, troubleshooting.

19

Classical McCabe-Thiele Approach

FLUOR

- Teaches interplay of process variables: number of stages, optimum feed location, reflux ratio, feed thermal state, tray efficiency.
  (e.g., feed thermal state influences optimum feed location.)
- Develops intuitive feel for distillation.
- Primary educational tool for teaching ChE students about cascaded operations such as distillation, absorption and extraction.
- Benefit: develop "distillation craftsmanship or tacit knowledge."

20

Hybrid McCabe-Thiele Approach☞ – Design, Operability, Troubleshooting

FLUOR

Ohe	Energy-saving new technology
Zygula	Improved stream-stripping systems
Olsson and Roberson	Improve energy efficiency
Love, Shivler and Pierce	Evaluate revamp improvements
Bellner, Ege and Kister	Explain counter-intuitive observations in multi-feed distillation tower
Lee, Hauan, Lien and Westerberg	Understand complex separations, like reactive distillation

☞**Hengstebeck for multicomponent systems**

21

Debottlenecking Demethanizer Column (1)

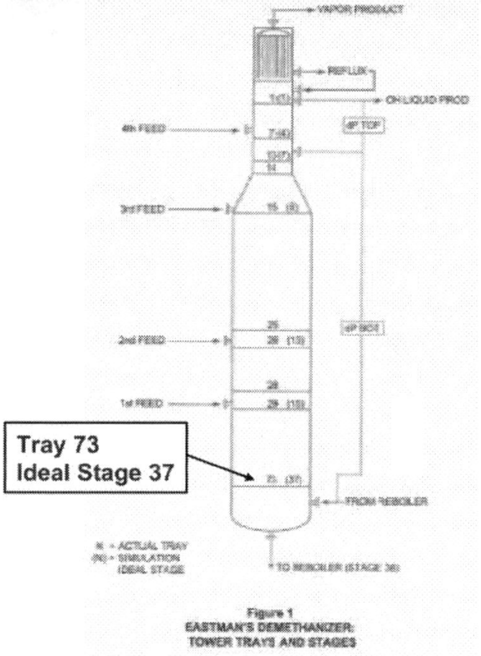

"Hydraulic Analysis Key to Effective, Low-Cost Demethanizer Debottlenecking"
Bellner, Ege and Kister
Eastman and Fluor
AIChE Spring Meeting, April 2004

- Tower operated well below bottleneck point.
- Normal tower $\Delta P \approx$ 5-6 psia.
- With tower feed rates above bottleneck point, flooding observed in top section, with $\Delta P \approx$ 8-12 psia.
- Corrective action was to reduce reflux rate, but result was increased ethylene losses in overhead.
- How to diagnose the problem and debottleneck the column?

22

Debottlenecking Demethanizer Column (2)

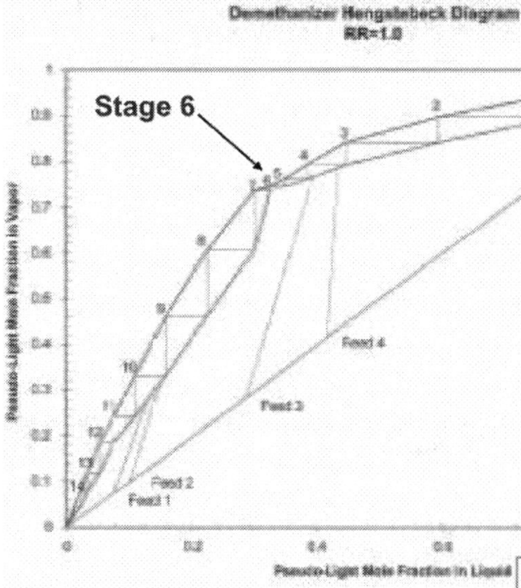

Bellner, Ege and Kister
AIChE Spring Meeting, April 2004

- Simulation model built and validated.
- Hengstebeck diagram revealed pinch near stage 6 (real tray 12).
- Hydraulic analysis revealed downcomer backup flooding in lower section.
- Solution: increase hole areas in trays 15-73 (ideal stages 8-37).

Understanding of equipment limitations, hydraulic analysis and Hengstebeck diagram correctly pinpointed bottleneck.

23

Still with Reflux, Dephlegmator and Condenser (1573)

FLUOR

Figure 3—Still with Reflux, Dephlegmator, and Condenser

(Lonicer, *Kunstliche Conterfeytunge*, Frankfurt, 1573)

Eglof and Lowry

"Distillation Methods, Ancient and Modern"

Ind. Eng. Chem. (1929)

24

Still with Cooled Head and Preheater (1606)

FLUOR

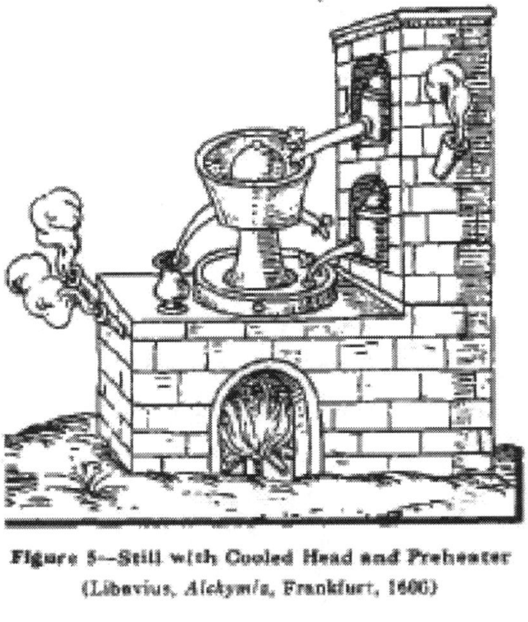

Figure 5—Still with Cooled Head and Preheater

(Libavius, *Alchymia*, Frankfurt, 1606)

Eglof and Lowry

"Distillation Methods, Ancient and Modern"

Ind. Eng. Chem. (1929)

25

Examples

- Physical Properties
- The McCabe-Thiele "circle"
- CO_2 Capture
- Solubility of Water as hydrate in Compressed CO_2
- Water Dew Point of Thermal Cracker Product

26

CO_2 Capture R&D is Enormous – and Growing

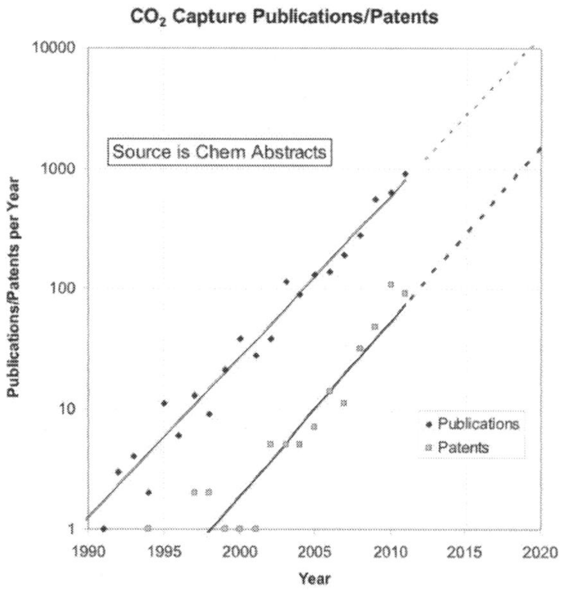

- Research in the field of CO_2 capture is enormous.
- Nearly one thousand papers were published in journals in 2011!
- The trend for patents is similar.
- Claims of dramatic breakthroughs are often made.
- Many breakthroughs have fizzled out.

27

Claimed Advantages of Chilled-Ammonia CO_2-Capture Process

FLUOR

- Low solvent cost
- High stability
- High capacity for CO_2
- Ability to capture other acid gases: SO_2, NO_2, etc.
- Low heat of regeneration
  - 800 Btu/lb CO_2 for MEA Versus 260 Btu/lb CO_2 for NH_3
- Regeneration flexibility: Low temperature or high pressure
- Low slip of NH_3 in the absorber

Process simulation, using a rigorous thermodynamic model, provides a powerful tool for objective analysis and process optimization.

28

Heat of Solution – CO_2 in Aqueous Ammonia at 100˚F

FLUOR

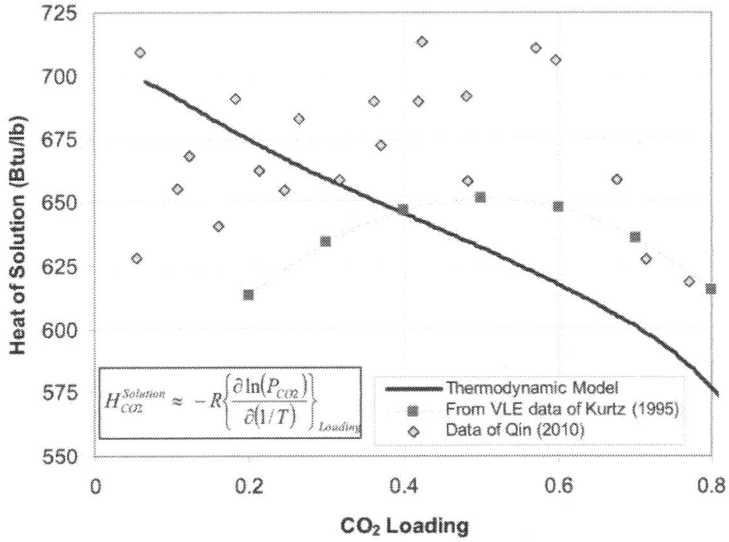

Data of Qin came after Fluor presentations and publications.

29

Thermodynamic Analysis of Chilled-Ammonia Process

- Process simulation with rigorous thermodynamic model provides a powerful, reliable tool to analyze CO_2-capture process.
- Comparison between chilled-ammonia process and alkanolamine processes:
  - LP steam requirements of chilled-ammonia process are similar to MEA, but rich solvent leaving cross exchanger will have solids for temperatures below 50°F.
  - Chilled-ammonia process requires refrigeration energy, which exceeds the benefits of relatively high-pressure CO_2 product.
  - Chilled-ammonia process has additional equipment (refrigeration system, NH_3 abatement) and process complexity (solids handling).

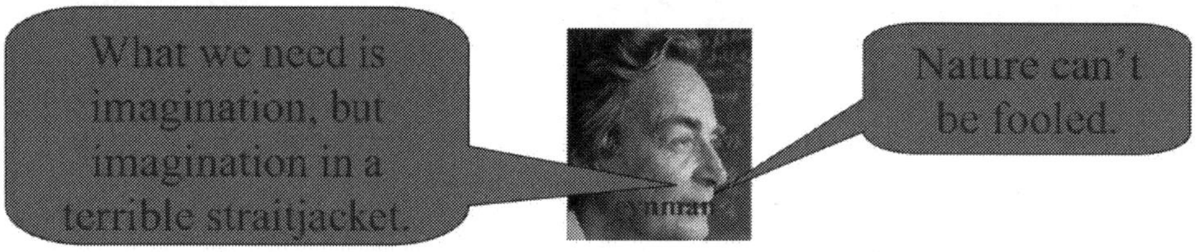

30

Examples

- Physical Properties
- The McCabe-Thiele "circle"
- CO_2 Capture
- Solubility of Water as hydrate in Compressed CO_2
- Water Dew Point of Thermal Cracker Product

31

What is the Solubility of Water as Hydrate in Liquid CO_2? Timeframe: 1986 to 2011.

FLUOR

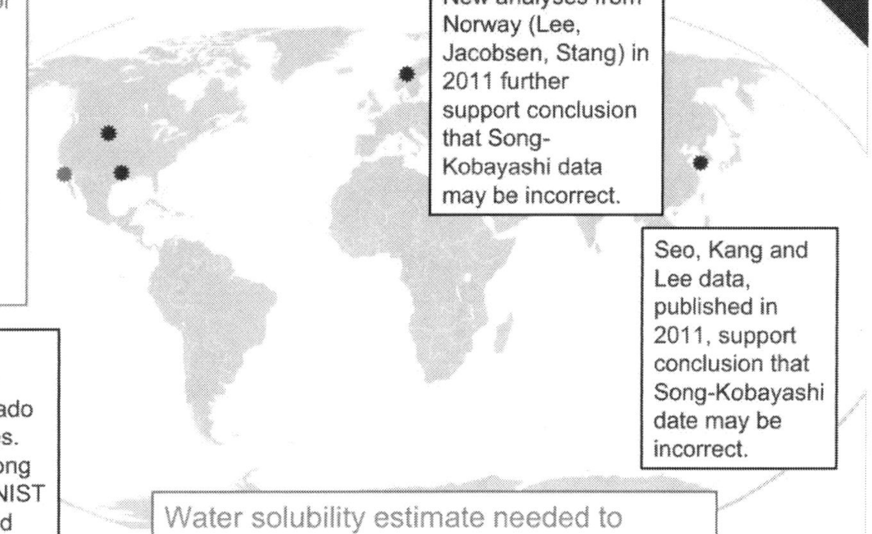

Mathias at Fluor in California needs this estimate for in 2010 for CO_2 pipeline in Alaska. Data from Song and Kobayashi (1986) are unexpected!

Mathias calls hydrate expert Sloan at Colorado School of Mines. Sloan says Jeong Kang, visiting NIST from Korea, told him about same issue last week.

New analyses from Norway (Lee, Jacobsen, Stang) in 2011 further support conclusion that Song-Kobayashi data may be incorrect.

Seo, Kang and Lee data, published in 2011, support conclusion that Song-Kobayashi date may be incorrect.

Water solubility estimate needed to determine how much to dehydrate the CO_2 stream to prevent hydrate formation.

32

Moisture Content of Liquid CO_2

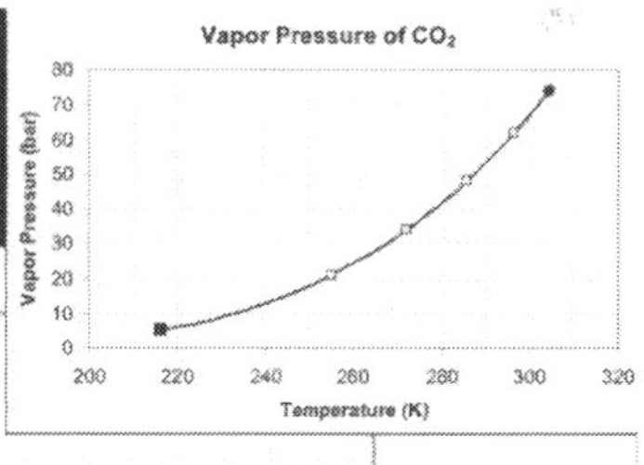

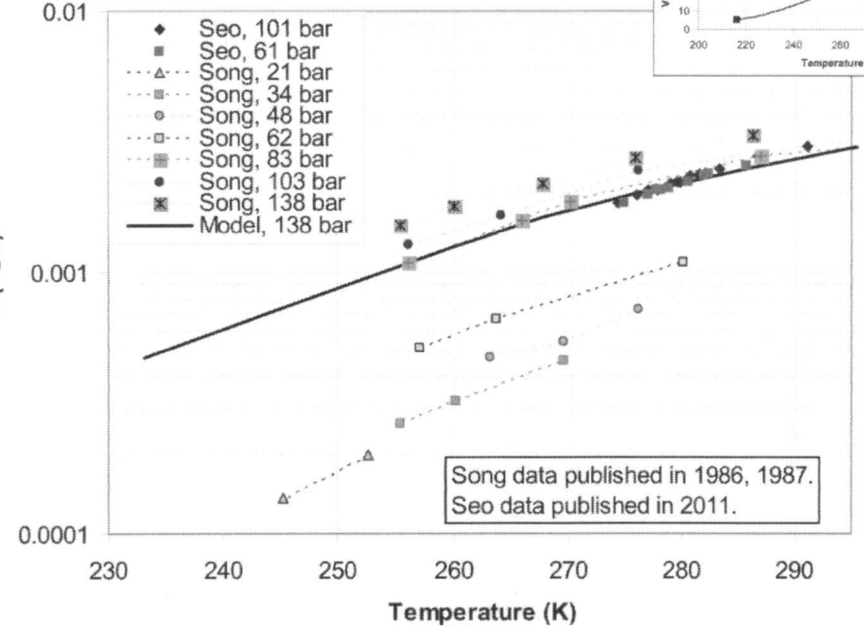

33

Moisture Content of Liquid CO_2 at 255 K

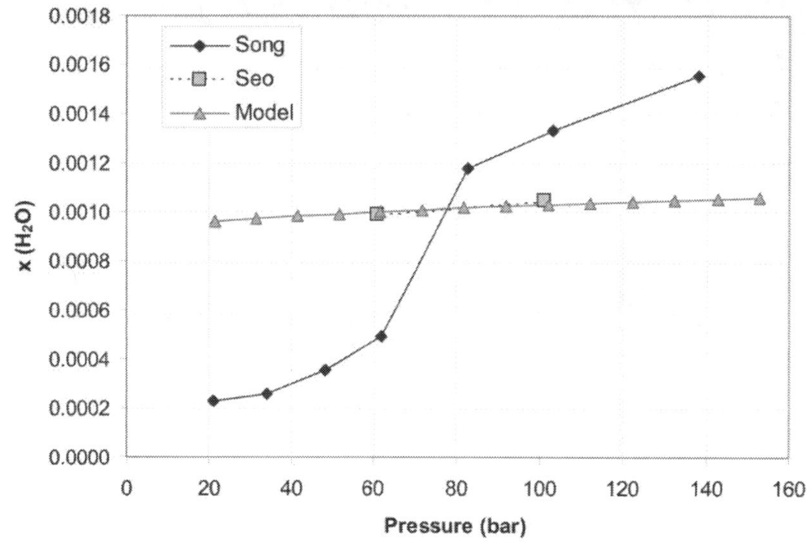

Song data are probably incorrect.

34

Density of Liquid CO_2

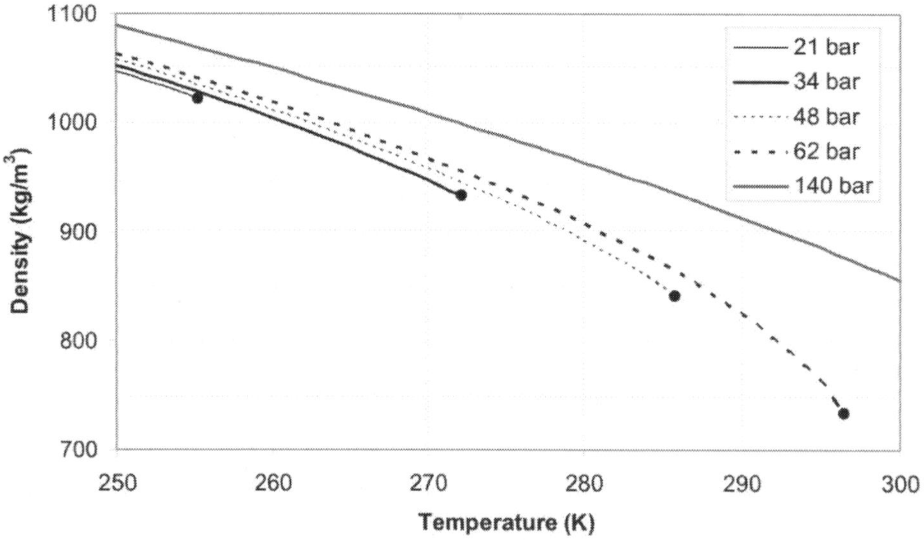

35

Examples

FLUOR

- Physical Properties
- The McCabe-Thiele "circle"
- CO_2 Capture
- Solubility of Water as hydrate in Compressed CO_2
- Water Dew Point of Thermal Cracker Product

36

Aqueous Phase in Thermal Cracker Product

FLUOR

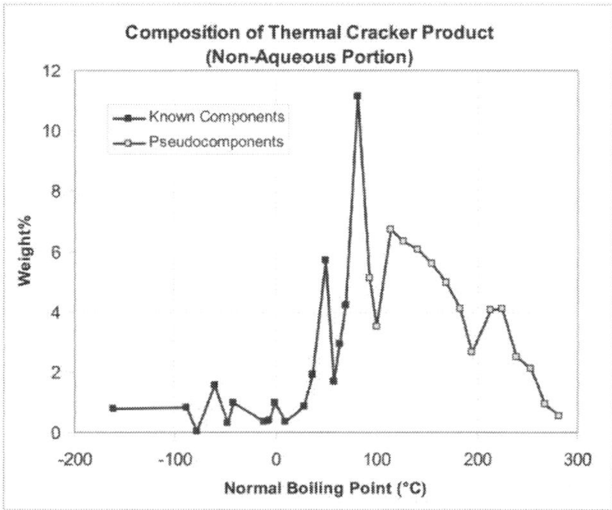

5.5 Wt% H_2O
Impurities as $SO_4^{-} \approx$ 5 ppmw

Pressure ≈ 475 kPa

Troubleshooting Issues:

- **Aqueous phase causes corrosion problems.**
- **What is the water dew point?**
- **How does it vary with impurity (H_2SO_4) level?**
- **What is our confidence in the model predictions?**

37

Aqueous Phase in Thermal Cracker Product

FLUOR

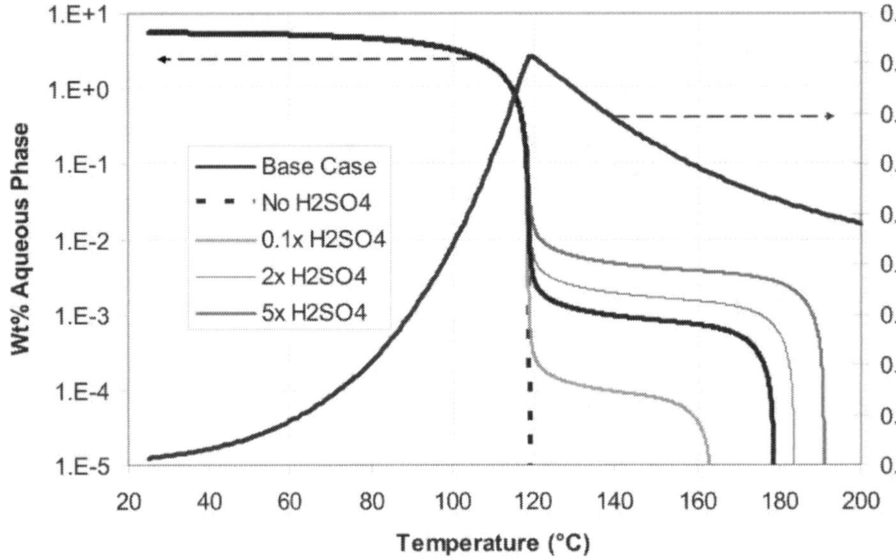

38

pH of Aqueous Phase – Thermal Cracker Product

FLUOR

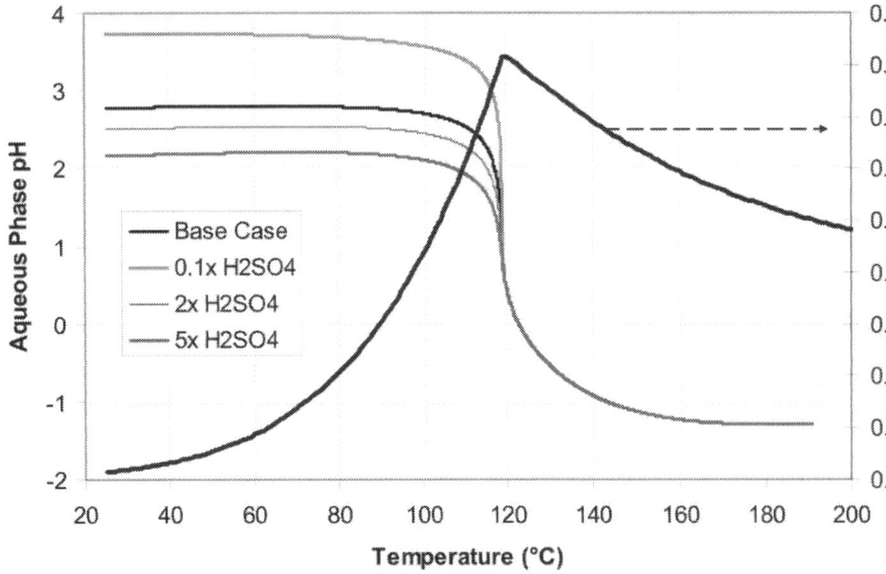

39

Aqueous Phase in Thermal Cracker Product

FLUOR

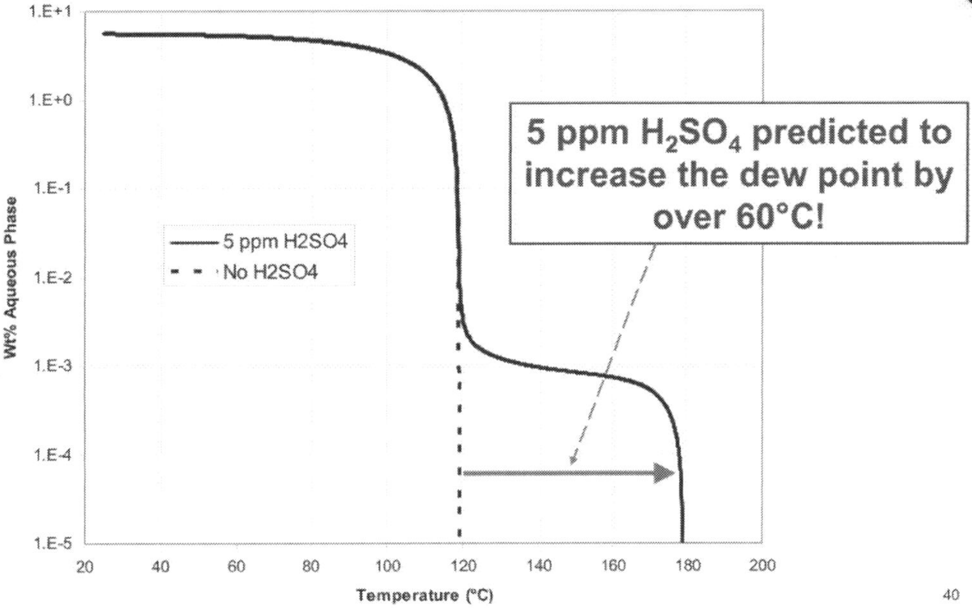

40

Aqueous Phase in Thermal Cracker Product - Discussion

FLUOR

- Validity and accuracy of model:
  - Calculations have expected trends, and hence are at least qualitatively correct.
  - Water dew point (no H_2SO_4 impurities) is accurate.
  - Effect of H_2SO_4 on increasing the dew-point temperature probably accurate to ±10-25°C.
  - pH calculations reinforce VLLE calculations, and provide an indicator of H_2SO_4 level.
- Problem mitigation:
  - Ensure sufficient temperature margin.
  - Eliminate organic oxygen that causes formation of H_2SO_4 – but note that even very low levels of H_2SO_4 cause significant increase in dew point.

41

Tacit Knowledge Should Lead and Direct Explicit Knowledge – Both are Needed

FLUOR

- The rational mind uses explicit knowledge - strongly emphasized in our education and research.
- The intuitive mind comes naturally to our evolved brains, and is honed by years of thinking, experience and problem solving.
- Excellence in process modeling requires both minds.

The intuitive mind is a sacred gift, and the rational mind is a faithful servant.

René Descartes 1596-1650

As for logic, its syllogisms and the majority of its other precepts are of avail rather in the communication of what we already know … than in the investigation of the unknown.

42

AICHE Regional Conference

Model based Real time performance monitoring for Operational Excellence

October 5, 2012

Vibhu Sharma

CEO, Ingenious Inc

www.ingen[illegible]

NGENIOUS

1

Industry Challenge

- Manufacturing industries face a constant challenge of maximizing productivity while minimizing costs
- TMI: Too much information and data, with no clarity: DCS, Historian, SAP/Oracle, SQL server, LIMS, models etc
- I think I know what I am producing, but I want to know how and what I should be producing!
- But Excel is my best friend!
- Overarching Objective: Achieve "Operational Excellence"

2

Operational Excellence is a philosophy of leadership, teamwork and problem solving resulting in continuous improvement throughout the organization by focusing on optimizing existing activities in the process.

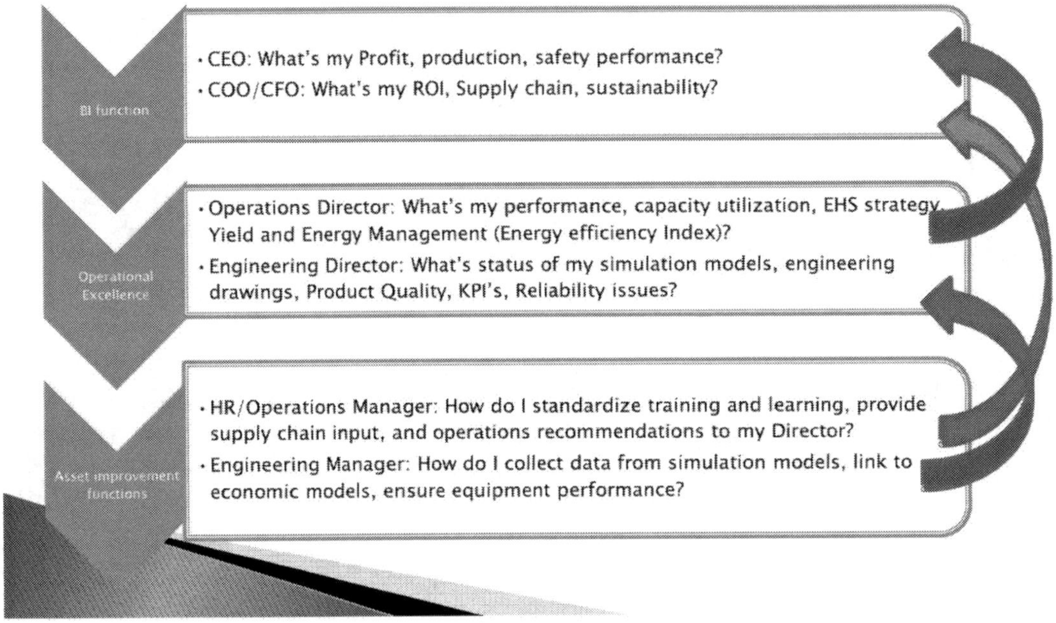

Solution & Benefits: RPM

- Real-time performance monitoring (RPM) is focused on collecting data, in real time, from various data sources, analyzing them and then making recommendations for performance optimization.
- **Know how your assets are performing, and how they SHOULD be performing**
- What features to expect?
  - Continuously monitor key performance indicators (KPI's) and performance
  - Perform "what-if" analysis and debottlenecking to improve product yields and quality.
  - Equipment condition monitoring
  - Google Maps (GIS) for tracking multiple sites anywhere in the world
  - Accurate Material Balance and loss analysis
  - Profit & Operations Analytics through dashboards
  - Work flow management and root cause analysis

4

How does RPM work?

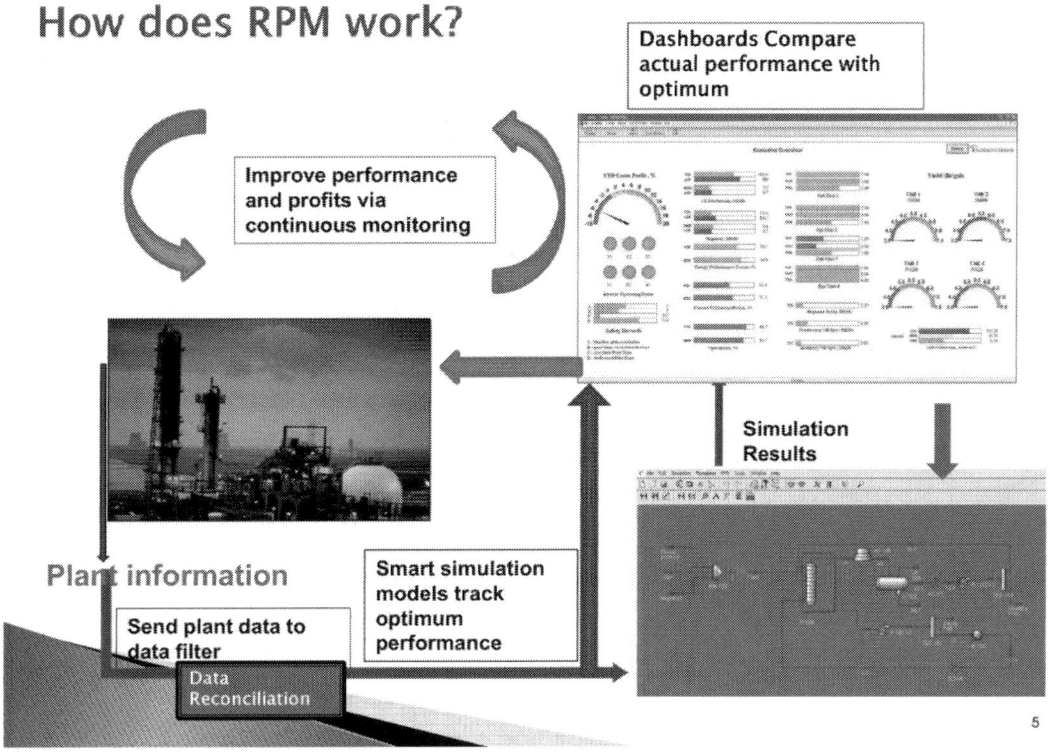

5

5 Steps to RPM

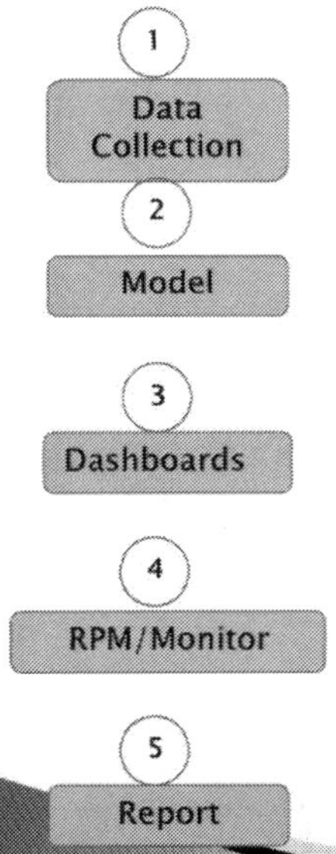

1. **Data Collection**: Data from the plant servers (PI/SQL server/Excel) is collected, Key performance indicators are identified and an objective function is established.
2. **Model**: Build a simulation or optimization model of the plant using commercial modeling & simulation tools
3. **Dashboards**: Based on the model, real data and the KPIs
4. **RPM/Monitor**: Engineers/operators improve and monitor performance parameters.
5. **Reports**: Reports are sent based on the performance of the plant. Reports are also sent if any critical alarm sets off.

6

Case Study: Continental Carbon, USA (1 of 2)

- **Info**: Leading producer of Carbon black for tire production. Part of $16 billion conglomerate. 3 US plants, 7 globally. www.continentalcarbon.com
- **Problem**: Benchmarking 3 US based plants with real time information, optimize performance, & reduce energy costs
- **Data**: Integrate monitoring software with CHEMCAD simulation, OSI-PI, DeltaV DCS, SQL server.
- **Special**: Neural network model for high Temperature reactor performance, EPA handbook calculations for Greenhouse Gas Emissions (GHG).

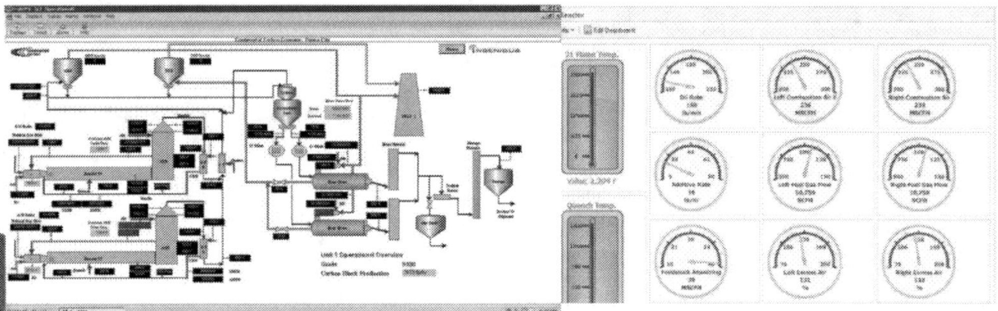

7

Benefits: Continental Carbon, USA (2 of 2)

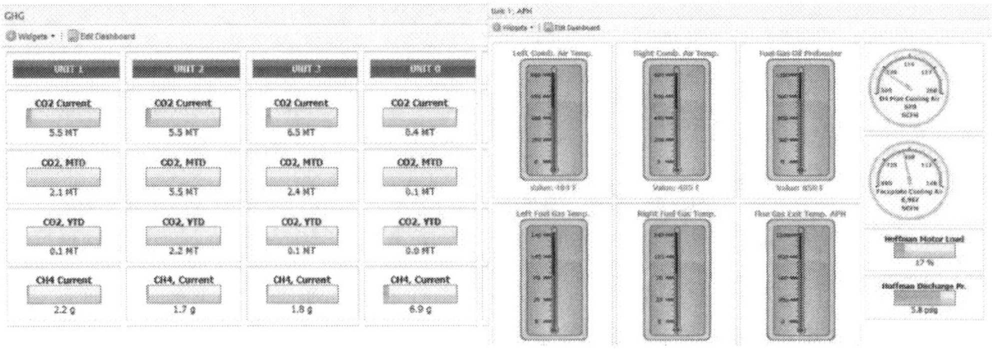

- Video Wall to monitor all assets, all reactors and capacity utilization in real time.
- Optimization of reactors, pre-heaters, compressors and automatic tracking of EPA Greenhouse emissions
- Benchmark various reactors, equipments and locations against each other to derive best in class performance
- Integrate dynamic training simulators, Smart P&IDs, e-learning etc to same real time monitoring platform for integrated Operational Excellence initiative

8

Case Study – MicroSeismic Inc

- **Info**: Leading geophysical services provider of micro seismic interpretations from reservoirs during hydraulic fracturing stimulation (www.microseismic.com)
- **Problem**: Analyze and visualize thousands of well and reservoir data points, with trends, predictive analytics, historical trends, & data repository
- **Solution**: ProRPM software implemented to integrate with custom spreadsheets, with 3D data visualization, embedded excel and PowerPoint info
- MicroSeismic data hosted on the CLOUD. Client Rollout planned shortly with direct CLOUD access to major oil & gas clients globally

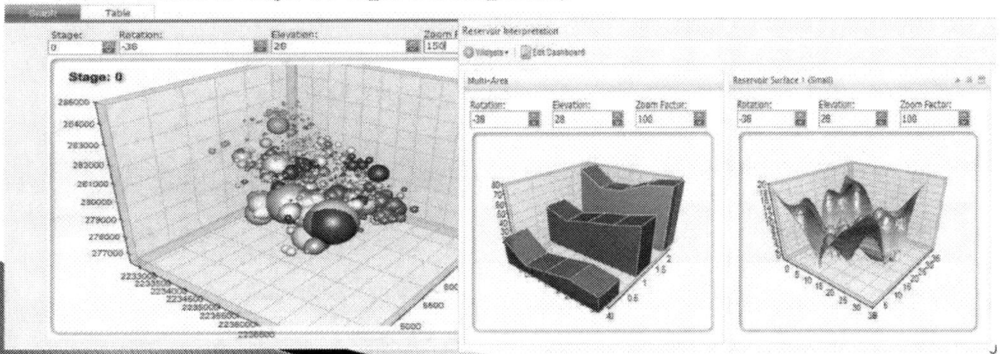

Benefits– MicroSeismic Inc

- **Benefits**: Real time access on the CLOUD to major oil & gas clients globally along with access to multiple wells, fields and reservoirs
- Accessible via ipad, iphone and any smart device
- Eliminates huge paperwork, and data transfer back and forth with clients for revisions and updates. Interactive access and updates directly on the CLOUD
- Solution: Software implemented to integrate with custom spreadsheets, with 3D seismic interpretation to make more informed decisions on investment strategy for billion dollar investments

10

Case Study: Energy Solutions International: Pipeline Dashboard

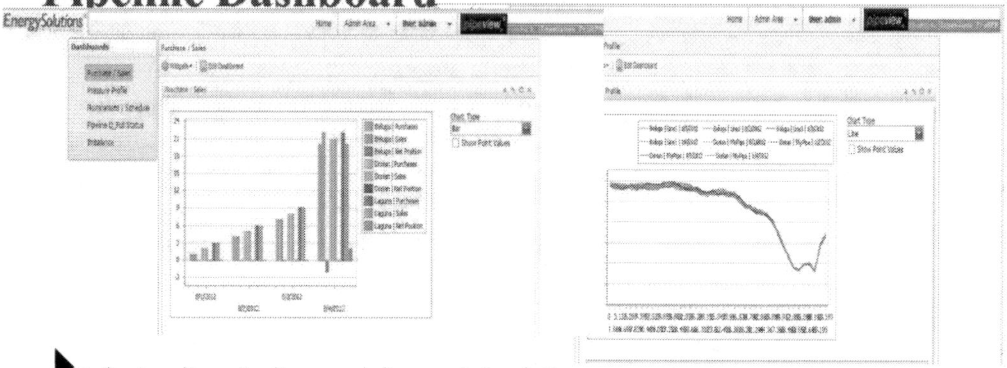

- **Info**: Leading pipeline modeling and simulation company, www.energy-solutions.com
- **Problem**: Real time monitoring , allocations/nominations, and leak detection for large & remote pipeline systems.
- **Data**: Integrate monitoring software with pipeline modeling software, data historian, Various SCADA, SQL server.
- **Benefits**: Real time visualization of pipeline pressure, allocations, nominations and invoicing with user integration and access via CLOUD

Case Study: Equipment Condition Monitoring

E-101

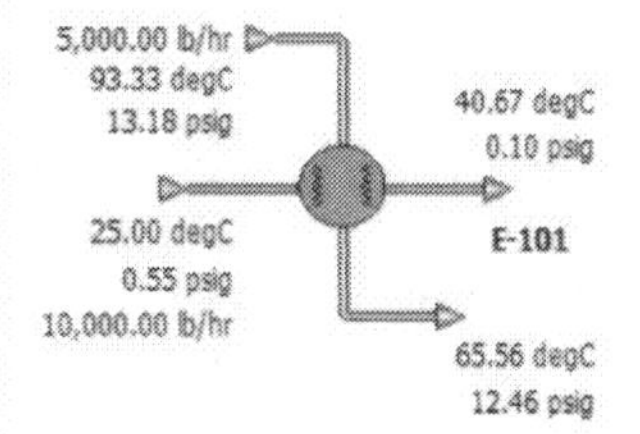

Description	Value	Units
E-101 Duty	266.5	kBtu/hr
E-101 Heat Transfer Coefficient	4.92	Btu/hr-ft2-F
E-101 UA	3195	Btu/F-hr
E-101 Log Mean Temperature Difference	28.56667	degC
E-101 Tube Differential Pressure	0.4351	psi
E-101 Shell Differential Pressure	0.7252	psi
E-101 Tube Differential Temperature	15.67	degC
E-101 Shell Differential Temperature	-27.77	degC

- **Problem**: Track exchanger fouling, catalyst deactivation, coker drum performance, and major equipment condition monitoring, with predictive analytics
- How/When will my equipment fail?
- **Data**: Custom algorithms to calculate equipment performance, statistical analytics using real-time data to predict future performance. Historical trends and extrapolation
- **Benefits**: Correct problems before they occur and influence efficiency and equipment life with operational changes and insights.

Digital Plant: The Future is here

- Business Intelligence: SAP, Oracle and SQL links
- Operational Intelligence: Performance of asset, equipment, Yields, Energy Management, Emissions etc
- Advisory tools, Predictive Analytics: Simulation and modeling engines, KPI calculations, equipment performance templates: feedback to Business level
- Real-time Training Simulators: Understand reason for variability or safety issues and respond instantly
- Real-time Learning: Learn at the click of a button with real plant info, Standard Operating Procedures, Quality manuals etc

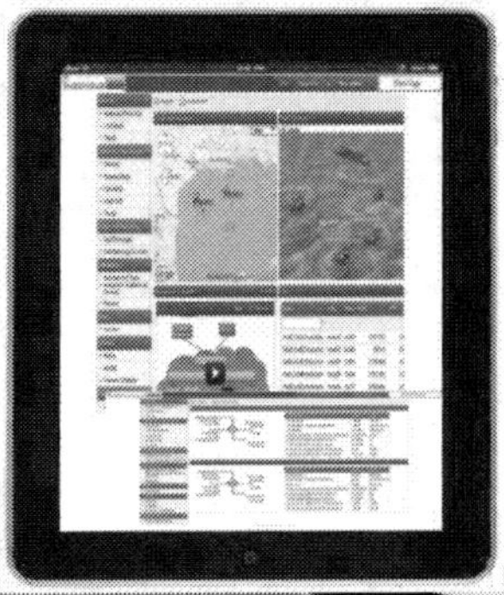

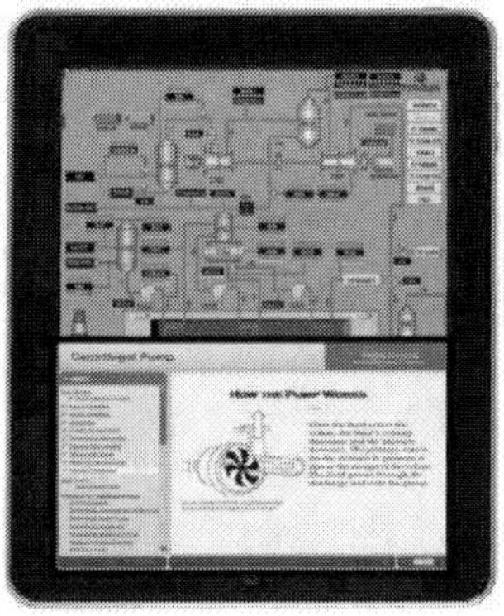

13

Path Forward

- Manufacturing industries have an opportunity to embrace new technologies and influence a new era of productivity.
- BIG DATA can be a good thing, if one can harness the power of information
- Sweat the assets, all day, everyday with the help of RPM and simulation and modeling tools.
- In an era of consolidation, each location operates in its own sphere with its own standards. RPM brings those worlds together in a synergistic manner
- Achieve "Operational Excellence" in combination with real time monitoring software, modeling tools, and plant information systems

14

Model based Real time performance monitoring for Operational Excellence

Ingenious Inc **For more information, please contact**

Vibhu Sharma, CEO

Vibhu.sharma@ingeniousinc.com

OR

Bharat Kamdar, President

Bharat.kamdar@ingeniousinc.com

Visit: www.ingeni

INGENIOUS

15

Chemical Engineering Disciplines

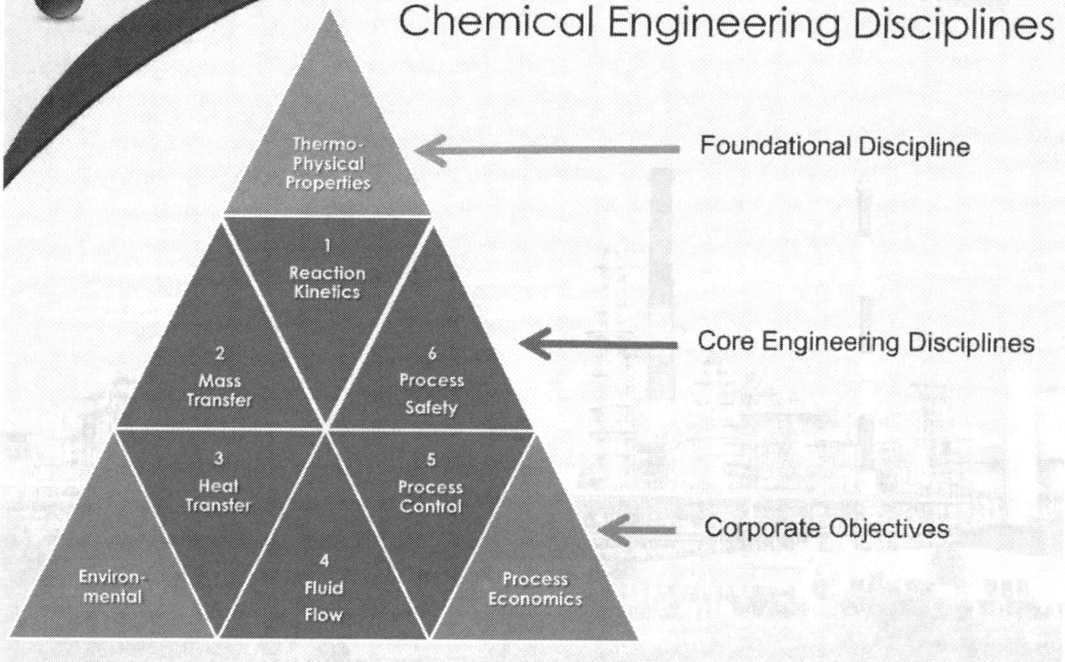

API, GPA, DDB & DIPPr Software & Data

The API Technical Data Book and GPA Data Bank have been developed to achieve accurate flash results for use in the oil, gas, chemical and petrochemical industries

- Pure Component Data
  - EPCON's THERMA™ Flash is included in both products including the complete DIPPr® Database of critically reviewed pure component data (approx. 2000 components)
  - DDB provides all GPA VLE and thermophysical property data from over 30 years of research in the GPA Data Bank
  - DDB ARTIST property estimation software and methods provides coverage for any component not available in DDB or DIPPr's pure component databases
- Petro Component Data
  - API provides accurate petro and characterization properties of petroleum streams when required for petroleum streams
- VLE Data and Binary Parameters
  - DDB provides accurate VLE data in the GPA Data Bank
  - DDB provide accurate fitting of binary parameters in the GPA Data Bank
- Accurate Flash Results
  - EPCON's THERMA 3-phase flash – in API and GPA thermophysical property products
  - EPCON's THERMA Mixture properties – in API and GPA thermophysical property products
  - Achieve accurate flash results using API, DDB, DIPPr and GPA methods and data
  - Benchmarked by the API Tech Data Committee to outperform major simulators in key areas

(DIPPr is a registered trademark of AIChE and is used by permission)

Pure Component Properties

energy API GPA DDBST DORTMUND DATA BANK SOFTWARE & SEPARATION TECHNOLOGY EPCONSOFTWARE It's time to upgrade to the latest technology

API Data Book Pure Components

Property Constants Group:

- Critical Properties
- Thermal Properties
- Characterization Properties
- Safety Related
- Other Properties
- Using DIPPr Database

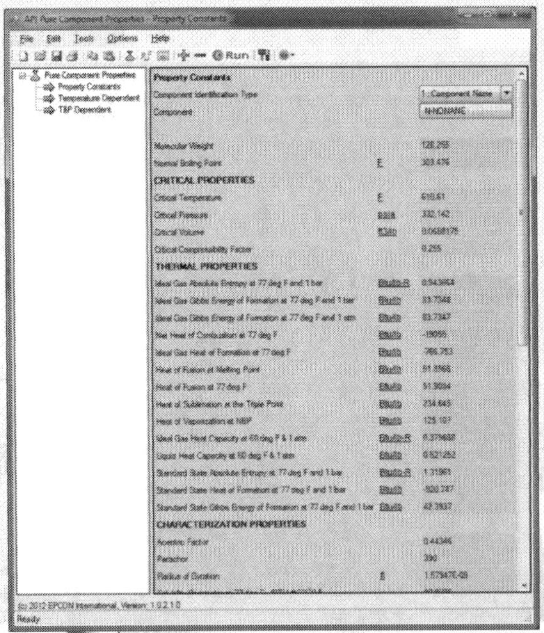

energy API GPA DDBST DORTMUND DATA BANK SOFTWARE & SEPARATION TECHNOLOGY EPCONSOFTWARE It's time to upgrade to the latest technology

API Data Book Pure Components

Temperature Dependent Properties

Liquid Properties not displayed if Input Temperature > Tc

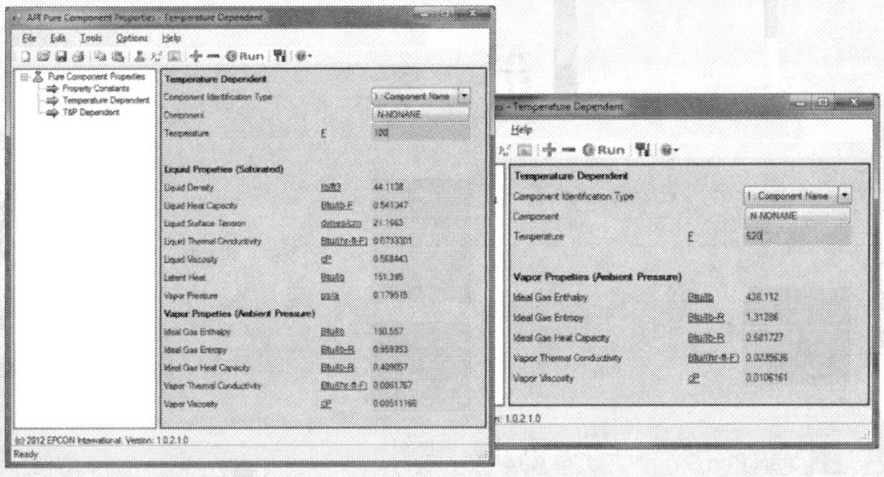

energy API GPA

DDBST DORTMUND DATA BANK SOFTWARE & SEPARATION TECHNOLOGY

EPCONSOFTWARE It's time to upgrade to the latest technology

API Data Book Pure Components

Temperature – Pressure Dependent Properties

- Phase Specific Properties Displayed
- Phase Calculated using Vapor Pressure

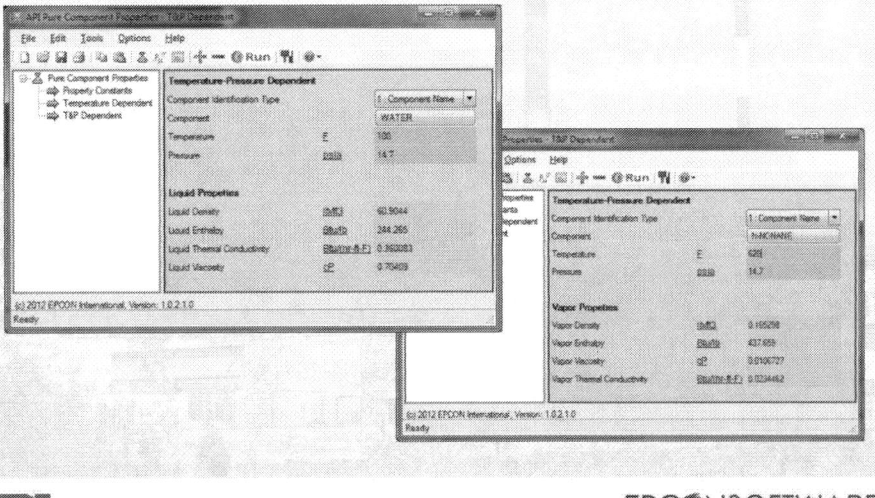

energy API TECHNICAL DATA BOOK 9th Edition

EPCONSOFTWARE It's time to upgrade to the latest technology

DDB ARTIST Property Estimation

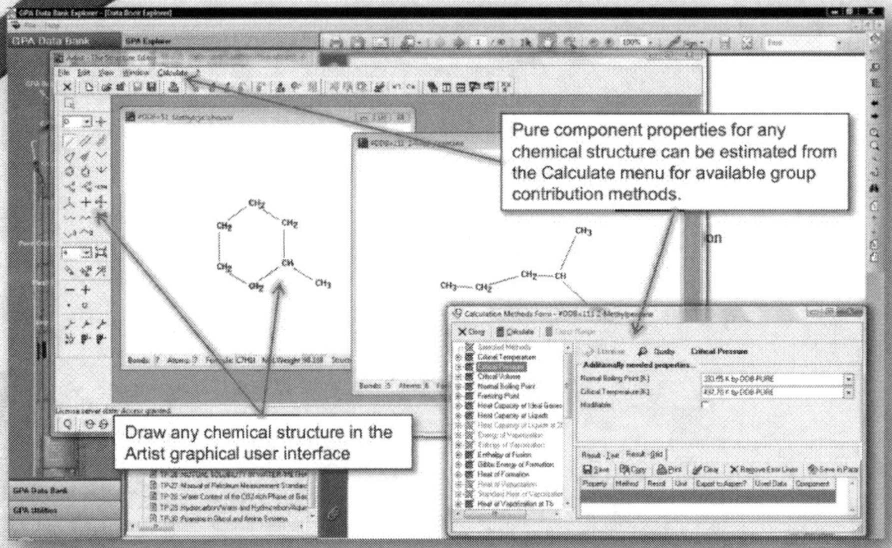

The DDB Artist graphical user interface allows for the drawing of any chemical structure and estimation of pure component properties. Group contribution methods are provided for basic estimation capabilities, and extended methods are available separately.

energy API GPA DDBST DORTMUND DATA BANK SOFTWARE & SEPARATION TECHNOLOGY EPCONSOFTWARE It's time to upgrade to the latest technology

DDB ARTIST Property Estimation

2009	Estimation of pure component properties. Part 4: Estimation of the saturated liquid viscosity of non-electrolyte organic compounds via group contributions and group interactions	Nannoolal Y., Rarey J., Ramjugernath D.	Fluid Phase Equilib., 281 (2), 97 119 (2009)
2008	Estimation of the vapour pressure of non-electrolyte organic compounds via group contributions and group interactions	Moller B., Rarey J., Ramjugernath D.	J.Mol.Liq., 143, 1, 52 63 (2008)
2008	Development of an Improved Group Contribution Method for the Prediction of Vapour Pressures of Organic Compounds, Bruce Moller, MSc. Thesis, 2008		MSc. Thesis, 2008
2008	Estimation of pure component properties: Part 3. Estimation of the vapor pressure of non-electrolyte organic compounds via group contributions and group interactions	Nannoolal Y., Rarey J., Ramjugernath D.	Fluid Phase Equilib., 269(1-2), 117-133 (2008)
2007	Estimation of pure component properties Part 2. Estimation of critical property data by group contribution	Nannoolal Y., Rarey J., Ramjugernath J.	Fluid Phase Equilib., 252, 1-2, 1 27 (2007)
2007	Development and Critical Evaluation of Group Contribution Methods for the Estimation of Critical Properties, Liquid Vapour Pressure and Liquid Viscosity of Organic Compounds, Yash Nannoolal, PhD Thesis, 2006		PhD Thesis, 2006
2004	Estimation of Pure Component Properties. Part 1. Estimation of the Normal Boiling Point of Non-Electrolyte Organic Compounds via Group Contributions and Group Interactions	Nannoolal Y., Rarey J., Ramjugernath D., Cordes W.	Fluid Phase Equilib., 226, 1, 45 63 (2004)

energy API GPA DDBST DORTMUND DATA BANK SOFTWARE & SEPARATION TECHNOLOGY EPCONSOFTWARE It's time to upgrade to the latest technology

Petro-Fraction Properties

energy API | GPA | DDBST DORTMUND DATA BANK SOFTWARE & SEPARATION TECHNOLOGY | EPCONSOFTWARE It's time to upgrade to the latest technology

API Data Book Petro Fractions

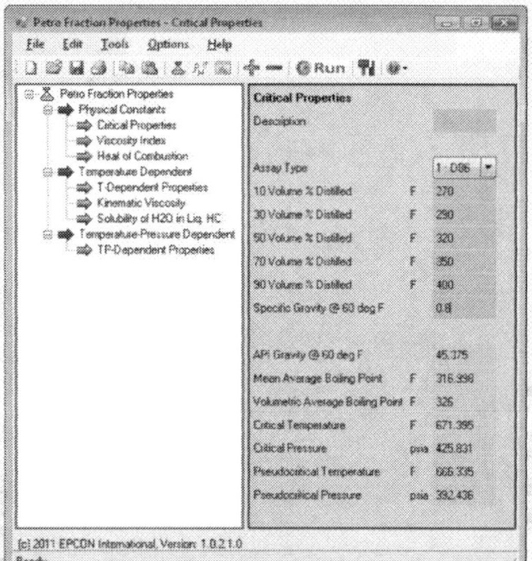

Critical Properties

energy API | GPA | DDBST DORTMUND DATA BANK SOFTWARE & SEPARATION TECHNOLOGY | EPCONSOFTWARE It's time to upgrade to the latest technology

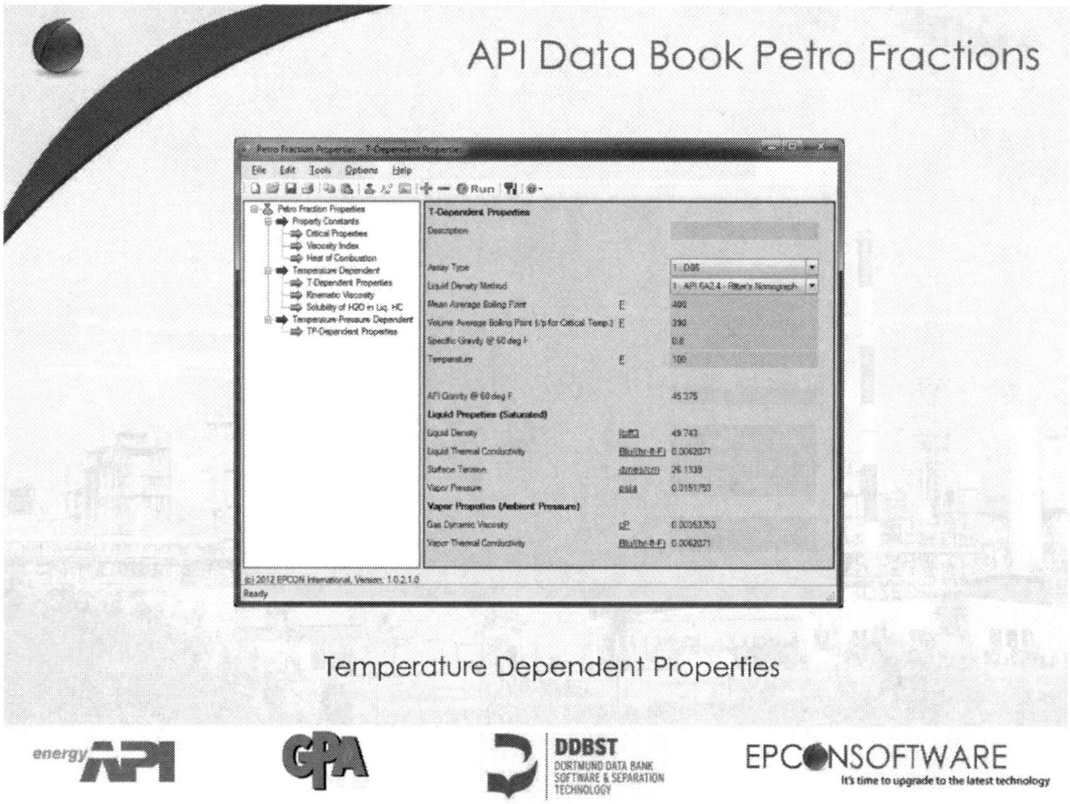
API Data Book Petro Fractions
Petro Fraction Properties - T-Dependent Properties
File Edit Tools Options Help
Run
Petro Fraction Properties
Property Constants
Critical Properties
Viscosity Index
Heat of Combustion
Temperature Dependent
T-Dependent Properties
Kinematic Viscosity
Solubility of H2O in Liq. HC
Temperature-Pressure Dependent
TP-Dependent Properties
T-Dependent Properties
Description
Assay Type
1 - D86
Liquid Density Method
Mean Average Boiling Point
F
400
Specific Gravity @ 60 deg F
0.8
Temperature
F
100
API Gravity @ 60 deg F
45.375
Liquid Properties (Saturated)
Liquid Density
Liquid Thermal Conductivity
Surface Tension
Vapor Pressure
Vapor Properties (Ambient Pressure)
Gas Dynamic Viscosity
Vapor Thermal Conductivity
Ready
Temperature Dependent Properties
energy API
GPA
DDBST
DORTMUND DATA BANK
SOFTWARE & SEPARATION
TECHNOLOGY
EPCONSOFTWARE
It's time to upgrade to the latest technology

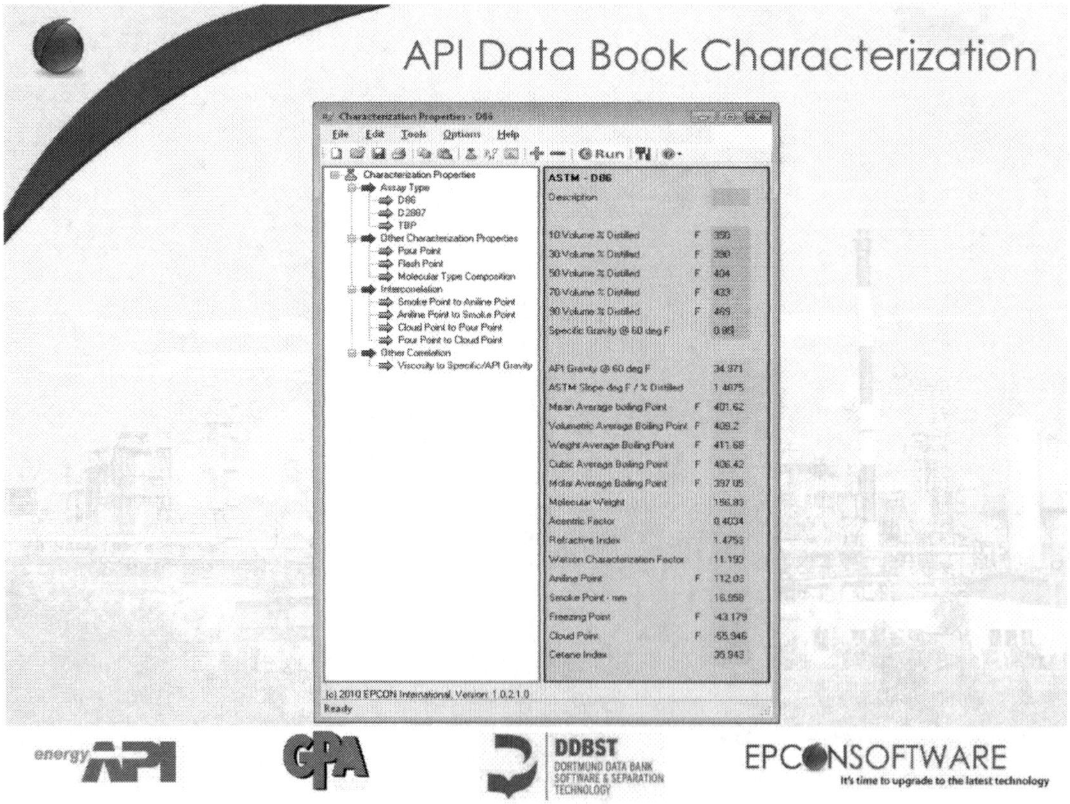
API Data Book Characterization
Characterization Properties - D86
File Edit Tools Options Help
Run
Characterization Properties
Assay Type
D86
D2887
TBP
Other Characterization Properties
Pour Point
Flash Point
Molecular Type Composition
Interconelation
Smoke Point to Aniline Point
Aniline Point to Smoke Point
Cloud Point to Pour Point
Pour Point to Cloud Point
Other Correlation
Viscosity to Specific/API Gravity
ASTM - D86
Description
10 Volume % Distilled
30 Volume % Distilled
50 Volume % Distilled
70 Volume % Distilled
90 Volume % Distilled
Specific Gravity @ 60 deg F
API Gravity @ 60 deg F
ASTM Slope deg F / % Distilled
Mean Average boiling Point
Volumetric Average Boiling Point
Weight Average Boiling Point
Cubic Average Boiling Point
Molal Average Boiling Point
Molecular Weight
Acentric Factor
Refractive Index
Watson Characterization Factor
Aniline Point
Smoke Point - mm
Freezing Point
Cloud Point
Cetane Index
Ready
energy API
GPA
DDBST
DORTMUND DATA BANK
SOFTWARE & SEPARATION
TECHNOLOGY
EPCONSOFTWARE
It's time to upgrade to the latest technology

API Pseudo Component Generation

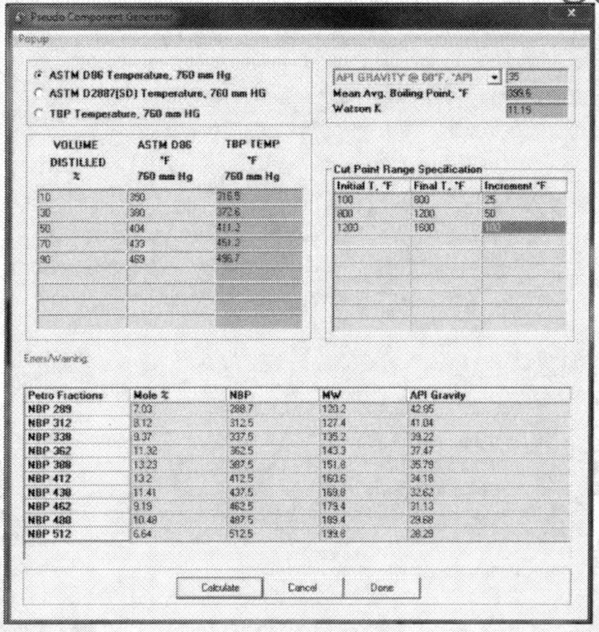

energy API GPA DDBST DORTMUND DATA BANK SOFTWARE & SEPARATION TECHNOLOGY EPCONSOFTWARE It's time to upgrade to the latest technology

VLE Data and Binary Parameters

energy API GPA DDBST DORTMUND DATA BANK SOFTWARE & SEPARATION TECHNOLOGY EPCONSOFTWARE It's time to upgrade to the latest technology

DDB Component Selection

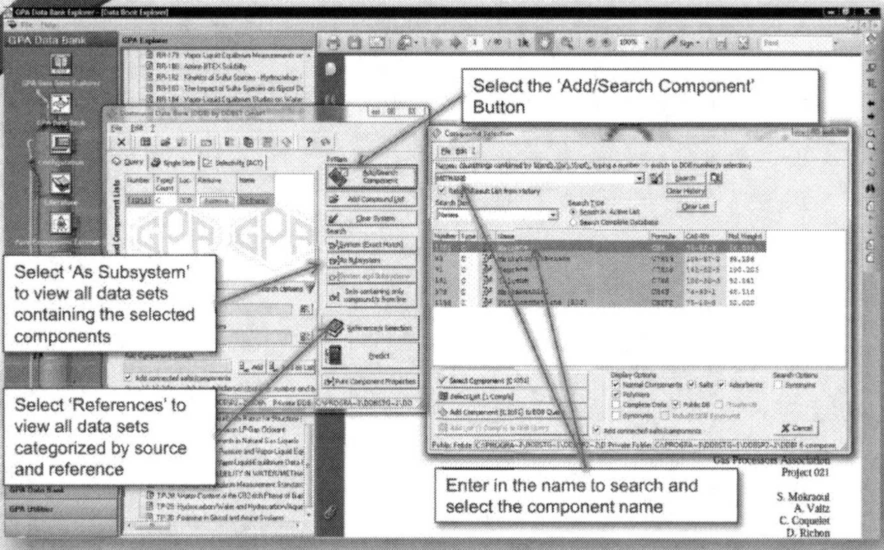

All DDB components can be searched in a number of ways for retrieval of pure and mixture data. Selection of 'System (Exact Match)' will display all systems where only the selected components appear while selection of 'As Subsystem' will display all systems where the selected components and all others appear. VLE data for these systems is then automatically selected and graphed for review.

energy API GPA DDBST DORTMUND DATA BANK SOFTWARE & SEPARATION TECHNOLOGY EPCONSOFTWARE It's time to upgrade to the latest technology

DDB Experimental Data Sets

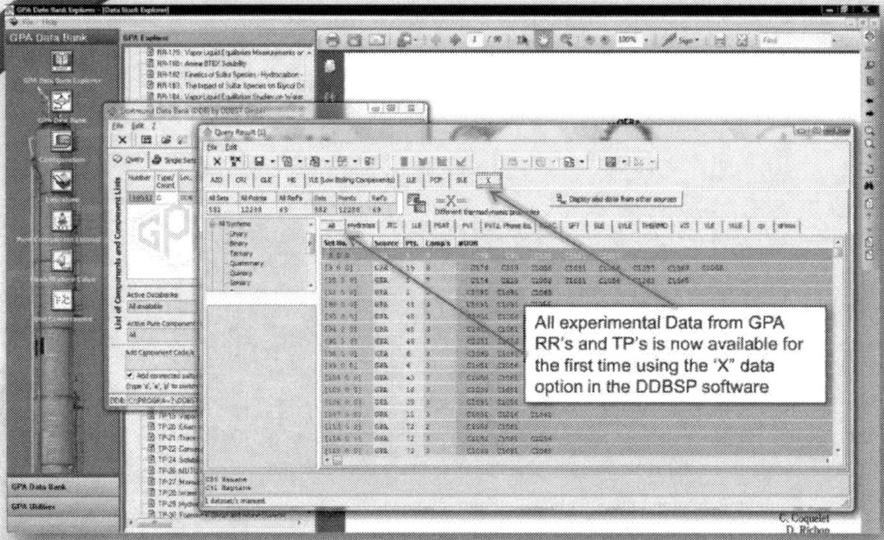

The DDB data types were extended with the 'X' data type so that all GPA experimental data could be captured into the GPA Data Bank. Selection of the 'X' data type provides a secondary set of property tabs that may be chosen to view the data sets available.

energy API GPA DDBST DORTMUND DATA BANK SOFTWARE & SEPARATION TECHNOLOGY EPCONSOFTWARE It's time to upgrade to the latest technology

DDB Data Displays and Graphs

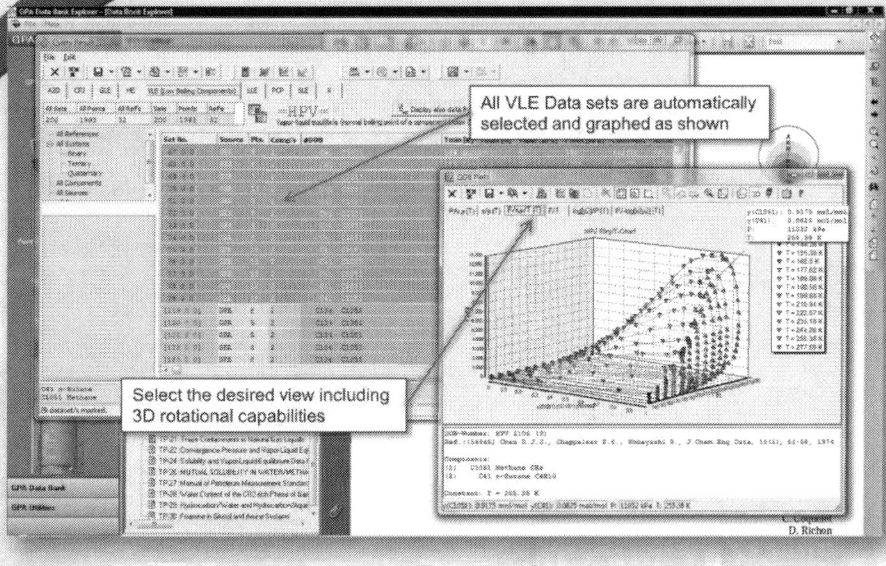

The Query Results window shows all data sets for each data type as shown on the property tabs (the default selection is VLE data). The graphing of VLE data is done automatically and other data types can also be graphed. As the pointer is moved over a data point on the graph, the numeric data is displayed in a floating window as shown. Data can be exported in a several file formats.

DDB VLE Prediction using UNIFAC

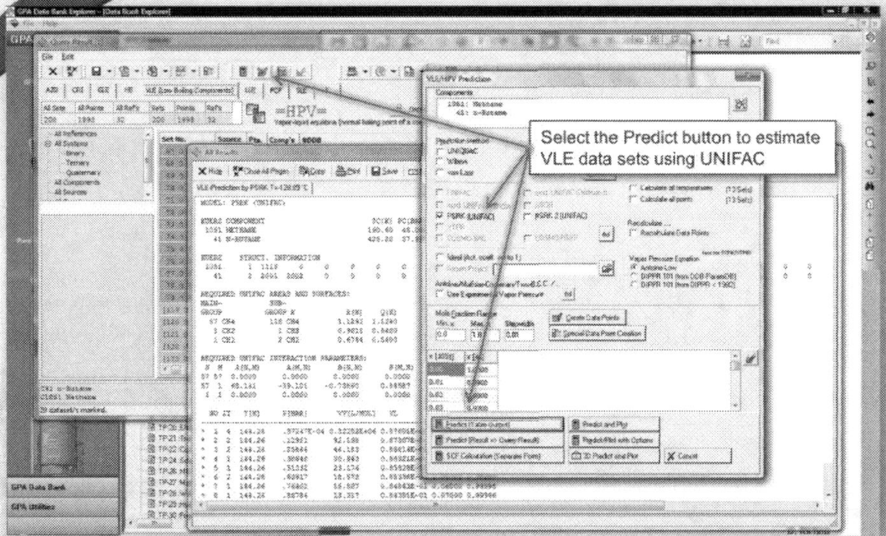

Prediction of VLE data sets using PSRK and UNIFAC can also be provided by the Dortmund Data Bank. The VLE data sets can then be used for prediction of binary parameters.

DDB Binary Parameter Fitting

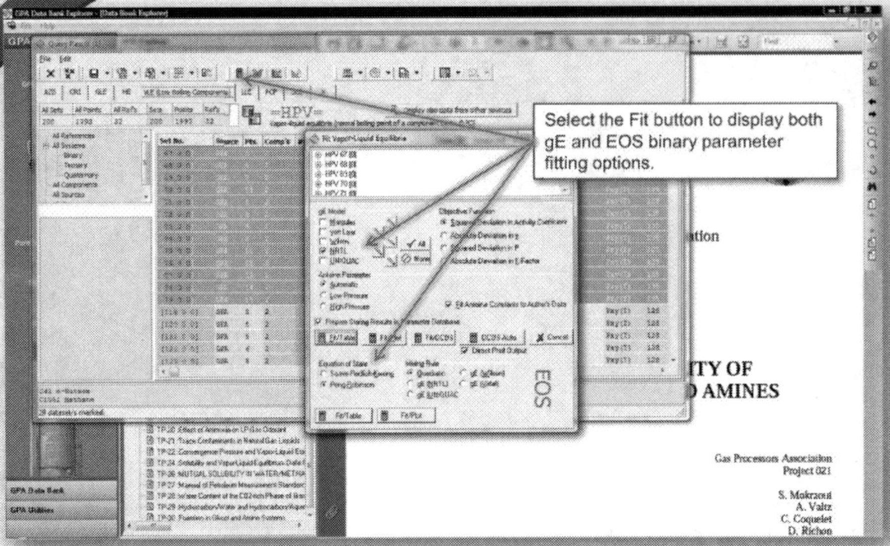

Automated fitting of binary parameters for either gE and EOS models is provided for selected data sets. View the fit graphically or numerically and save parameters to the Parameter database for use in process simulation.

energy API GPA DDBST DORTMUND DATA BANK SOFTWARE & SEPARATION TECHNOLOGY EPCONSOFTWARE It's time to upgrade to the latest technology

Accurate Flash Results

energy API GPA DDBST DORTMUND DATA BANK SOFTWARE & SEPARATION TECHNOLOGY EPCONSOFTWARE It's time to upgrade to the latest technology

EPCON Flash/Mixture Calculations

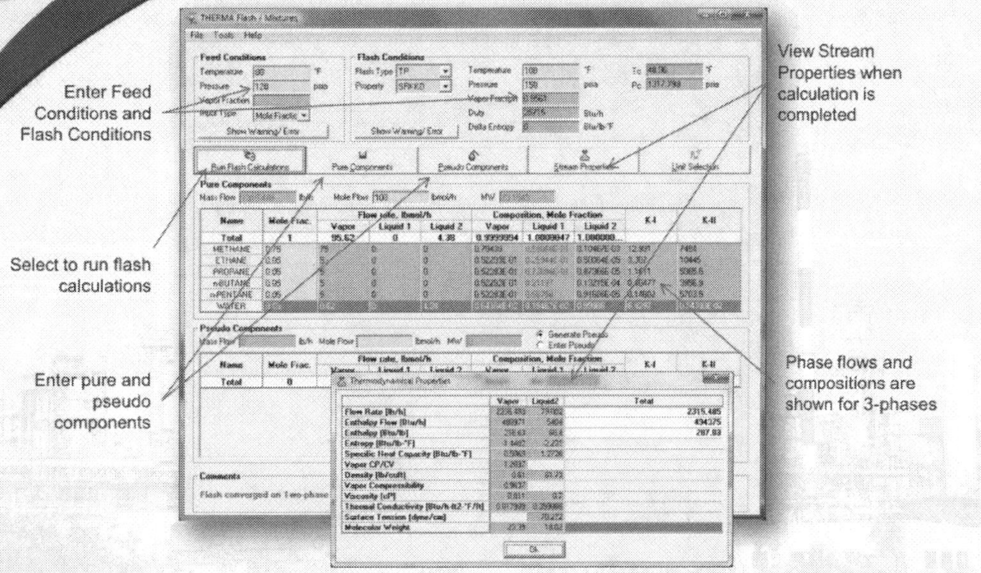

The Flash/Mixture Calculations program provides phase equilibrium and physical properties for each phase. The prediction of the second liquid (aqueous) phase is highly accurate based on its development and use in the API Technical Data Book. Use of binary parameters developed in the GPA Data Bank allow for highly accurate flash results.

energy API GPA DDBST DORTMUND DATA BANK SOFTWARE & SEPARATION TECHNOLOGY EPCONSOFTWARE It's time to upgrade to the latest technology

API Data Book Explorer

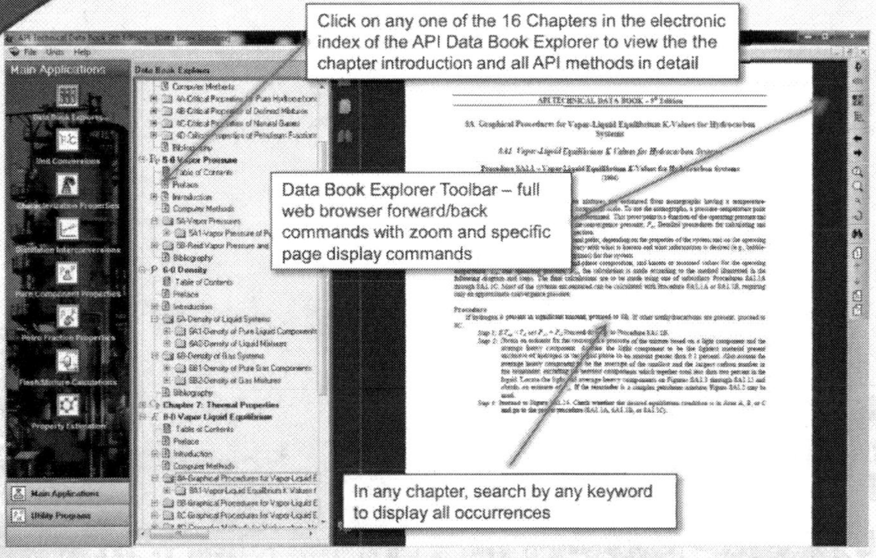

The API Data Book provides all of the API research on thermophysical properties using the Data Book Explorer interface. This provides quick access to over 45 years of API research work – providing the latest API data to insure accurate flash results

energy API GPA DDBST DORTMUND DATA BANK SOFTWARE & SEPARATION TECHNOLOGY EPCONSOFTWARE It's time to upgrade to the latest technology

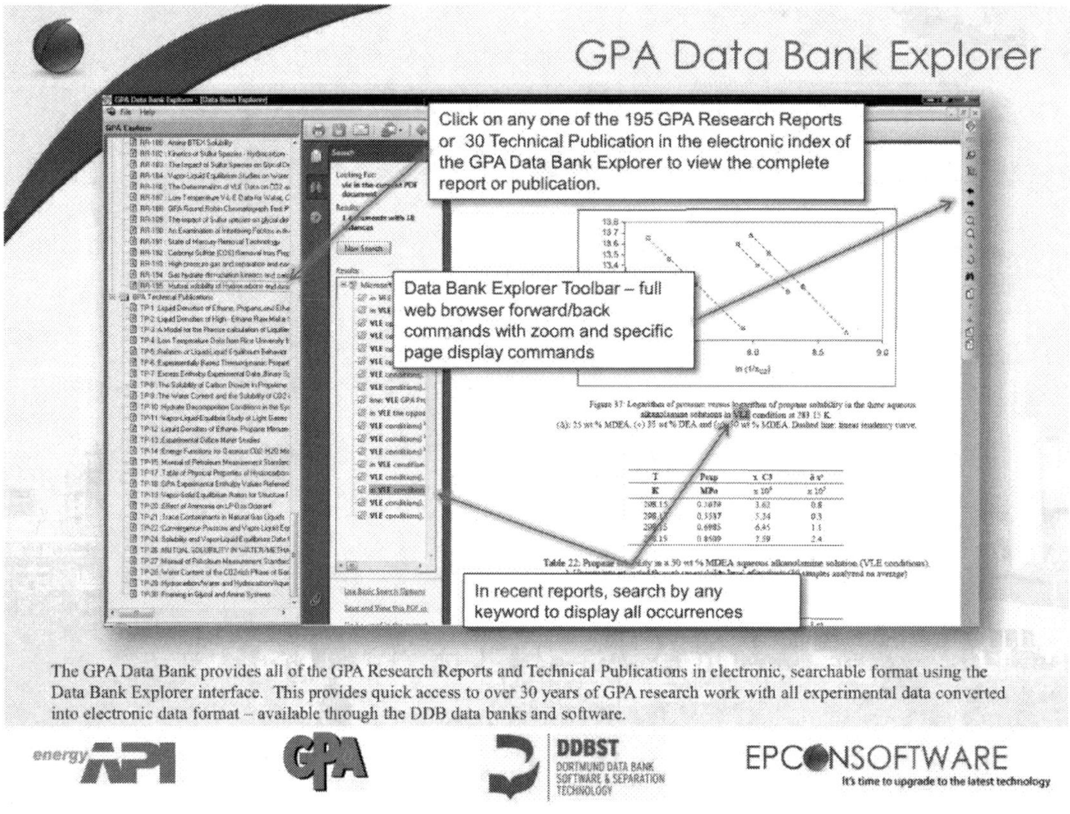
GPA Data Bank Explorer
Click on any one of the 195 GPA Research Reports or 30 Technical Publication in the electronic index of the GPA Data Bank Explorer to view the complete report or publication.
Data Bank Explorer Toolbar – full web browser forward/back commands with zoom and specific page display commands
In recent reports, search by any keyword to display all occurrences
The GPA Data Bank provides all of the GPA Research Reports and Technical Publications in electronic, searchable format using the Data Bank Explorer interface. This provides quick access to over 30 years of GPA research work with all experimental data converted into electronic data format – available through the DDB data banks and software.
energy API
GPA
DDBST
DORTMUND DATA BANK
SOFTWARE & SEPARATION TECHNOLOGY
EPCONSOFTWARE
It's time to upgrade to the latest technology

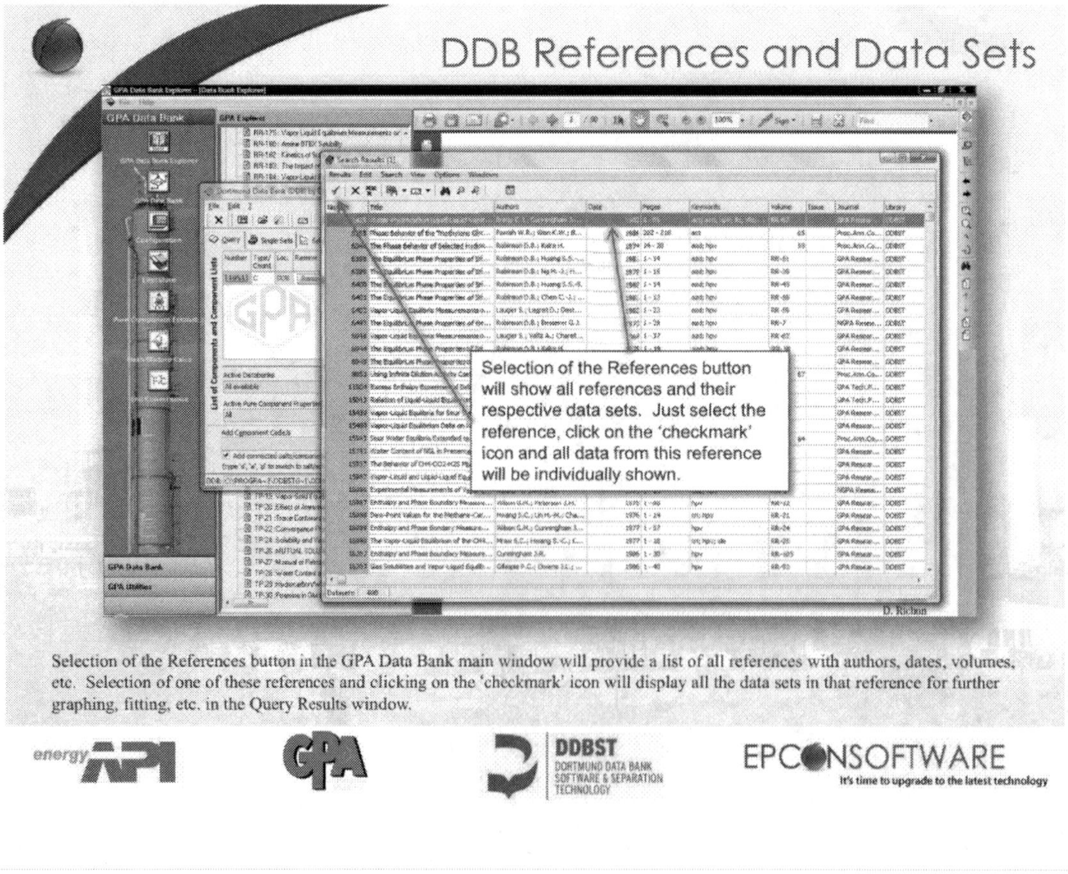
DDB References and Data Sets
Selection of the References button will show all references and their respective data sets. Just select the reference, click on the 'checkmark' icon and all data from this reference will be individually shown.
Selection of the References button in the GPA Data Bank main window will provide a list of all references with authors, dates, volumes, etc. Selection of one of these references and clicking on the 'checkmark' icon will display all the data sets in that reference for further graphing, fitting, etc. in the Query Results window.
energy API
GPA
DDBST
DORTMUND DATA BANK
SOFTWARE & SEPARATION TECHNOLOGY
EPCONSOFTWARE
It's time to upgrade to the latest technology

Practical Application of Thermodynamics

CHAPTERS

1. Introduction
2. PvT Behavior of Pure Components
3. Correlation and Estimation of Pure Component Properties
4. Properties of Mixtures
5. Phase Equilibria of Fluid Systems
6. Caloric Properties
7. Electrolyte Solutions
8. Solid-Liquid Equilibria
9. Membrane Processes
10. Polymer Thermodynamics
11. Application of Thermodynamics in Separation Technology
12. Enthalpy of Reaction and Chemical Equilibria
13. Special Applications
14. Practical Applications
15. Example Problems (MathCAD, DDB, Excel and VBA)

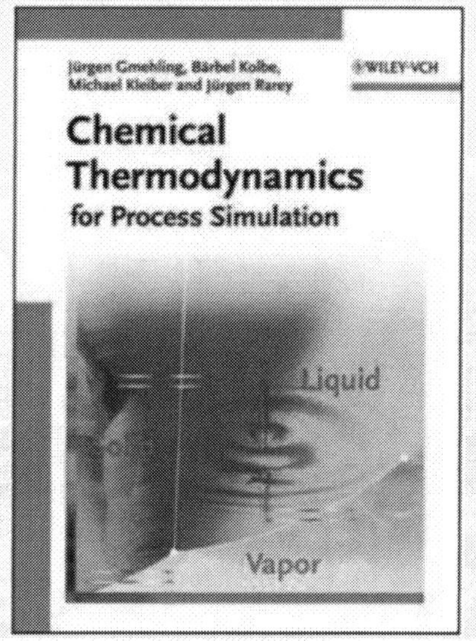

APPENDICES

A. Pure Component Parameters
B. Coefficients for High Precision Equations of State
C. Useful Derivations
D. Standard Thermodynamic Properties for Selected Electrolyte Compounds
E. Regression Technique for Pure Component Data
F. Regression Techniques for Binary Parameters
G. Ideal Gas Heat Capacity Polynomial Coefficients for Selected Compounds
H. UNIFAC Parameters
I. Modified UNIFAC Parameters
J. PSRK Parameters
K. VTPR Parameters

energy API GPA DDBST DORTMUND DATA BANK SOFTWARE & SEPARATION TECHNOLOGY EPCONSOFTWARE It's time to upgrade to the latest technology

EPCONSOFTWARE

Prediction and Mitigation of CO2 Freeze during operation and blowdown

Mark Beyleveld,
Virtual Materials Group Inc., Houston
Marco Satyro,
Virtual Materials Group Inc., Calgary

AIChE Regional Process Technology Conference
Oct 4-5, 2012

Outline

- Introduction
- Thermodynamic model
- Column Operation
- Column Blowdown
- Results/Discussion
- Conclusions

virtualmaterialsgroup™

Introduction

- The occurrence of processes with high CO_2 concentration is becoming more common
- Formation of solid CO2 is a problem both during normal operation, and during blowdown
- Process simulation can assist with prediction of CO2 freeze
- Models can be used to evaluate alternate operating points and procedures
- Simulations in presentation performed with VMGSim

virtual**materials**group™

Thermodynamic Model

- APR Solid CO2 property package predicts formation of solid CO_2 (pure CO_2 and mixtures) based on the APR model for vapor/liquid and a model for the fugacity of solid CO_2 where equilibrium is described:

$$f_{CO_2}^{v} = f_{CO_2}^{l} = f_{CO_2}^{s}$$

virtual**materials**group™

Rigorous VLS Calculations

- Need good model for vapour and liquid fluid phases (APR)
- Need good model for solid phase equilibrium (NIST/Refprop)
- Need good model for solid phase H, S and V (for engineering calculations)

High CO2 Column

- Feed is 70% CO_2, 30% CH_4, with trace heavier HC
- Remove as much pure CO_2 as possible
  - CO_2 from bottoms
  - CO_2/CH_4 mixture overhead
- Operates at 600 psia

virtualmaterialsgroup™

Column Profiles

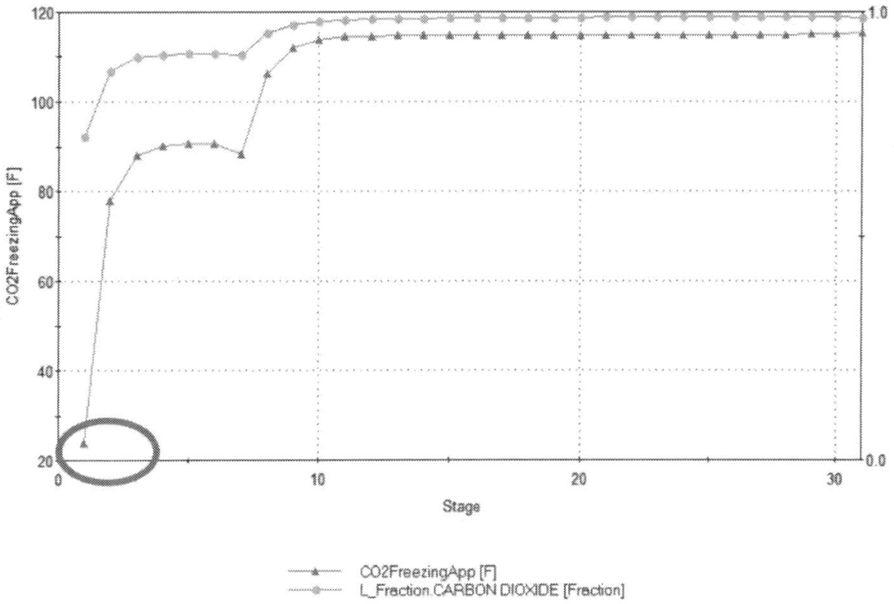

virtualmaterialsgroup™

Improve Operation

- Overhead still contains about 30% CO2
- Freeze margin is about 20 deg
- Analyze by looking at T-X-Y diagram
- Consider higher pressure
  - Freezing temperature is weak function of pressure

virtualmaterialsgroup™

Improve Operation?

T-X-Y at 600 psia

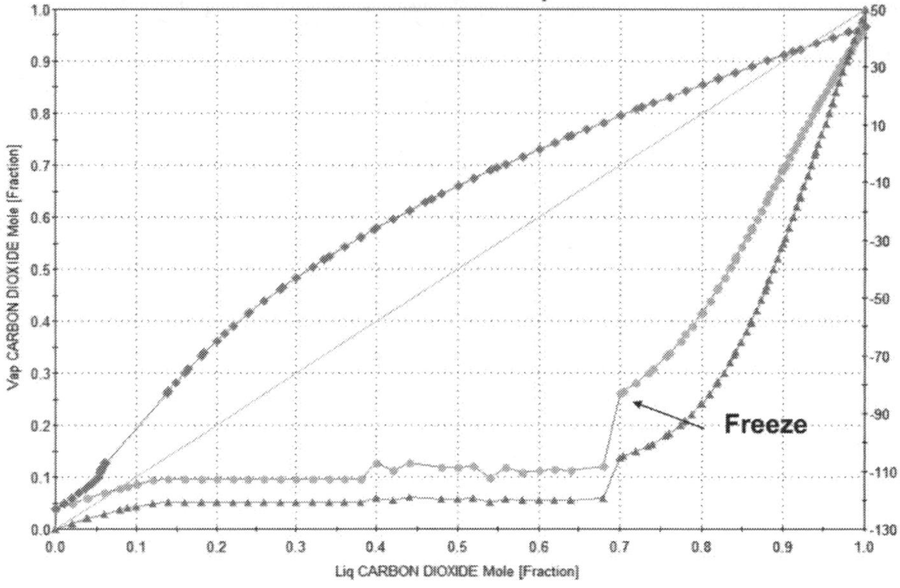

virtual**materials**group™

Improve Operation?

T-X-Y at 900 psia

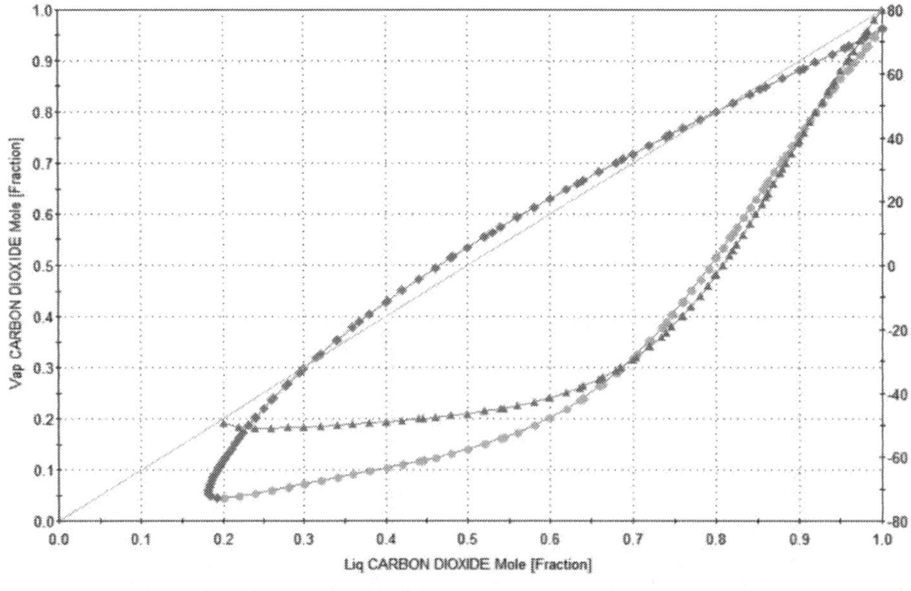

virtual**materials**group™

Phase Envelope – 10% CO2

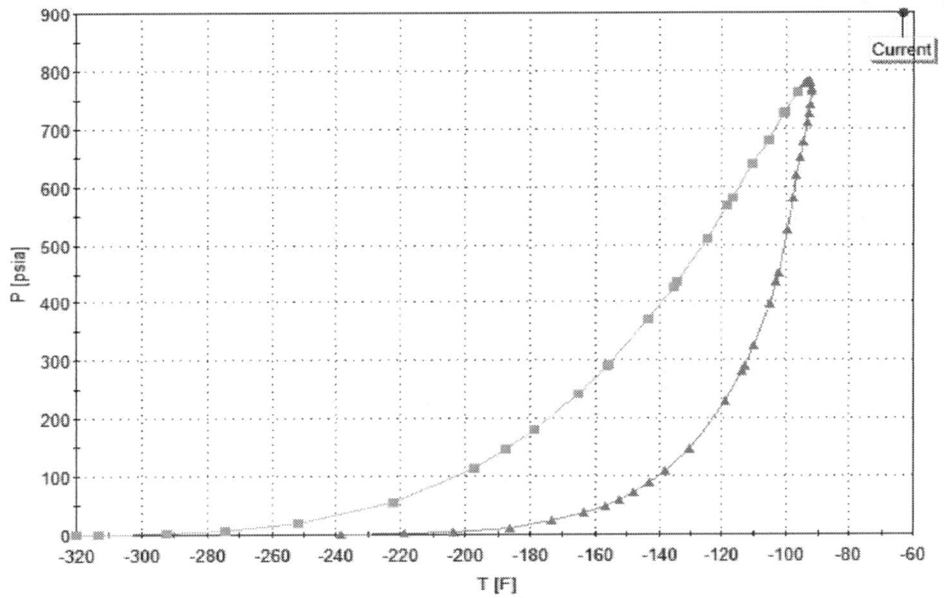

Higher Pressure Infeasible

- Separation cannot be improved
  - At 900 psia and low CO2 conc we are above critical point

Dynamic Model

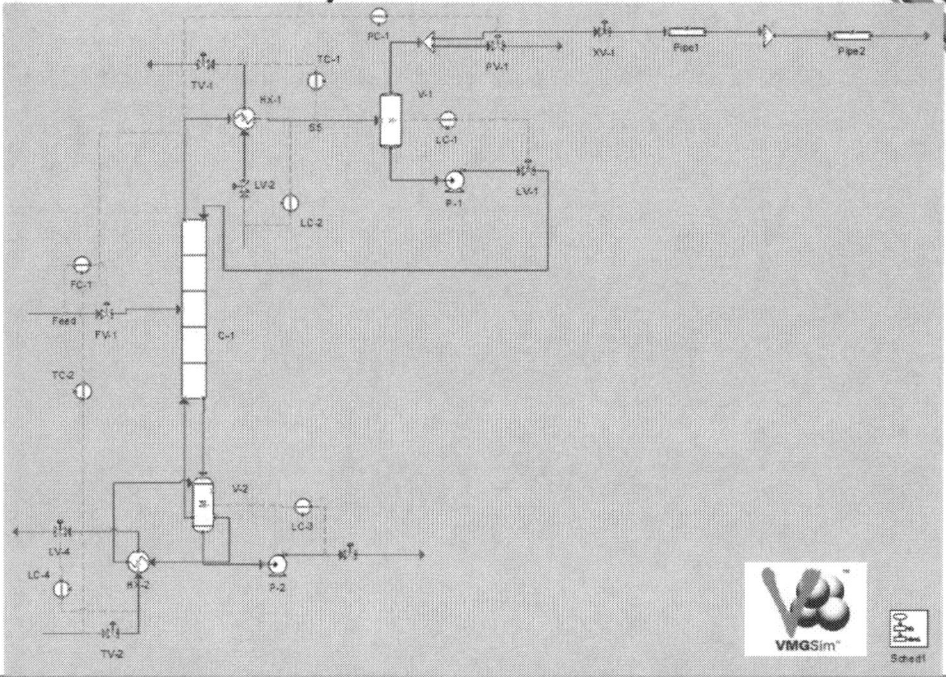

virtual**materials**group™

Column Operation

- Dynamic model mirrors plant control scheme
- Can assess control quality
- Can explore alternate control schemes

virtual**materials**group™

Column Blowdown

- Cut refrigeration and steam
- Shut normal operating valves
- Open blowdown from overhead accumulator
- 2" Ball Full Port
- 6" line discharging into 18" flare header

Blowdown

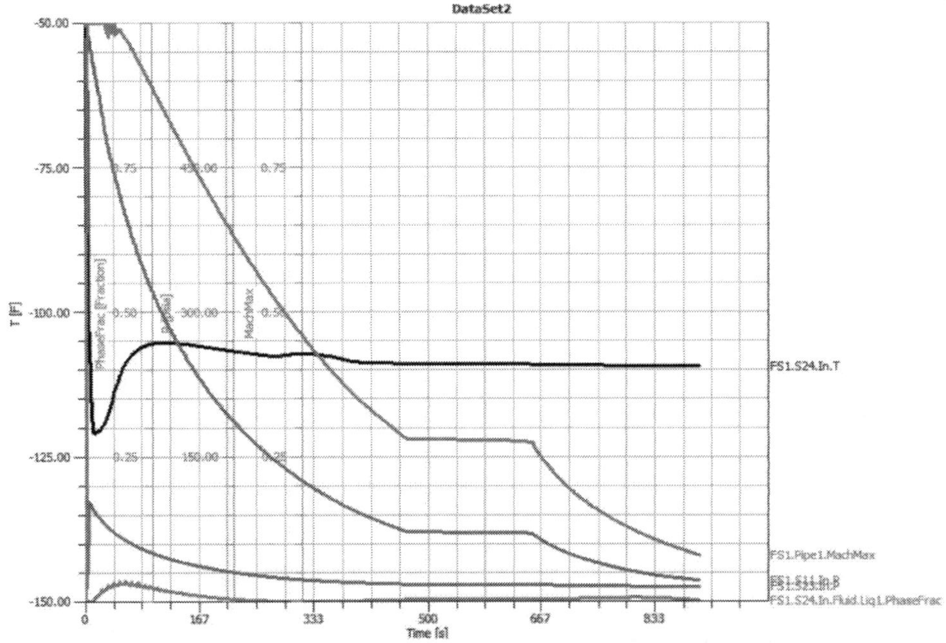

Blowdown - Analysis

- Blows down within 15 min
- Solid CO2 forms almost from outset
- What is impact of partial pipe blockage?
- Reduce diameter in section of pipe to 2”
  - 11% of previous area

Blowdown with Restriction

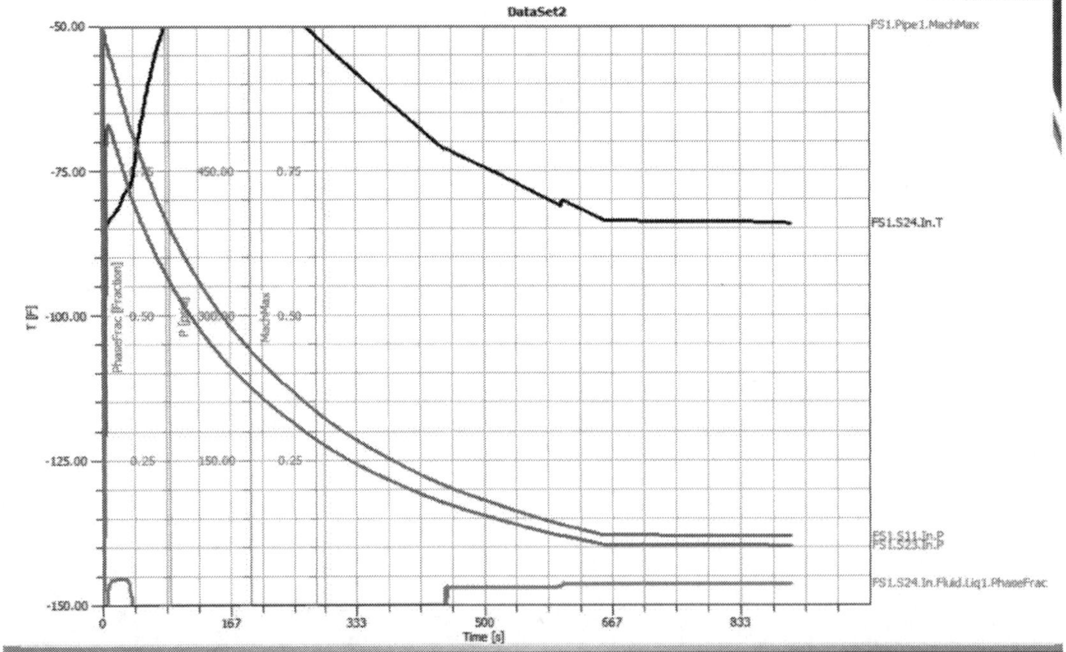

Blowdown - Analysis

- Backpressure in pipe much higher
- Gas is warmer because less expansion
  - (Upstream of choke/blockage)
- Reduced solid CO2 formation
  - Partial blockage will reach balance point?
- Time is longer
  - Less than 100 psi within 15 mins
- Analysis based on 90% blockage

Conclusions

- Simulation assists with understanding process and phase behavior
- Good thermodynamics permits accurate prediction of solid formation
- Dynamic simulation can analyze transient behavior such as blowdown

virtualmaterialsgroup™

Insights into the Phase Behavior and Interfacial Structure of Complex Fluids

Chapman, W. G., Rice University
Feng, Z., Rice University
Gong, K., Rice University
Ballal, D., Rice University
Marshall, B. D. II, Rice University
Cox, K. R., Rice University
Originally presented on: 10/5/2012 13:40:00 - 14:05:00

To design and optimize processes and products in the energy, chemicals, and materials industries, engineers need computational tools to predict bulk and interfacial equilibrium properties for complex fluids in terms of temperature, pressure, composition, and molecular geometry. Systems that have been particularly challenging to model include associating or polar molecules as well as polymers and amphiphiles. Such molecules are important in a range of applications from flow assurance and enhanced oil recovery in the energy industry to paints and coatings to self-assembly of polymer / nanoparticle systems. Advances in statistical mechanics have led to successful molecular models for bulk fluids such as the statistical associating fluid theory (SAFT). SAFT enables engineers to predict the effects of molecular weight, polydispersity, polarity, association, and compressibility on the bulk phase behavior of mixtures containing solvents, monomers, and polymers. Further extensions of this approach using density functional theory show great promise in modeling interfacial properties and molecular structure, including the behavior of fluids at hydrophilic and hydrophobic surfaces, segregation in polymer blends, self assembly of surfactants and block copolymers, and polymer depletion forces in polymer/colloid systems.

In this presentation, we briefly review the physical basis for the SAFT approach and the extension to interfacial systems. An advantage of theoretically based models is that the theory can be systematically improved to address weaknesses or to extend the theory to more complex fluid systems. Some recent extensions and applications of the SAFT approach for bulk and interfacial systems are presented. These points are illustrated through comparisons with molecular simulation as well as phase equilibria data for monomers, solvents, and copolymer systems.

Creating Opportunities from Challenges: *Maximizing Propylene Yields from Your FCC*

Bob Mehlberg
Jeff Knight
UOP LLC, A Honeywell Company

2012 AIChE Regional Process Technology Conference
October 4-5
League City, Texas

UOP
A Honeywell Company

UOP 5747-1

Creating Opportunities from Challenges: *Maximizing Propylene Yields from Your FCC*

UOP
A Honeywell Company

- Markets
- Propylene Keys
- Propylene Revamps
- RxPro^{TM} Technology

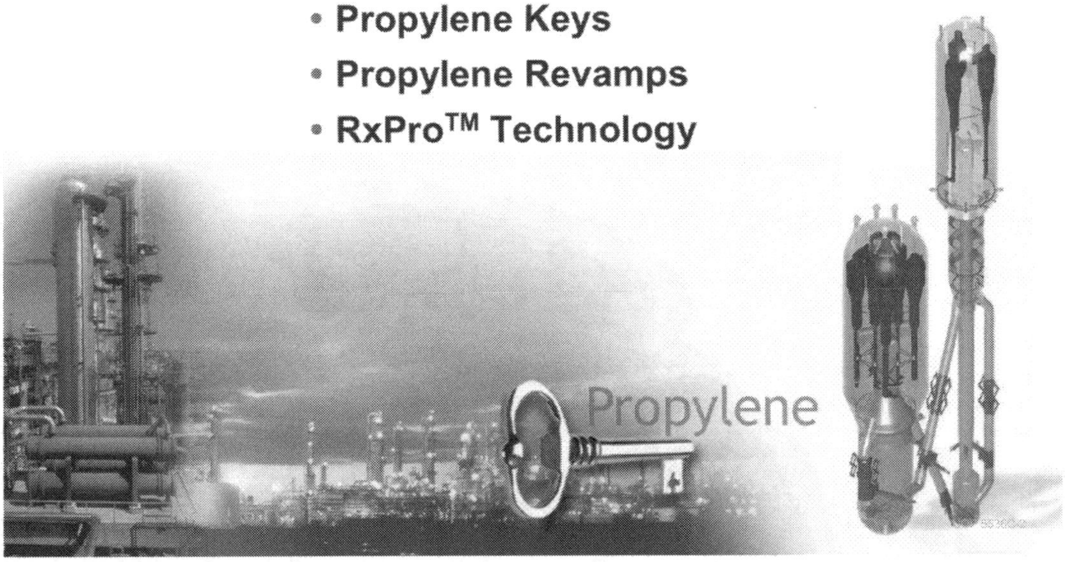

U.S. Gasoline Demand Dropping

UOP
A Honeywell Company

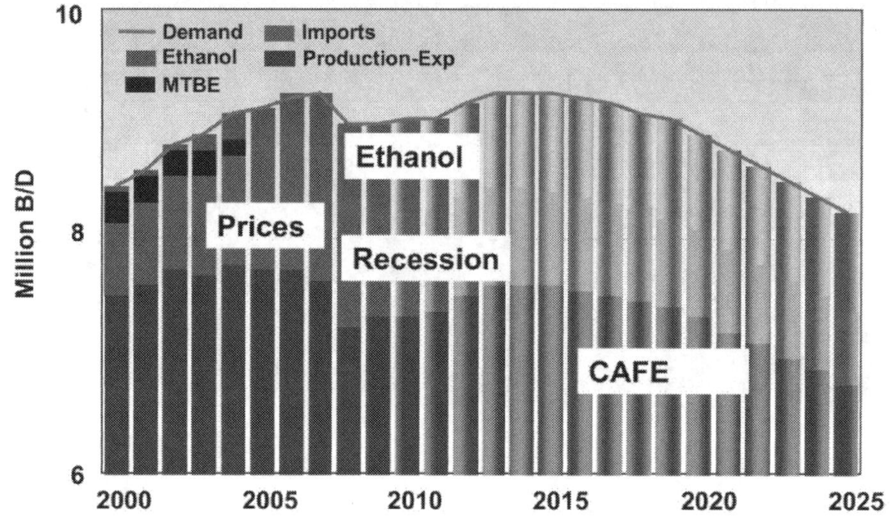

Data Source: ***Global Petroleum Market Outlook 2011, Purvin and Gertz***

Demand Is Weakening for FCC's Primary Product

UOP 5536C-3

Propylene Demand Growing

UOP
A Honeywell Company

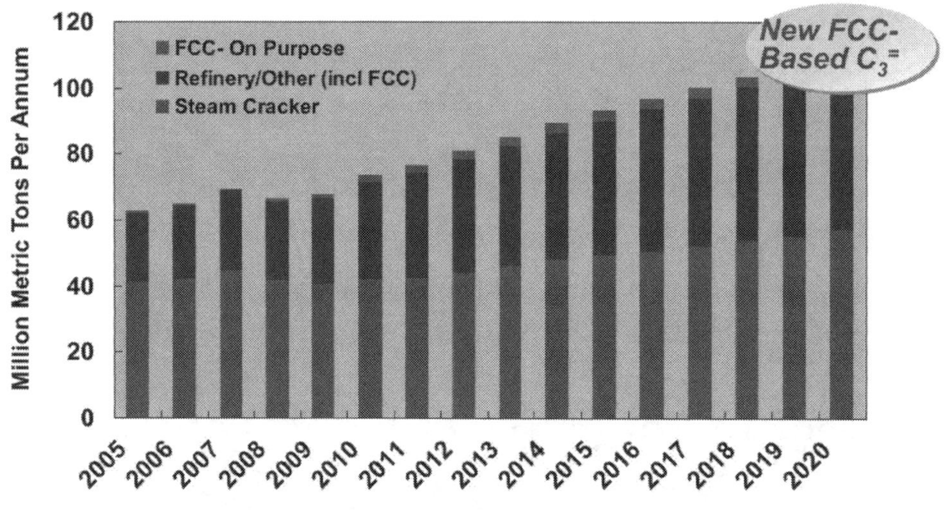

Data Source: ***CMAI 2010 World Light Olefins Analysis (WLOA)***

Utilize the FCC's Open Capacity to Close Propylene Gap

UOP 5536C-4

Unlocking Propylene

UOP
A Honeywell Company

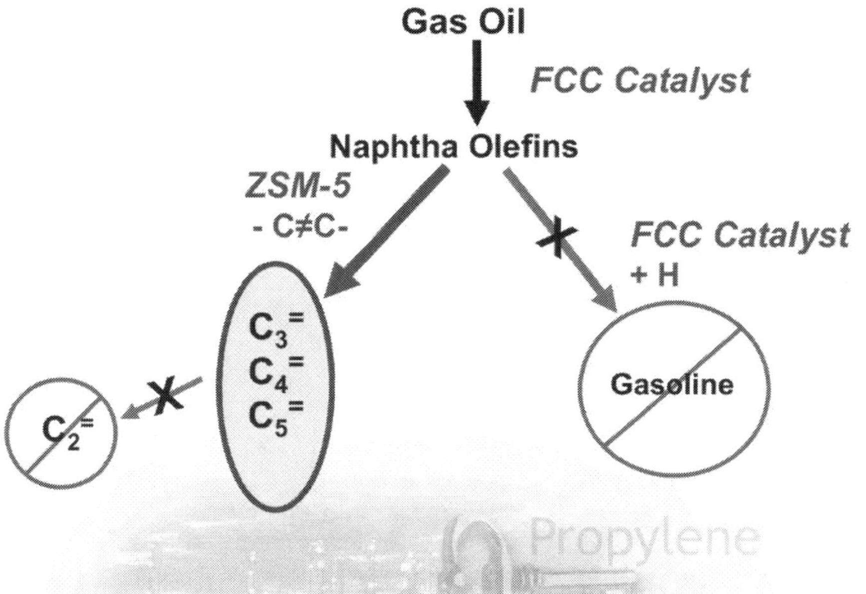

Reaction Management is Critical to Increasing Propylene

UOP 5536C-5

Propylene Equilibrium

UOP
A Honeywell Company

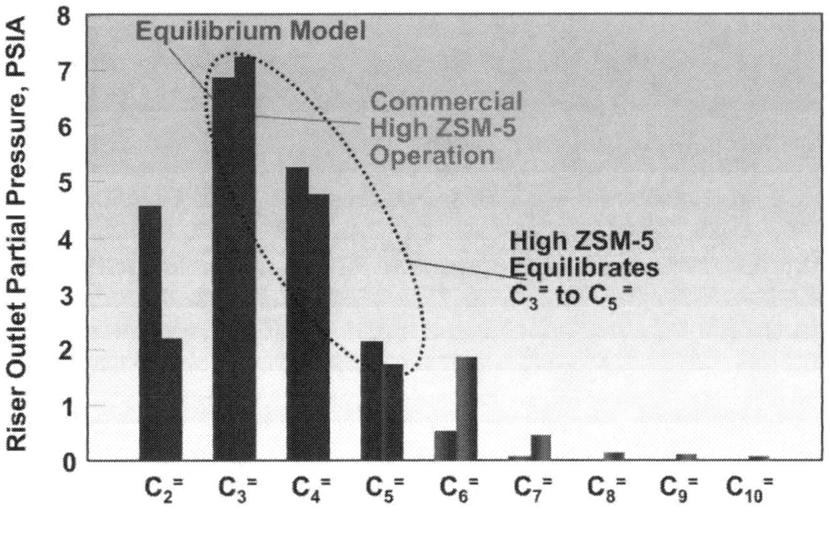

FCC Propylene Yield Is Equilibrium-Limited

UOP 5536C-8

FCC Propylene from Butenes

UOP
A Honeywell Company

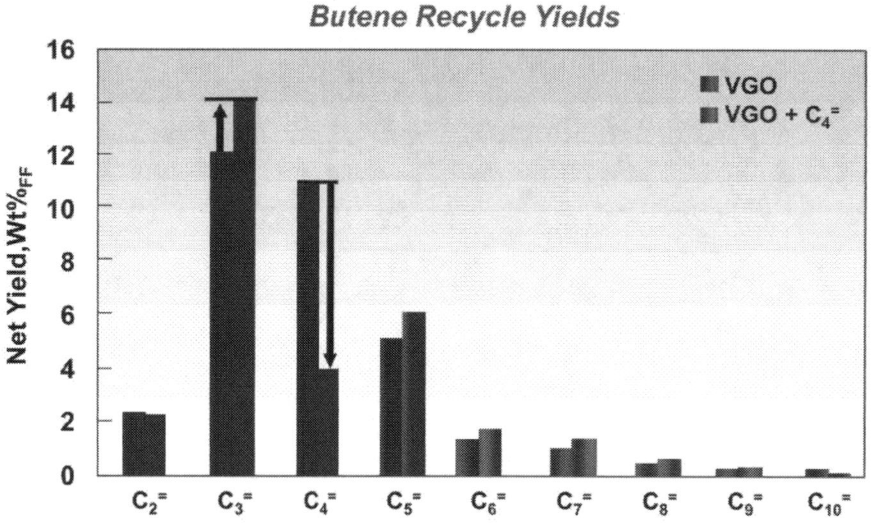

ZSM-5 Converts Butenes to Propylene

UOP 5536C-7

Stepping Up Propylene

UOP
A Honeywell Company

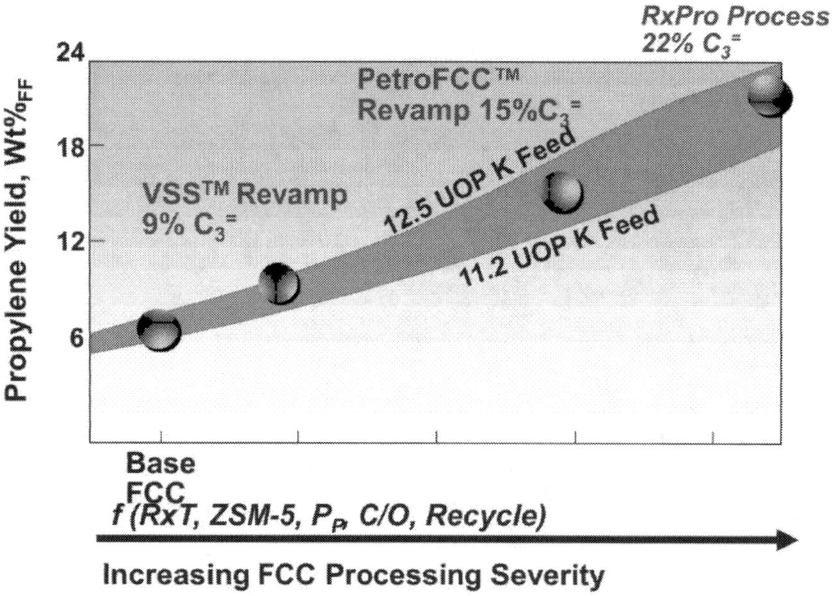

Leverage Operating Conditions with Technology Upgrades

UOP 5536C-8

VSS Technology- The First Step

UOP
A Honeywell Company

- Revamp Scope
  - Feed quality Improvement
  - High Severity
  - New Gas Concentration Unit
  - Optimix™ Feed Distributors
  - Reactor Revamp to VSS
- UOP Vortex™ Separation System
  - Primary separation system
  - Highest reactor containment
  - Greatly reduced hydrogen transfer
- Results
  - $C_3^=$ doubled from 4.5 to 9.0 Wt-%
- Future
  - Possible increase to 13 Wt-% $C_3^=$

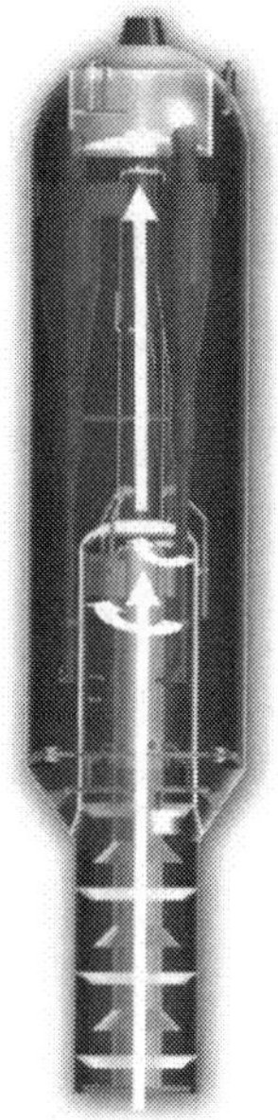

VSS is a Propylene Enabler

UOP 5536C-9

9% Propylene Revamp - *Results*

UOP
A Honeywell Company

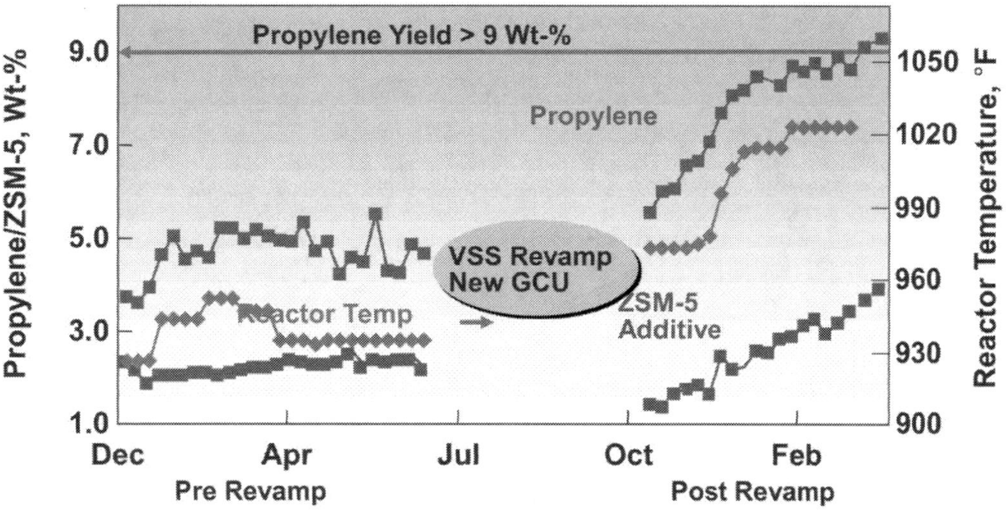

UOP 5536C-10

PetroFCC Revamp – 15% $C_3^=$

UOP
A Honeywell Company

- Revamp Scope
  - Feed Expansion
  - Maximum Propylene
  - New PetroFCC Reactor
  - New Combustor Regenerator
  - Product Recovery Expansion
- Results
  - 80-85 vol% Conversion
  - 15^+ Wt-% $C_3^=$

Leverage Operating Conditions with Technology Upgrades

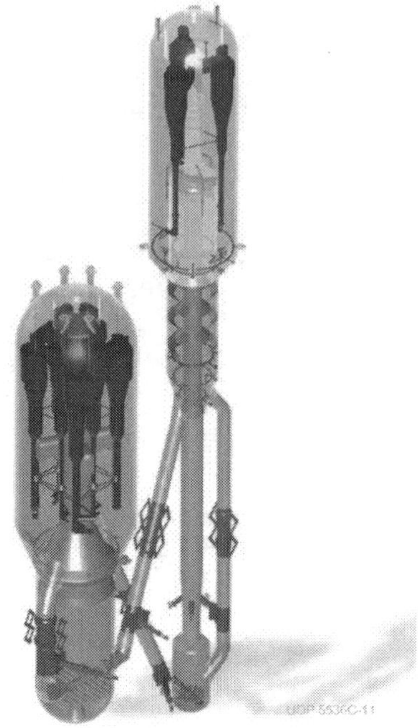

UOP 5536C-11

RxCat Technology for Propylene

UOP
A Honeywell Company

RxCat Technology Enables High $C_3^=$ Yields

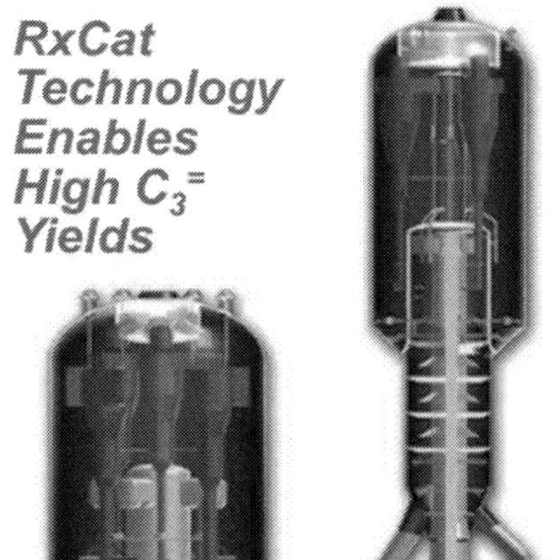

1) Heat balance neutral recycle → Tunable riser catalyst to oil
2) Catalyst and oil contact temp decreases → Tempers dry gas production
3) Riser catalyst to oil increases → Higher conversion
4) Catalyst coke increases → Warms Regenerator

Commercial Experience List

- Refiner #1 - Americas: Apr '05
  - Regenerator temp control
- Refiner #2 – Asia: Mar '08
  - Regenerator temp control
- Refiner #3 – Asia: Feb '08
  - Propylene Production
- Refiner #4 – Asia: Apr '09
  - Propylene Production
- Refiner #5 – Asia: Nov '10
  - Propylene Production

+5 Others in Construction

UOP 5536C-12

PetroFCC Revamp - *Results*

Commercial Propylene vs. Partial Pressure

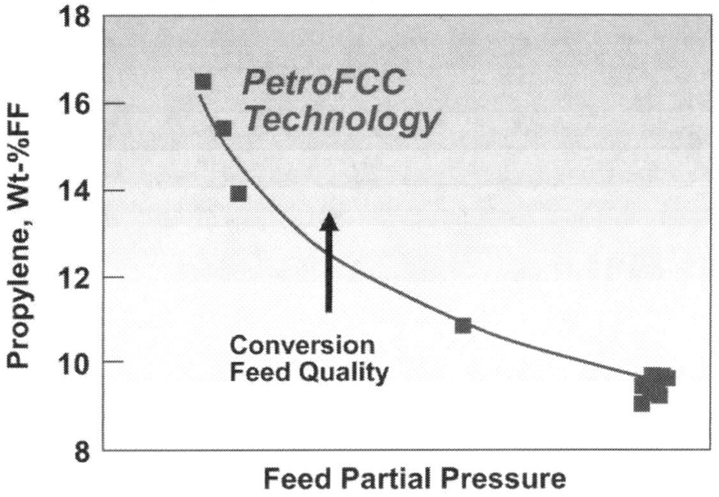

PetroFCC Technology Extends C_3= vs. Pressure Experience

UOP 5536C-13

RxPro Reaction Section

Features

- **Proven Reactor Technologies**
- **Targeted Olefin Recycle**
- **Two Reactor Stages**
  - **Maximum $C_3^=$ Yield**
  - **High Selectivity**
  - **Flexibility**
- **UOP RxCat Technology**
  - **Dry Gas Control**
  - **Recycle Conversion**

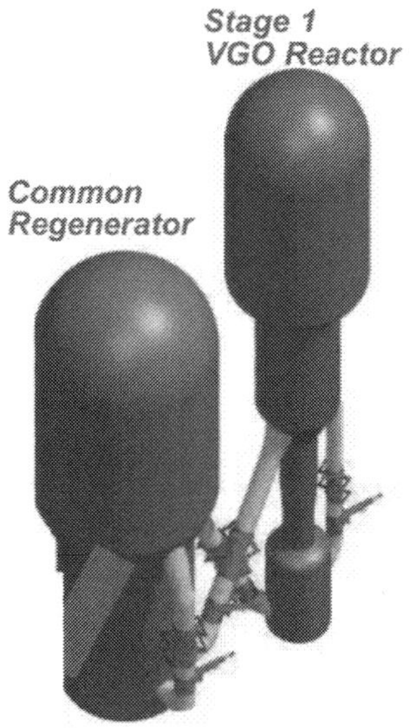

UOP 5536C-14

RxCat Butene Recracking

- **Butene Conversion**
  - Required for high $C_3^=$
  - High ZSM-5
  - High Cat/Oil

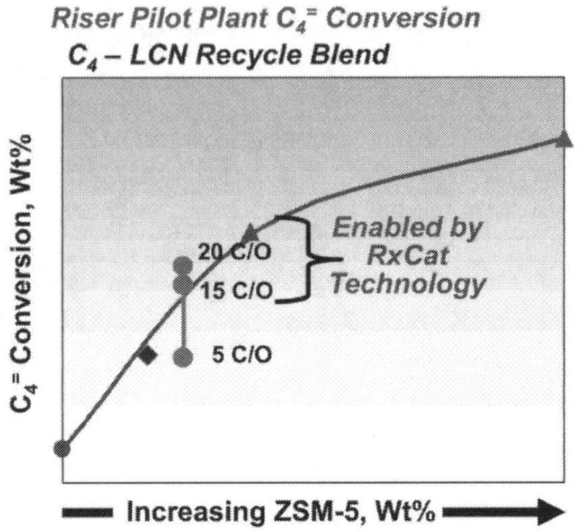

RxCat Technology Doubles $C_4^=$ Conversion

UOP 5536C-15

Creating the Targeted Recycle

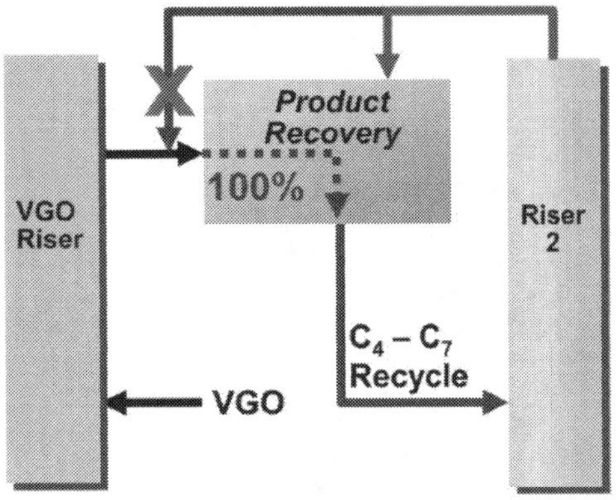

Targeted Olefin Recycle

- **No co-mingling**
- **100% containment of VGO Riser $C_4 - C_7$ olefins**
- **Minimizes paraffin dilution**
- **Reduces recycle 30%**

Targeted Recycle Results in 9% Lower Cost

UOP 5536C-16

FCC Yields vs. Severity

UOP
A Honeywell Company

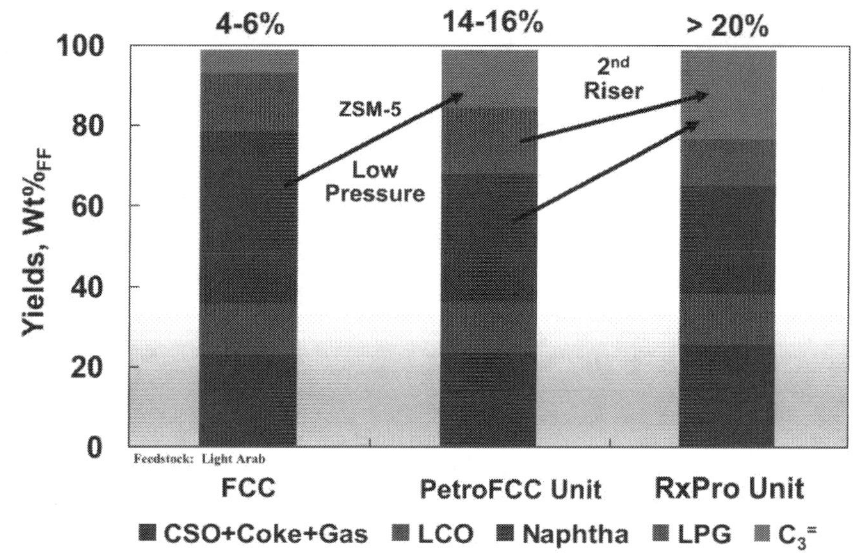

UOP 5536C-17

High FCC Propylene Economics

UOP
A Honeywell Company

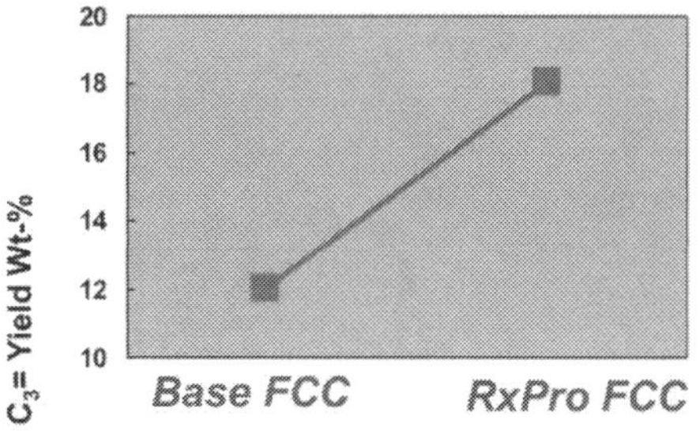

Refinery Study

- Base FCC - 12% $C_3^=$
- RxPro Technology - 18% $C_3^=$
- $400/MT Propylene/Gasoline

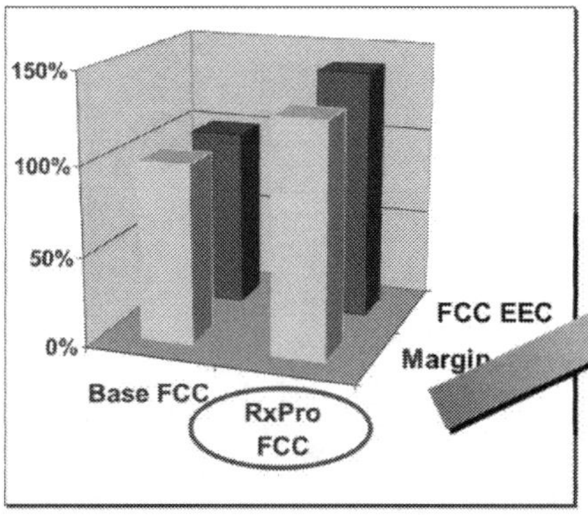

28% IRR

RxPro Technology Most Capital Efficient

UOP 5536C-18

Comparing Propylene Technologies

UOP
A Honeywell Company

Parameter	Other $C_3^=$ Technology	UOP RxPro Technology
$C_3^=$ Yield, Wt-%	>20	>20
Feed Hydrogen	> 13.2%	12.4
Dry Gas, Wt-%	12.9	6.4
Coke, Wt-%	9.4	7.5
Steam, Wt-%	20 – 30	10 - 15

UOP Advantages

- Broad feed quality
- Less coke and gas
- Energy efficient
- Capital efficient

UOP 5536C-19

Concluding Remarks

UOP
A Honeywell Company

- UOP has more than 500 kBSD of 8-15 Wt-% FCC propylene expansion experience
- UOP offers fundamentally sound, selective and proven technologies to enhance or maximize FCC propylene
- Propylene markets present opportunities to repurpose FCC to higher valued products

UOP
A Honeywell Company

Q&A

UOP 5536C-21

Alternatives to Rare Earth - Commercial Evaluation of REpLaCeR® FCC Catalysts at Montana Refining Company

Benoit, A.,
Colwell, R., Montana Refining Company
Hunt, D., Grace
Jale, S.,
Jergenson, D., Montana Refining Company
Udvari, E.,
Originally presented on: 10/5/2012 12:25:00 - 12:50:00

After the sweeping reduction of Chinese export quotas, rare earth metals experienced price hyperinflation in the range of 2700% between August 2010 and July 2011. FCC catalyst has experienced unprecedented inflation as a result. Rare earth is used to stabilize FCC catalyst zeolite in order to provide high FCC catalyst activity, liquid selectivities and superior coke selectivity. Removing rare earth from FCC catalyst will provide relief in catalyst expenses; however, it is not an economical solution in most FCC operations due to a lower activity and product value.

Alternate materials and processing must be used to stabilize the FCC zeolite to ensure a profitable yield slate without rare earth. Grace has recently developed and commercialized its new REpLaCeR® family of catalysts, a portfolio of low and zero rare earth catalysts applicable to a broad range of FCC applications. Presently, over 50 units worldwide use REpLaCeR® FCC technology.

Montana Refining Company (MRC) partnered with Grace to commercialize a rare earth-free REMEDY™ catalyst from Grace's new REpLaCeR® family of catalysts. MRC has successfully applied a zero rare earth REMEDY™ over a moderate rare earth GENESIS® FCC catalyst. Commercial results demonstrate similar unit conversion at similar catalyst additions. Coke selectivity and slurry cracking were maintained with REMEDY™ while gasoline selectivity increased.

Presentation to:

AIChE

Regional Process Technology Conference

October 5, 2012

AFPM
American Fuel & Petrochemical Manufacturers
The new NPRA.

Coker Furnace On-Line Spalling

Safe, Clean, Proven and Profitable

Presented By:

Jack Adams, PE
Adams Project Managers, Inc.
(APMI.Jack@gmail.com)

Gary C. Hughes,
Bechtel Hydrocarbon Technology Solutions, Inc.
(GHughes1@Bechtel.com)

Agenda

- Background
- Benefits and Applications of On-Line Spalling
- On-Line Spalling Overview
- On-Line Spalling Procedure and Uses
- Coker Double Fired Furnaces
- Summary / Questions

Coker Furnace On-Line Spalling
Safe, Clean, Proven and Profitable

AIChE
October 2012

Background - Delayed Coker Units (DCU) Operation:

- Thermal crack residue to increase refined product yields
- Heat residue to crack at 875 - 940°F
- Designs are to heat residue in the furnace(s) and to crack residue in the coke drums, not in the furnace
- Steam, velocity, tube metallurgy, advanced designs are used to minimize carbon lay down in furnace tubes
- Carbon lay-down (coke) does still occurs in furnaces
- Furnace tubes are decoked to recover furnace efficiency
- On-Line Spalling, Steam-Air Decoking, and Pigging are acceptable practices for decoking coker furnace(s) tubes

(Paper presented to AFPM, March 2012, # AM-12-72) Page 3 AIChE - Regional Process Technology Conference – October 2012

Coker Furnace On-Line Spalling
Safe, Clean, Proven and Profitable

AIChE
October 2012

Benefits of On-Line Spalling (OLS)

- OLS reduces annual shutdown days needed to decoke the coker furnace(s) tubes, thereby improving stream factor and profitability.
- OLS is proven. – In use in many refineries.
- OLS is safest procedure. – Minimizes shutdowns and startups.
- OLS is the cleanest method to decoke the inside of furnace tubes. – Environmentally friendly.
- A carefully developed coker design and OLS procedure are essential.

(Paper presented to AFPM, March 2012, # AM-12-72) Page 4 AIChE - Regional Process Technology Conference – October 2012

Coker Furnace On-Line Spalling
Safe, Clean, Proven and Profitable

AIChE
October 2012

Applicable Coker Furnaces

- On-Line Spalling (OLS) is used in many Delayed Coker Units (DCU) to remove coke from the inside of the coker furnace tubes, in lieu of pigging or steam-air decoking.
- OLS applications:
  - DCU's with two or more furnaces or cells
  - Furnace having multiple tube passes
  - Furnaces with adequate design temperature and pressure

Coker Furnace On-Line Spalling
Safe, Clean, Proven and Profitable

AIChE
October 2012

On-Line Spalling Overview

- Maintain furnace tube velocity with steam
- Remove all Hydrocarbon flows to operating drum
- Double block oil
- Increase steam rate
- Manage tube velocity
- Manage temperature
- Tubes Contract/Expand
- Manage Antifoam

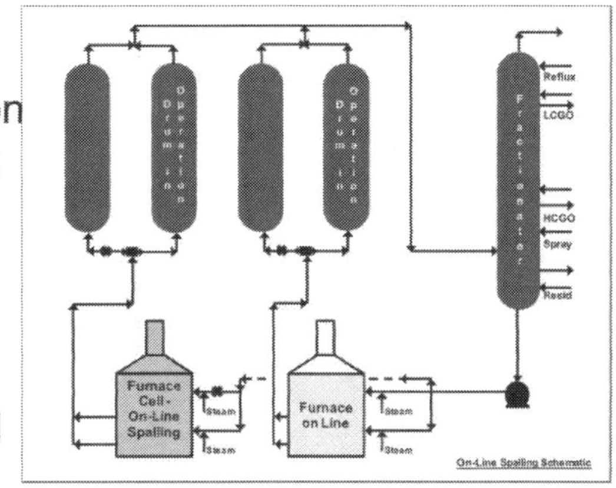

On-Line Spalling Schematic

On-Line Spalling Procedure

General steps:

1. Review OLS procedure and prior reports and deviations.
2. OLS begins with normal residue and steam rates.
3. Residue rate is reduced.
4. Double block the flow of residue to the cell to be spalled.
5. High pressure steam is increased to maintain roughly constant velocity through the furnace tubes.
6. Steam rate is increased within velocity limits.
7. Tube spalling temperatures are increased and monitored.

(continued...)

Coker Furnace On-Line Spalling
Safe, Clean, Proven and Profitable

AIChE
October 2012

OLS Procedure (continued):

8. Maximum tube temperature is set based on mechanical limitations of the tubes during spalling.
9. For best OLS, steam temperature set at about 25 to 50 F above the maximum end of run skin temperature for the furnace tubes.
10. The intent is to slough off coke gradually.
11. Steam velocity should not exceed 350 - 375 feet per second.
12. Thermal shocks of the tubes is usually completed based on experience of the coker operator.
13. Antifoam injection to prevent a foamover of the coke drum.

(continued...)

Coker Furnace On-Line Spalling
Safe, Clean, Proven and Profitable

AIChE
October 2012

OLS Procedure – Other considerations:

14. Spalling should begin three to four hours after drum switch.
15. Due to the relatively low temperatures of steam (and/or Boiler Feed Water) and hydrocarbons going into the coke drum, there will be a higher foaming tendency of the material entering drum.
16. If OLS is not completed before it is time to switch drums again, delay bringing the spalled furnace on-line, unless this can be done with about three to four hours left before the coke drum switch.
17. Drum switch should occur while the spalling temperatures are at their hottest.

(continued...)

Coker Furnace On-Line Spalling
Safe, Clean, Proven and Profitable

AIChE
October 2012

OLS Procedure – Other considerations:

18. The initial switch from residue to steam through the furnace tubes occurs rapidly and in large increments. Maintain temperature of steam as steam pushes the residue out of the tubes.
19. Fuel gas rates to the furnace being spalled are much lower than normal. During On-Line Spalling, expect the fuel gas rate to the furnace cell being spalled to be 10-20% of the normal fuel gas rate.
20. As necessary, take burners off-line to control furnace temperature. But it is important to fire as many burners as possible at low rate to provide uniform heat distribution throughout the firebox.

Coker Furnace On-Line Spalling
Safe, Clean, Proven and Profitable

AIChE
October 2012

Where OLS cannot be used

1. Furnaces with tube passes that cannot be separated such that one or two passes can be spalled while the other passes continue in operation.
2. Significant levels of inorganic compounds
3. Soft coke tube plugging
4. Within 3 hours to drum switch
5. When the tube maximum design temperature and pressure will be exceeded by the targeted spalling conditions.

Coker Furnace On-Line Spalling
Safe, Clean, Proven and Profitable

AIChE
October 2012

Furnace Design Applicable to OLS

- Typical for 9Chr-1Mo DCU furnace tubes:
  - Skin temperatures will increase from 1050 F at start-of-run to a maximum tube wall temperature of about 1200 to 1275
  - At the tube design skin temperature, coke laid down must be removed or the furnace firing rate reduced
  - Desired procedure includes operating the tube skin temperatures 25 to 50 degrees F higher than the maximum operating skin temperature prior to the beginning of the spall
  - Steam-Air decoking, Pigging, and On-Line Spalling (OLS) are accepted procedures to decoke the furnace
  - New furnaces designs often consider only OLS and pigging

AIChE - Regional Process Technology Conference – October 2012

Coker Furnace On-Line Spalling
Safe, Clean, Proven and Profitable

AIChE
October 2012

Double Fired Furnaces

- Have lower peak to average flux profile
- Lower residence time of the hydrocarbons at high temperatures in the heater (30 percent less tubes)
- Increased run lengths between spalling events
- Lower costs (less tubes burners, box size)
- Modular construction

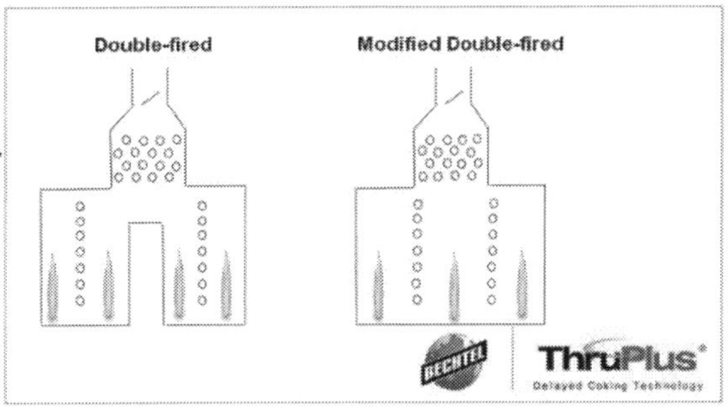

(Paper presented to AFPM, March 2012, # AM-12-72)

Page 13

AIChE - Regional Process Technology Conference – October 2012

Coker Furnace On-Line Spalling
Safe, Clean, Proven and Profitable

AIChE
October 2012

Furnace Design

- OLS is applicable to furnaces with multiple tube passes in separate cells
- OLS may not be applicable for older, smaller furnaces
- OLS is most efficient if it can be used
- Metallurgy and designs vary

(Paper presented to AFPM, March 2012, # AM-12-72)

Page 14

AIChE - Regional Process Technology Conference – October 2012

Coker Furnace On-Line Spalling
Safe, Clean, Proven and Profitable

AIChE
October 2012

Summary / Questions

- On-Line Spalling Benefits the refiner, because it is:
  - Improves stream factor which increases profitability
  - Proven
  - Safe
  - Clean
- On-Line Spalling procedure requires experience, careful design, and Operations training and "buy-in".

Coker Furnace On-Line Spalling
Safe, Clean, Proven and Profitable

AIChE
October 2012

Mission Statement

Adams Project Managers Inc. Deliver Proven, State-of-the Art Technology, Project Development, Project Management, and Engineering Services, which Captures Total Economic Value for our Refinery Clients;

- By becoming aligned with our client's goals and objectives
- By evaluating and installing proven, safe, and economically beneficial technology
- By providing single source responsibility for the Project Development, Program Management, Management of Engineering, Purchasing, and Construction of Delayed Coker Units, Drum Replacement, Unheading Device Installation, Revamp and Expansion Projects
- By focusing on completing projects on time and within budget

Mission Statement

Bechtel Deliver Proven, State-of-the Art Technology, Project Development, Project Management, Engineering, Purchasing, and Construction (EPC) Services;

- By becoming aligned with our client's goals and objectives
- By installing proven, safe, and economically beneficial technology
- By providing single source responsibility for the Project Development, Program Management, Management of Engineering, Purchasing, and Construction of Delayed Coker Units, Drum Replacement, Unheading Device Installation, Revamp and Expansion Projects
- By focusing on completing projects on time and within budget

Coker Furnace On-Line Spalling
Safe, Clean, Proven and Profitable

AIChE
October 2012

Jack Adams' CV – www.CokerDNA.com

- Jack has 38 years in refining as a process engineer, project manager, and business manager.
- Jack is a hands on project development and project management. Comfortable in the field.
- Master's Degree in Chemical Engineering from Arizona State University.
- PE in Texas.

Gary C. Hughes' CV

- **Gary C. Hughes is a** Coking Technology Specialist at Bechtel Hydrocarbon Technology Solutions, Inc. in Houston, Texas. Gary has 44+ years experience in the downstream refining business supporting both operations and process design efforts.
- Gary has co-authored patents and papers related to Delayed Coking and led the technical delayed coking group responsible for supporting all the fuel and anode cokers for Conoco Inc and ConocoPhillips Company (CoP).
- Gary holds B.S. in Chemical Engineer from Oklahoma State University, Stillwater, Oklahoma.

Processing Challenging Crudes Enabled with Increased Hydrotreating Performance

Jonathan Konen
Process Engineer
LyondellBasell Houston Refining

Matthew Goodman
Hydroprocessing Specialist
Criterion Catalysts & Technologies

lyondellbasell

Overview

- Facility Upgrades Enable Operational Flexibility
- The Challenges of Heavy/Bitumen/Upgraded Crude Processing
- Impacts Observed Taking Advantage of Available Feeds
- Evaluation of Options - Application of CENTERA Technology
- Utilizing Capability Allows for Further Flexibility
- Learning and Path Forward

Houston Refining
Process Overview

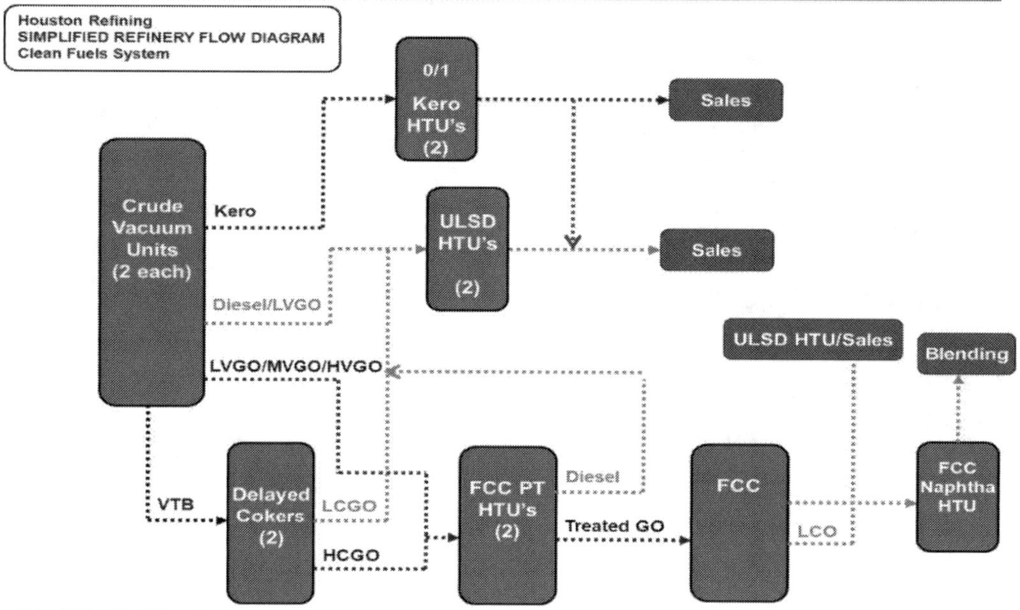

HTU 0 / 1 Summary

Unit 0	
Feed Rate, bpd	15000
H_2 consumption, scf/bbl	100
Treat strategy	HDN
Cycle length, mo	60

Unit 0

~500 psia
<90 mol% H_2 treat gas
>1000 scf/b treat gas
LHSV 2.0 hr^{-1}

Unit 1	
Feed Rate, bpd	15000
H_2 consumption, scf/bbl	150
Treat strategy	HDS/HDN
Cycle length, mo	24

Unit 1

Same operating conditions as unit 0

Hydrotreater 1
Key Past Issues

- Cycle lengths limited by:
  - Activity
  - Pressure drop in lead reactor
- Operating mode impacted by clean fuels strategy
- History in LCO, distillate, and naphtha processing
- Catalysts systems / technologies utilized

Challenges to Upgraded Crudes
Quality, Chemistry, Variability

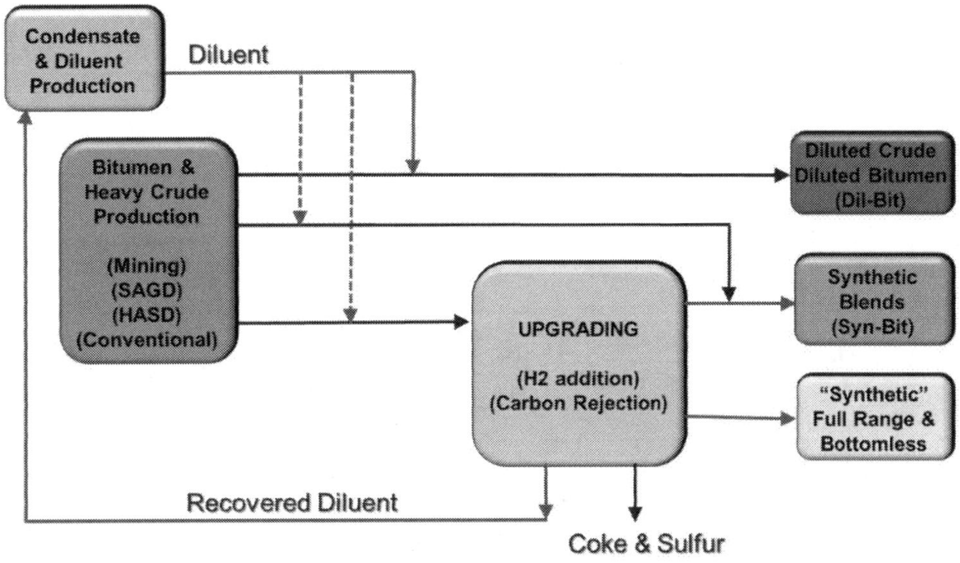

Challenges to Upgraded Crudes
Quality, Chemistry, Variability

- Required increased understanding of feed streams
  - Quantitative and Qualitative Analysis of feed streams across refinery
  - Segregation of feeds based on feed properties
  - Limited rates of specific crudes

- Houston Refining begins processing opportunity crudes
  - High sulfur
  - High nitrogen
  - Increased aromaticity

Houston Refining Operational Changes
Specific Changes to 1

Unit 0
Segregated to typical Heavy Sour Crude

Unit 1
Opportunity crudes including synthetics

Impact On Unit 1 Operations	Prior to 2010	Post 2010
Sulfur (wt %)	0.35	1.0-1.2
API gravity	37.5	35.5
Feed Nitrogen (ppm)	30-50	125-200
Reactor pressure (psig)	Same	Same
Treat gas purity (mol % H2)	Same	Same
Hydrogen ratio (scf/bbl)	Same	Same
LHSV (hr^{-1})	Same	Same
Chem H2 cons (scf/bbl)	150	225

Impacts Observed With Challenging Feeds
Exercising Facility Flexibility Provides Surprises

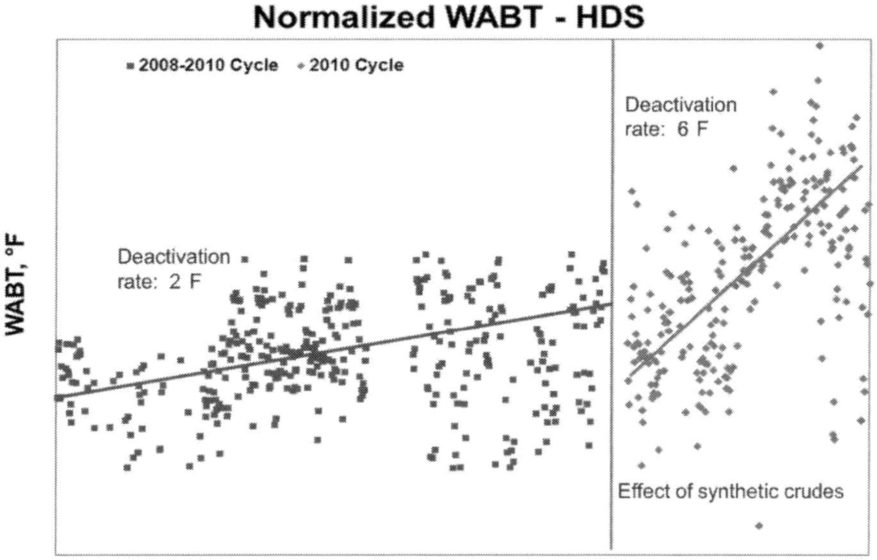

Catalyst for both cycles was DC-2533 ASCENT

Evaluation of Options
Regaining Unit Capability

- Duration of operations with opportunity crude uncertain
- Maximum cycle life requested with a minimum of 1 year
- Ability to treat to ultra-low sulfur product specifications
- Feed evolution compared and reviewed with Criterion
- Feed case for new cycle determined and modelled with various catalyst systems

CENTERA Key to Advancement in Capability
Early Adoption of Technology Aids Operations

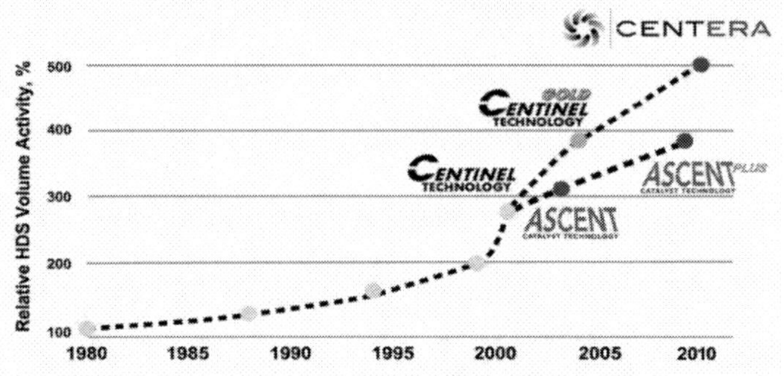

- Criterion's ULSD portfolio, basis HDS activity
- CENTERA DC-2618 is a top tier CoMo catalyst developed with Criterion's state-of-the-art CENTERA technology
- Proven ability to extend operational cycles in ULSD units
- Increased HDN Performance

lyondellbasell.com lyondellbasell 11

Evaluation of Options
Regaining Unit Capability

- Decision and Justification:
  - Process Team selected the CENTERA DC-2618
  - CENTERA DC-2618 expected to dramatically decrease nitrogen inhibition effect
  - Conservative prediction of 4 month cycle life improvement
  - Economic evaluation of CENTERA DC-2618 provided a similar fill cost in comparison to ASCENT DC-2533

lyondellbasell.com lyondellbasell 12

Observations of Initial Unit 1 Operations
Upgraded Performance with CENTERA DC-2618

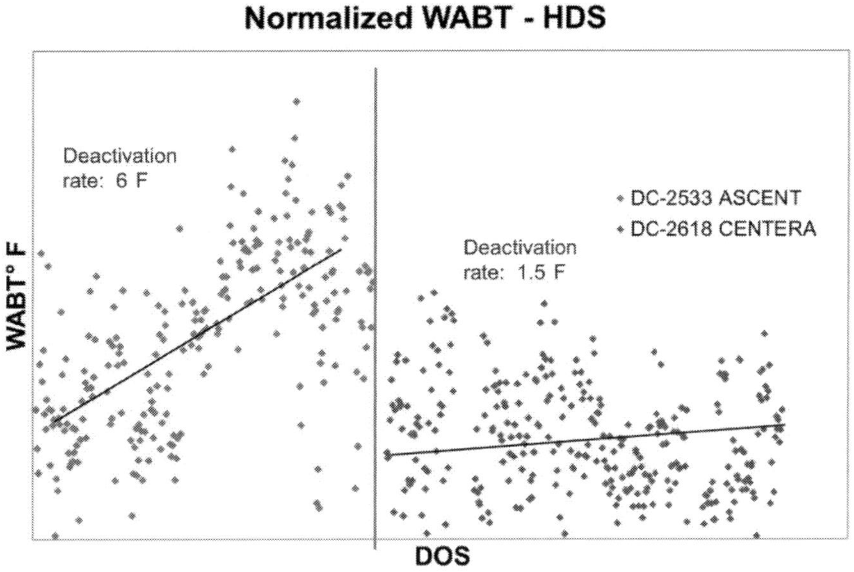

Observations of Initial Unit 1 Operations
Upgraded Performance with CENTERA DC-2618

- Enhanced catalyst stability
- Improved operational flexibility
  - Able to pull back stacked material from storage during catalyst outage at maximum unit rates and at predicted SOR WABT
- Reduced required WABT for HDS
- Increased EOR color stability
- Significantly improved HDN performance, consistently > 90%

Observations of Initial Unit 1 Operations
Upgraded Performance with CENTERA DC-2618

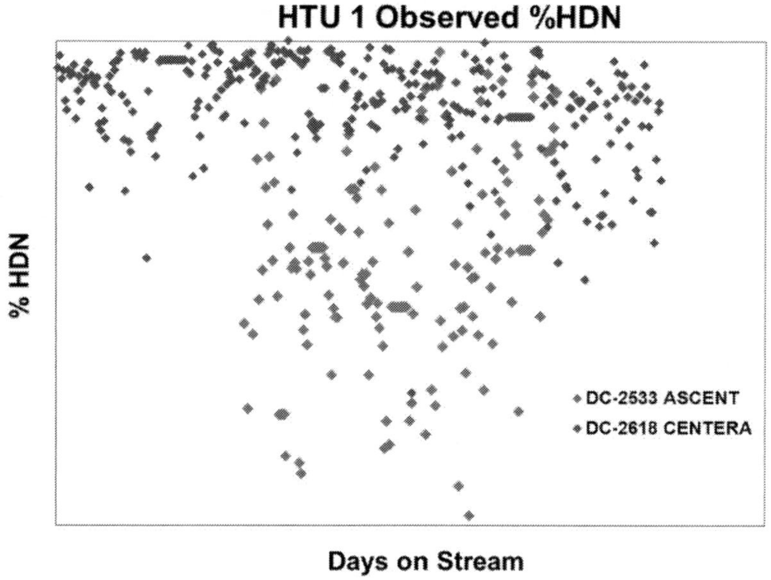

Enhanced Operational Flexibility Provided
Increased Jet Fuel With Difficult Feed

- Margins in May supported treating both kerosene products to jet specifications for sales
- Increase in feed nitrogen and increase in reactor outlet temperature with no change in product color stability
- Product nitrogen continued below 10 ppm

Enhanced Operational Flexibility Provided
Increased Jet Fuel With Difficult Feed

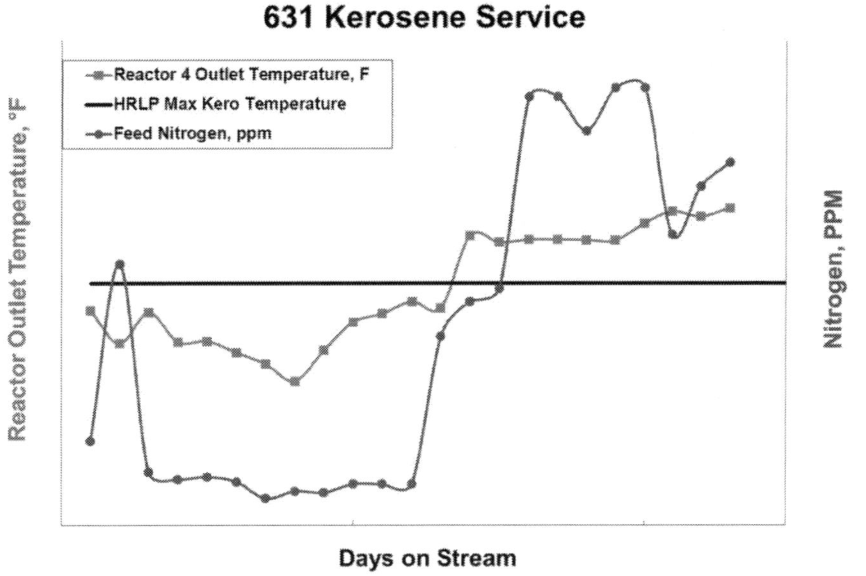

Making the Most of Improved Capability
Finding the Balance of Cycle Life & Performance

- A number of evaluations were performed with various feed/operational scenarios studied
- Cycle life projections show an expected cycle life of X months or more based on activity
  - This was double the original projections with the same feed and original DC-2533 catalyst system
- Increased flexibility through longer term operational planning and utilization of assets
- Ability to run in jet mode at elevated temperatures with cracked crude stocks

Making the Most of Improved Capability
Finding the Balance of Cycle Life & Performance

- Plant savings on catalyst cycle length
- Unit throughput not limited by reaching maximum budgeted WABT curve temperatures
- Economic incentive when kerosene/ULSD margin swings in favor of kerosene
- Improved color stability at higher bed temperatures in kerosene service due to improved HDN

Successes & Path Forward

- Ability to run more opportunity crude
- Enhanced jet production capability
- Activity gains of 75-100% observed after 1 year of operation
- Improved activity management on secondary HTU units
- Review specific cuts of available crudes to study specific cut's impacts
- Monitoring the impact of crude changes on individual process units

Learnings and Keys to Success

- Focus on Unit Interactions and System Economics
- Unit improvements:
  - –Optimization of unit feeds
  - –Optimization of Cycle Length vs. Treat Strategy
  - –Continued catalyst technology improvements

lyondellbasell.com

lyondellbasell 21

Challenges with Synthetic Crude Processing
Enabled with Increased Hydrotreating Performance

Questions?

lyondellbasell.com

lyondellbasell 22

By using any Technical Information contained herein, you agree that said technical information is given for convenience only, without any warranty or guarantee of any kind, and is accepted and used at your sole risk. You are encouraged to verify independently any such information to their reasonable satisfaction. As used in this paragraph, "Technical Information" includes any technical advice, recommendations, testing, or analysis, including, without limitation, information as it may relate to the selection of a product for a specific use and application.

AIChE Regional Process Technology Conference (RPTC) October 4-5, 2012:

New Technology for Enhanced Clean Diesel Production

RESEARCH | TECHNOLOGY | CATALYSTS

Henrik Rasmussen

HALDOR TOPSØE

Presentation Outline

- Implications for processing heavy crudes
  - increased conversion requirements
  - increased desulfurization requirements
- Staged Partial Conversion technology
  - incremental gas oil conversion to ULSD
  - incremental FCC feed pretreatment
- ULSD Reverse Poly-shift technology
  - make ULSD at substantially reduced hydrogen consumption
  - minimize cost of expanding diesel hydrotreaters
- Optimize key hydroprocessing facilities

HALDOR TOPSØE

Heavy crude oil opportunities

- Light sweet crudes are being utilized at fast pace
- Heavy sour crude oil represents 38% of current reserves
- Opportunities to take advantage of price differentials
- Market and product considerations
  - steady growth in demand for diesel
  - reduction of sulfur mandated in all products
- Technology considerations
  - need for extra conversion capacity
  - need for extra hydroprocessing capacity
  - tight margins constrain capital investment

HALDOR TOPSØE

Bottom of the barrel upgrading

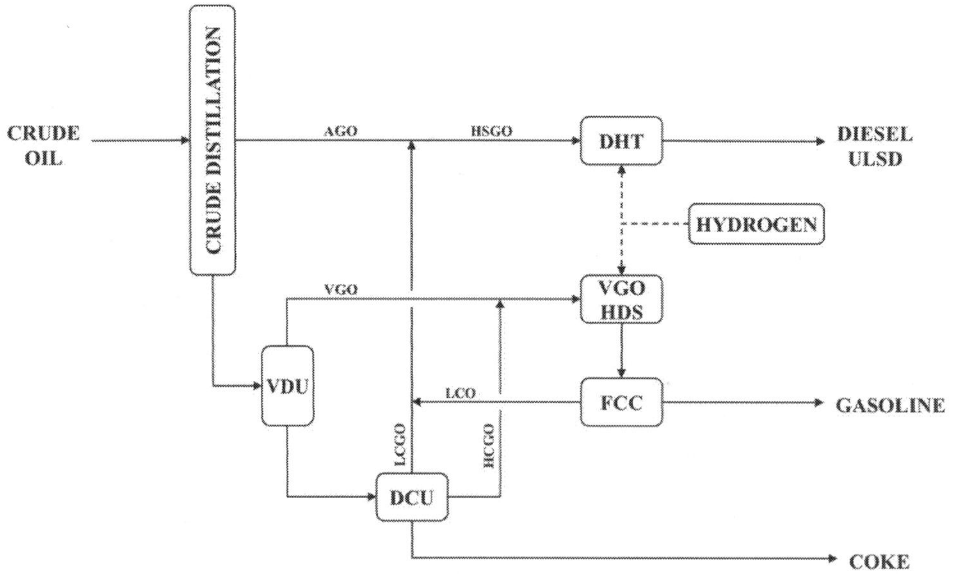

HALDOR TOPSØE

Incentives for hydrotreating FCC feed

- Sulfur reduction is mandated
- FCC feed HDS - proven process for ULSG
  - also reduces sulfur in light cycle oil
- Lowers SOx / NOx from FCC regenerator
- Benefits of aromatics and nitrogen reduction
  - Increased conversion
  - higher gasoline yields
- Profitable investment, not just stay in business

HALDOR TOPSØE

Value of FCC feed hydrocracking

- Increases refinery net conversion capacity
- Avoids FCC bottleneck, most FCC units are already full
- Hydrocracking increases diesel/gasoline ratio
  - flexibility to adjust G/D to market demands
  - diesel demand growth exceeds gasoline in many markets
- Low incremental cost relative to VGO HDS
  - MP partial conversion HDC lower cost than HP HDC
  - revamp existing VGO HDS unit to hydrocracker
  - use existing HDS for ULSD, build new HDC
- How to make high quality diesel under moderate pressure and low conversion conditions?

HALDOR TOPSØE

FCC feed pretreatment

Switch to heavy crude

Crude Source		Mid-continent VGO	Synthetic Crude VGO+HCGO
Capacity	BPSD	40,000	50,000
Hydrogen Consumption	SCF/B	450	600
Feed Properties			
Gravity	°API	21	16
Sulfur	wt.%	2.4	3.0
Nitrogen	wppm	1300	1440
Carbon Residue	wt.%	0.5	0.9

HALDOR TOPSØE

Impact of VGO HDS Severity on Product Sulfur

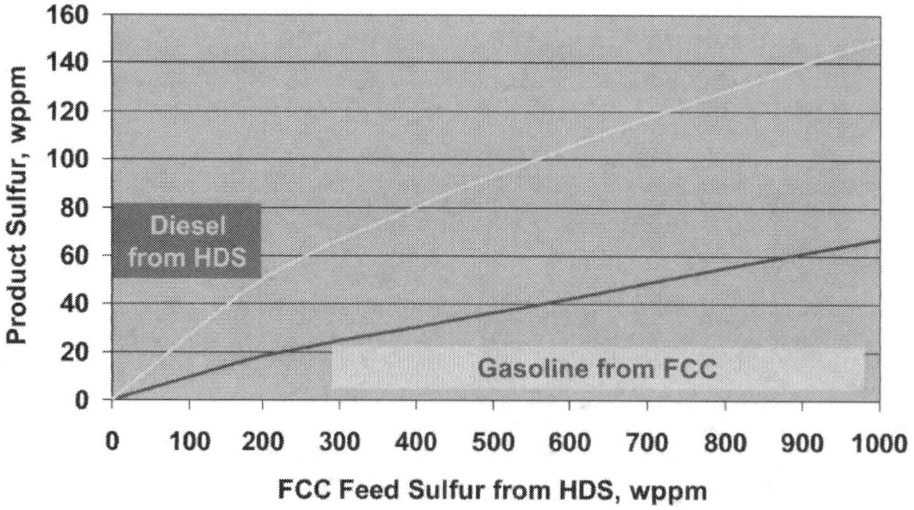

HALDOR TOPSØE

Impact of conversion on diesel quality

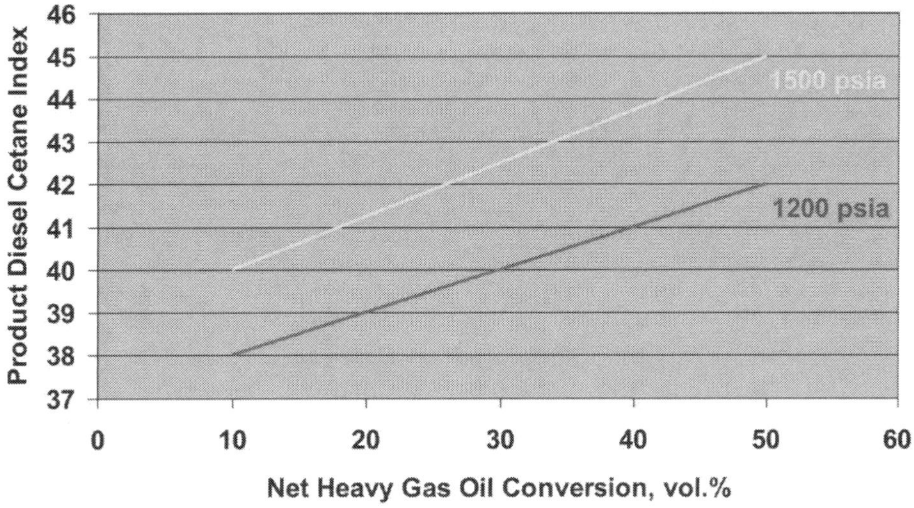

HALDOR TOPSØE

Simple VGO HDS, Cat Feed Hydrotreater

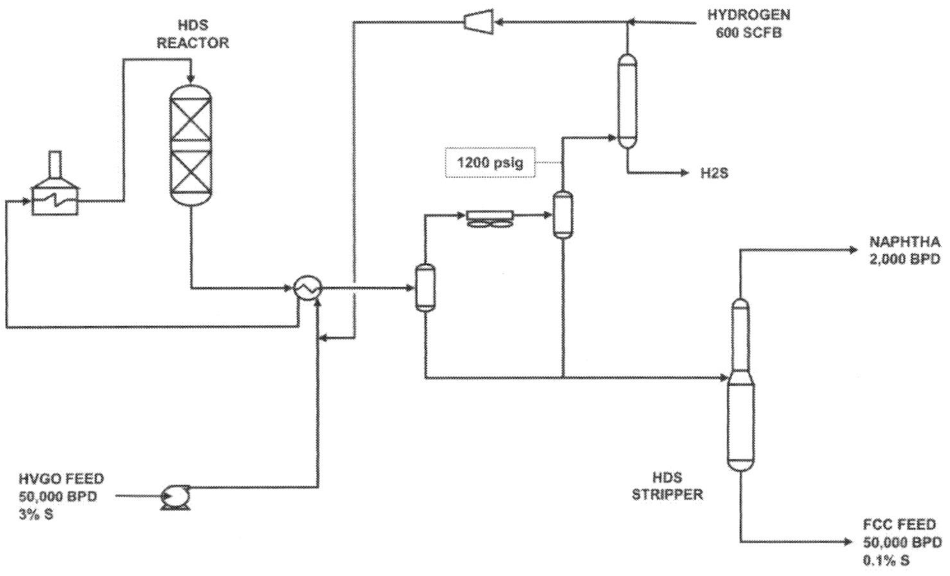

HALDOR TOPSØE

Revamp to Mild Hydrocracker "MHC"

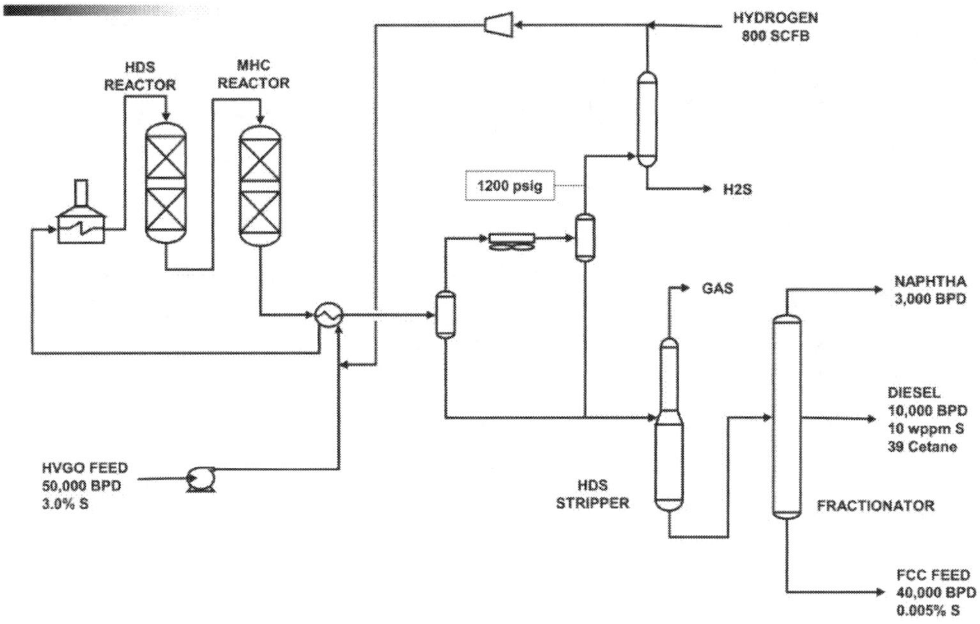

HALDOR TOPSØE

Staged Partial Conversion approach

- Alternative to conventional MHC
- High net conversion with low gross conversion
  - desulfurize the whole feed in a first reactor
  - bypass portion of desulfurized feed directly to FCC
  - hydrocrack remaining gas oil at elevated conversion severity
- Higher cracking severity allows for deeper HDS, density reduction and cetane uplift of diesel product
- Hydrogen is preferentially added to diesel product instead of FCC feed

HALDOR TOPSØE

Staged Partial Conversion Hydrocracker

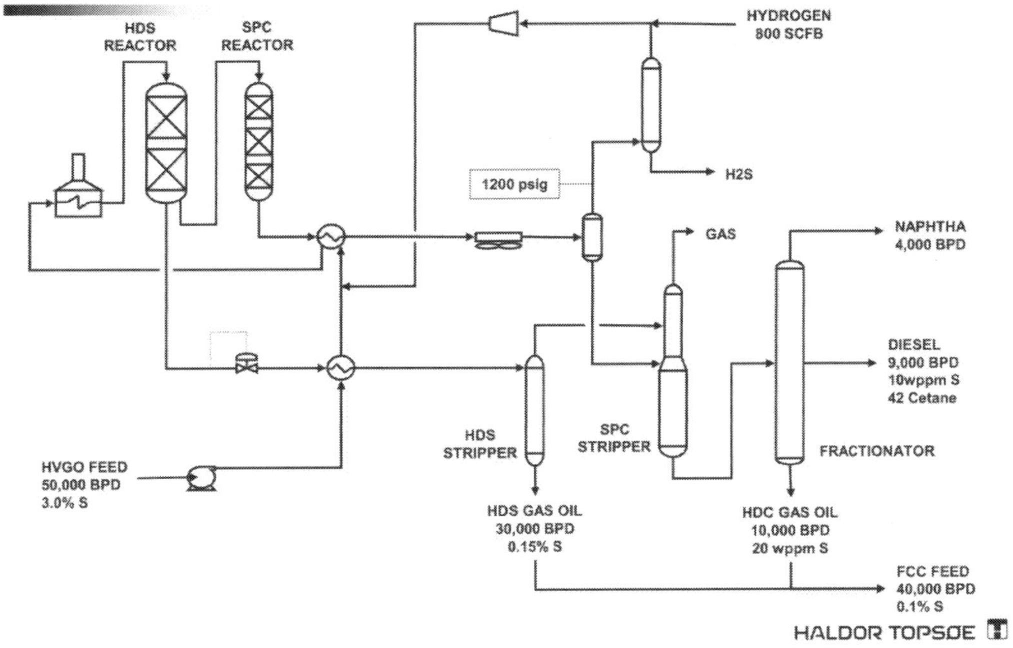

Integral separation in bottom head

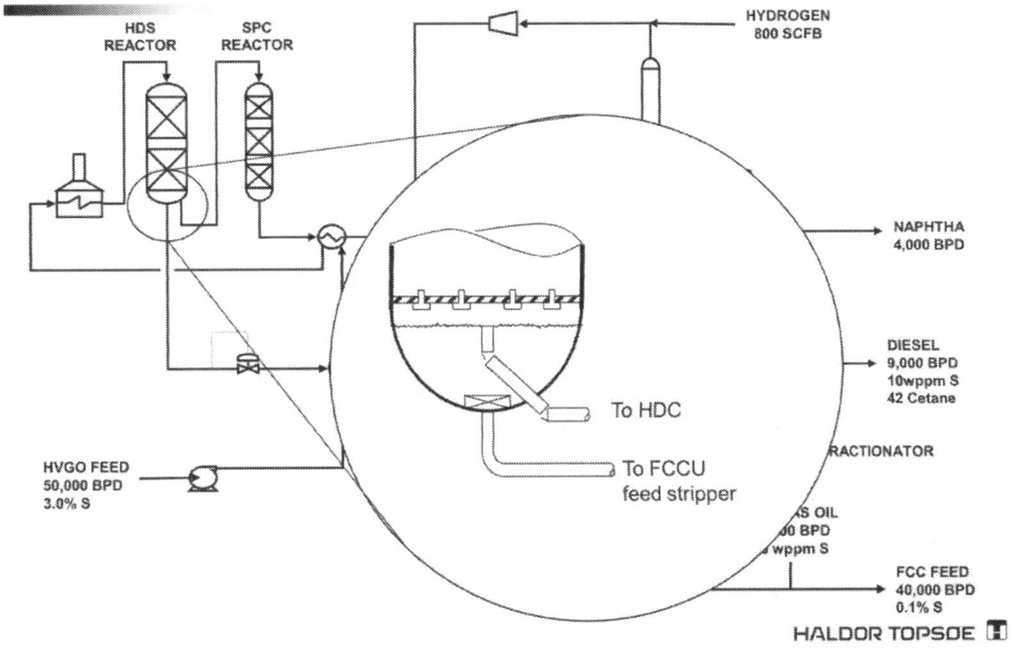

Revamp with flash drum

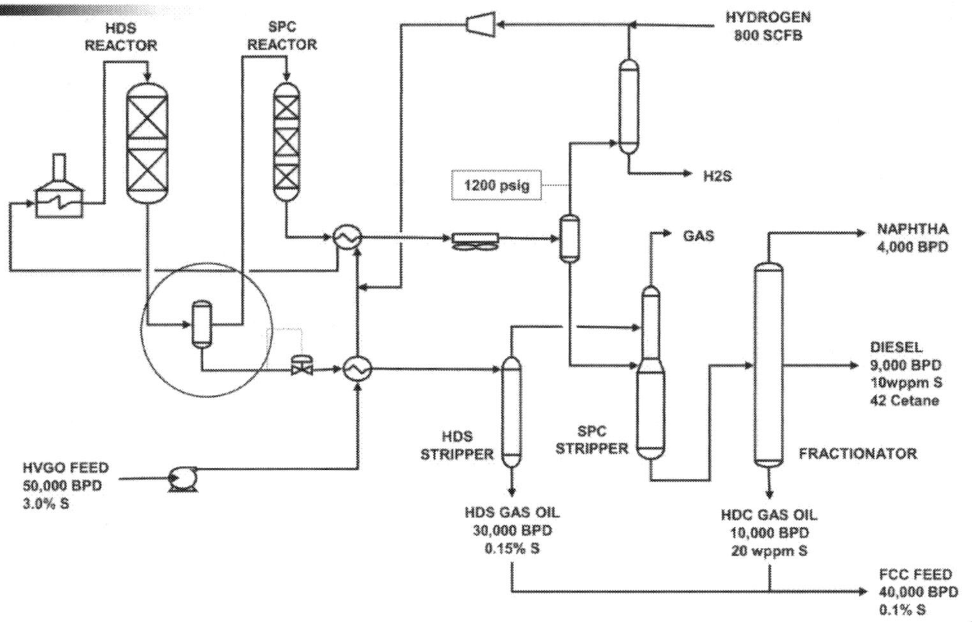

HALDOR TOPSØE

Feedstock segregation

Feed Type		Heavy	Light
Composition of Total FCC Feed	%vol.		
LVGO		0	35
HVGO		55	0
DAO		10	0
Feed Properties			
Gravity	°API	20	25
Sulfur	wt.%	2.7	1.5
Nitrogen	wppm	1050	650
Carbon Residue	wt.%	1.3	0.2
Metals, Ni+V	wppm	2.0	0.2
Asphaltenes	wppm	700	<100

HALDOR TOPSØE

Segregated feed configuration

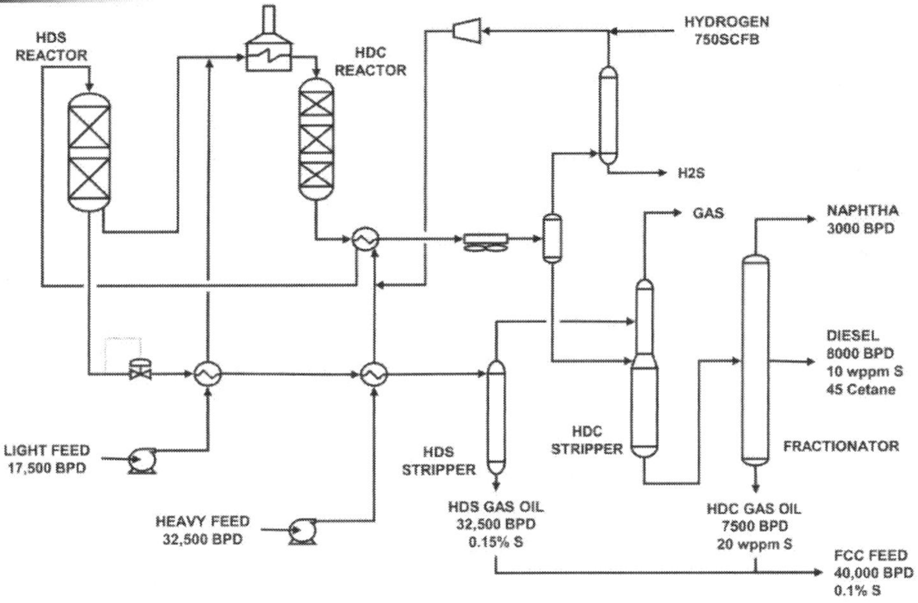

HALDOR TOPSOE

Summary of SPC key features

- Phase separation in bottom of lead reactor
  - minimizes extra separators (lower capex)
  - allows HDS product to bypass HDC
- HDS and HDC severities independently controlled
- HDS and HDC products segregated
- Much better distillate quality
- Lower H_2 consumption for same product quality
- Large product uplift at medium pressure

HALDOR TOPSOE

Production of ULSD

Factors impacting hydrogen consumption

- Feed properties / composition
- Hydrogen partial pressure
- LHSV
- HDS Severity (product sulfur)
- *Reactor temperature*
  - poly-aromatic saturation is equilibrium limited
  - H_2 consumption decreases with increasing temperature in almost all ULSD applications
  - large variation, start to end of run
- Temperature response is basis for Topsoe Poly-Shift technology

HALDOR TOPSØE

Poly-Aromatics Saturation

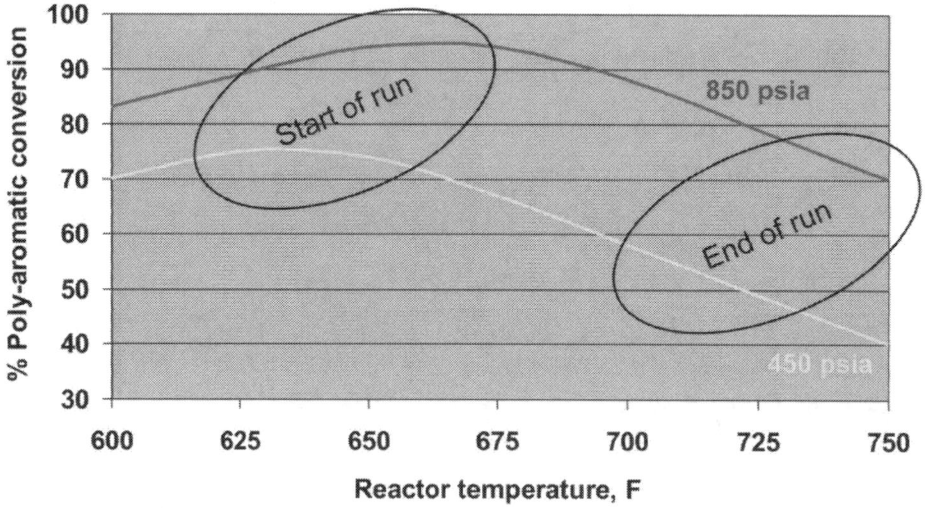

HALDOR TOPSØE

Commercial Example – Aromatics Shift

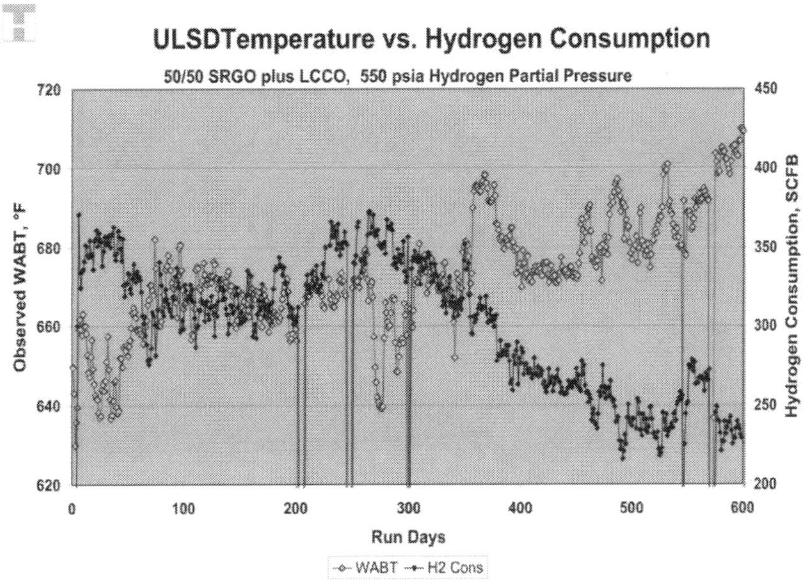

HALDOR TOPSØE

ULSD Reverse Poly-Shift Technology

- Operate trailing catalyst at constant EOR temperature based on product quality (density, aromatics, cetane, etc), ***high LHSV***
- Fix pressure and overall LHSV based on product uplift and cycle life
- Leading catalyst operates in conventional SOR/EOR deactivation mode
- Trailing reactor equilibrates aromatics and maintains constant H_2 consumption and product quality
- 20 to 50% savings in peak (SOR) hydrogen demand
- 10 to 25% savings in average cycle hydrogen consumption
- Corresponding savings in hydrogen production and compression capex / opex

HALDOR TOPSØE

Distillate ULSD hydrotreater

Switch to heavy crude

DHT Operation		Current Base	Future
Capacity	BPSD	17,000	18,500
Composition of DHT Feed	%vol.		
Straight run distillate		65	55
Cracked distillate (LCGO+LCCO)		35	45
Feed Properties			
Gravity	°API	31.2	29.6
Sulfur	wt.%	2.0	2.2
Aromatics	wt.%	37	40

HALDOR TOPSØE

Reverse Poly-Shift configuration

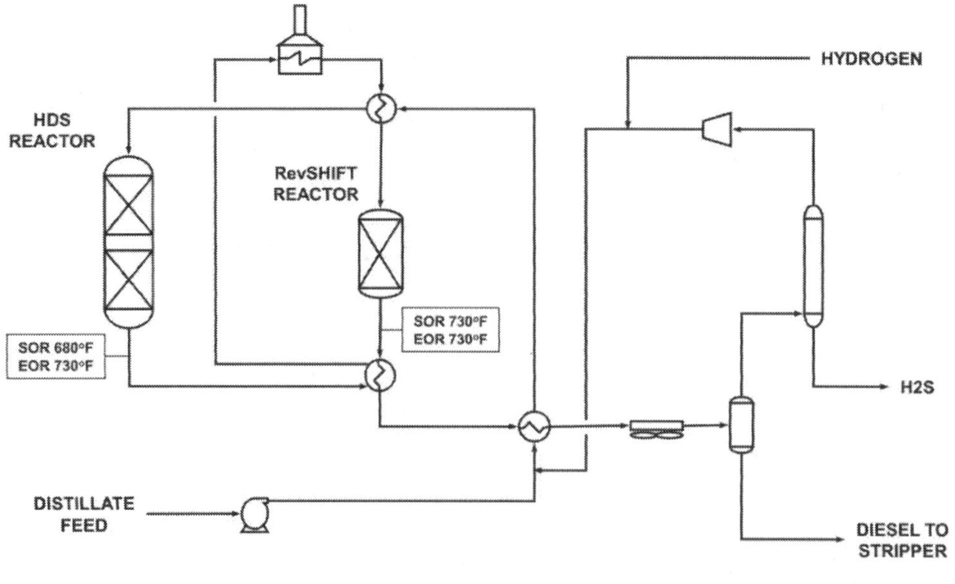

HALDOR TOPSØE

ULSD hydrotreater expansion study

DHT Operation		Current Base Configuration	Conventional Revamp	Reverse Poly-Shift Revamp
Capacity	BPSD	17000	18500	18,500
Reactor Pressure	psig	800	800	800
Reactor LHSV	v/v/hr			
Existing Reactor		0.65	0.70	0.70
New Reactor		-	3.2	3.2
Overall		0.65	0.57	0.57

HALDOR TOPSOE

ULSD hydrotreater expansion study

DHT Operation		Current Base Configuration	Conventional Revamp	Rev Poly-Shift Revamp
Average Temperature @ SOR	°F			
Existing Reactor		685	683	674
New Reactor		-	693	730
Overall		685	685	685
Recycle H_2/Oil Flow	SCFB	1630	1500	1500
Hydrogen Consumption SOR EOR	SCFB	 480 400	 550 440	 440 440
Excess Hydrogen Ratio(1)		3.4	2.7	3.4
Total Hydrogen Makeup	MMSCFD	8.2	10.2	8.2
Catalyst Cycle Life	Months	24	24	24

(1) Ratio of hydrogen recycle to hydrogen consumption at start of run

HALDOR TOPSOE

Reverse Poly-shift benefits

- Minimum <u>constant</u> hydrogen demand
- No product quality giveaway (SOR to EOR)
- Expand capacity of conventional designs without adding hydrogen generation or compression
- Simple process modifications, no new mechanical features
  - high space velocity, single bed shift reactor
  - two heat exchangers
- Small incremental capex vs. conventional design
- Chemistry demonstrated in multiple pilot plant tests (governed by thermodynamic equilibrium)
- Yields track overall WABT, not maximum shift temperature
- Benefits are not catalyst dependent

HALDOR TOPSOE

Conclusions

- Heavier crudes are coming
- Hydroprocessing technology key to upgrading
- New technology needed for clean fuels
- Topsoe introduces "Staged Partial Conversion"
  - improved diesel quality at moderate pressures
  - lower capex and H_2 consumption
  - unique but simple configuration
- Topsoe introduces "ULSD Reverse Poly-Shift"
  - minimizes H_2 consumption for diesel HDS
  - expand existing facilities with minimum hydrogen addition

HALDOR TOPSOE

DON'T FORGET ABOUT THE PRODUCT!

T.A. JURGENS-KOWAL
5 OCT 2012
SE TEXAS RPTC
SOUTH SHORE HARBOR

Global NP Solutions, LLC
www.globalnpsolutions.com 281-280-8717

Goals for Today's Talk

2

- Why should chemical engineers care about product development?
- New Product Innovation delivers profits

 www.globalnpsolutions.com

Why Should ChEs Care?

3

- How many products?

- <1,000
- 1,000 to 5,000
- 5,000 to 10,000
- 10,000 to 25,000
- 25,000 to 50,000
- >50,000

 www.globalnpsolutions.com

Commercializing a New Product

4

50% of New Products Fail

Why?

- Reasons why new products fail
  - Lack of up-front homework and customer understanding
  - Failed marketing

 www.globalnpsolutions.com

Who is Your Customer?

5

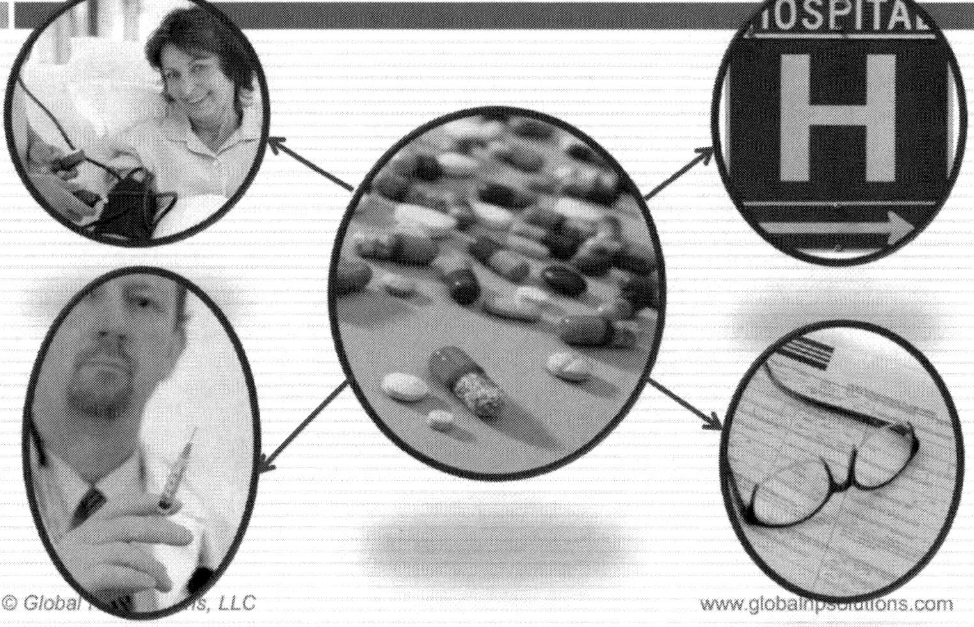

© Global … ns, LLC

www.globalnpsolutions.com

Commodity - Polyethylene

6

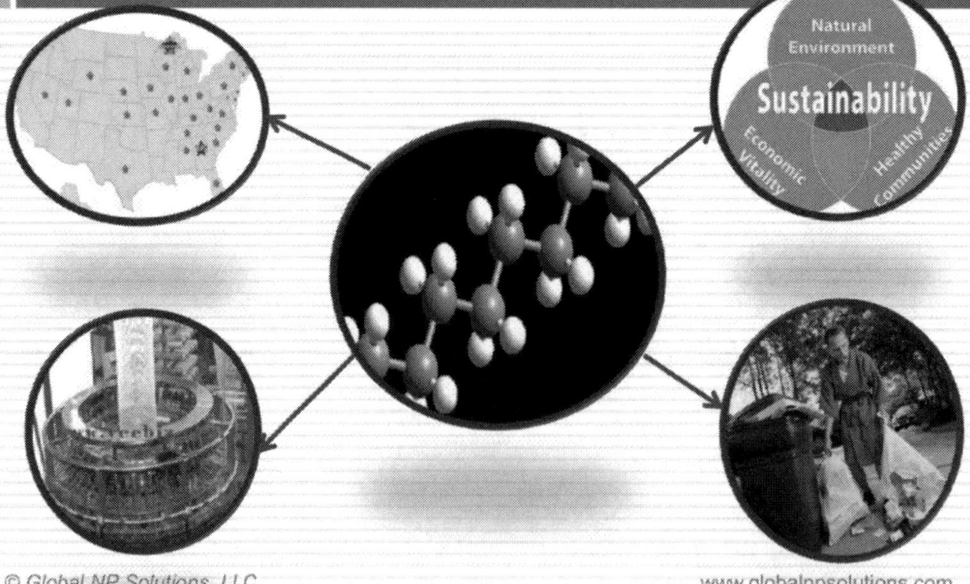

© Global NP Solutions, LLC

www.globalnpsolutions.com

Best Practices for Innovation

7

1. Link new product development to **Strategy**
   - Top 20% use **Strategy** to make project decisions

2. Improved **Time to Market**
   - 181 weeks to 104 weeks
3. Average **market share** of ~50%
4. Early selection of **best** ideas

 www.globalnpsolutions.com

Best Practices for Innovation

8

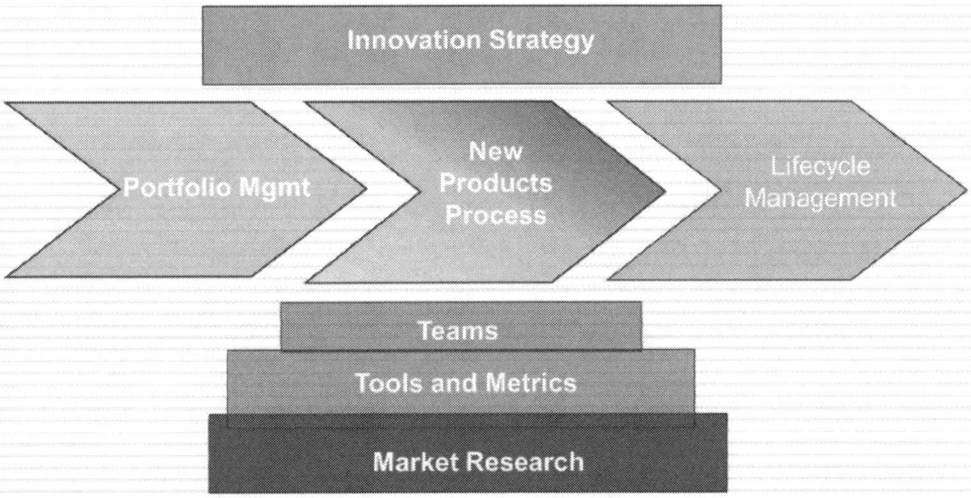

 www.globalnpsolutions.com

☑ Thanks for Participating !

9

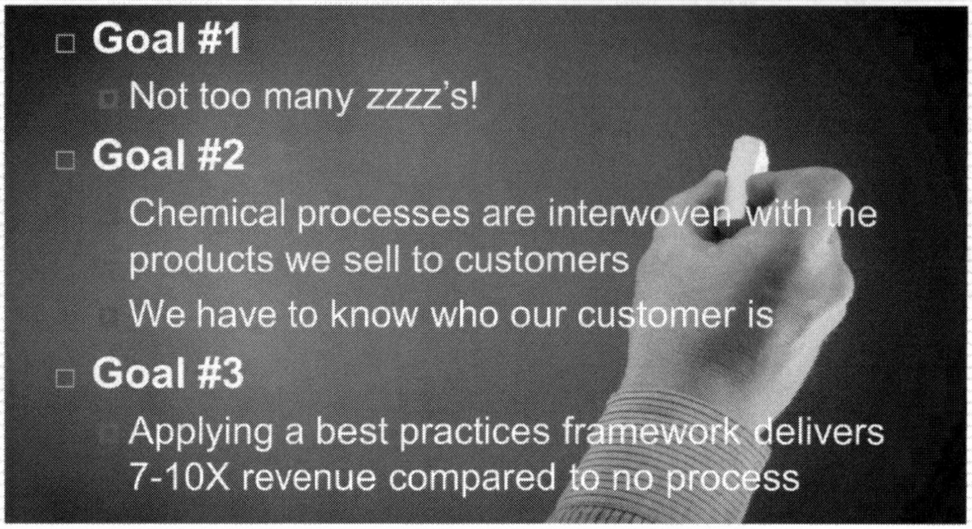

- **Goal #1**
  - Not too many zzzz's!
- **Goal #2**
  - Chemical processes are interwoven with the products we sell to customers
  - We have to know who our customer is
- **Goal #3**
  - Applying a best practices framework delivers 7-10X revenue compared to no process

 www.globalnpsolutions.com

For More Information

10

- **Lots of Books and Linked In Groups**
  - The Innovator's Book Club (Linked In)
- **New Product Development Training**
- **Project Management Workshops**
  - Facilitated on-line or face-to-face
  - Self-study….Coming Soon
- **Blogs and Websites**
  - Harvard Business Review
  - Idea Incubator blog

 www.globalnpsolutions.com

Nitrogen Free Retarder For Styrene Production - An Economic Comparison to DNBP

2012 AIChE Regional Process Technology Conference
Oct 4-5, 2012 League City Texas
Joe Lally

EVONIK INDUSTRIES
SiYPro™ Performance Additives

Overview

EVONIK INDUSTRIES

About us

- Who is Evonik Industries
- What is SiYPro™ Performance Additives

The current situation

- DNBP producer
- Hazardous topics
- Supply situation

Alternatives to DNBP?

- What is technically available
- What is the cost position of the alternative

Page 2

About Evonik Industries

EVONIK INDUSTRIES

- Evonik is one of the world's leading specialty chemicals companies.
- 2011 sales $18.7 billion with 33000 employees.
- EBITDA reached $3.6 billion for 2011.
- Active with production sites and sales offices in more than 100 countries.
- Strong technology platforms and integrated production structures.
- Balanced spectrum of business activities and end-markets.
- Close collaboration with customers.
- Market-oriented research and development.

Page 3

About Evonik Industries

EVONIK INDUSTRIES

We focus on economic megatrends

- resource efficiency
- health and nutrition
- globalization

Strong presence in petrochemical market:

- Monomers such as butadiene, acrylics and specialties
- Processing aids including inhibitors, antifoams, dispersants, ...

Page 4

SiYPro™ Performance Additives

This is our portfolio of additives to prevent fouling in vinyl monomer plants.

Designed in our Research Center in Germany to meet customer needs.

Implemented and supported by local process experts.

Applied in dozens of petrochemical plants around the world.

Current situation - DNBP

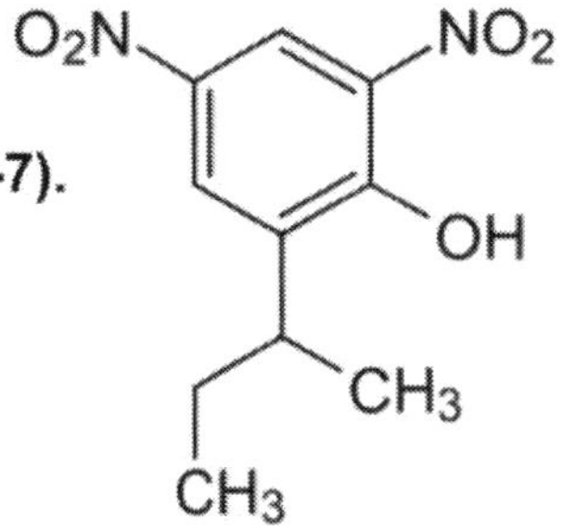

A widely used retarder for styrene production is DNBP (2-sec-butyl-4,6-dinitrophenol CAS: 88-85-7).

It is Acute - Toxic 

It is Repro - Toxic 

It is hazardous to the aquatic environment 

Supply situation - DNBP

ORTHO-SECONDARY-BUTYLPHENOL

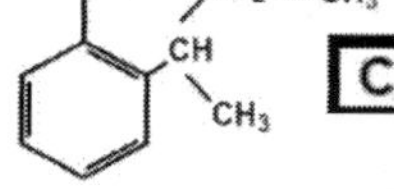

CAS 89-72-5

DNBP is produced in*

China: Three producers, rest are trader. No one registered under REACH. All rely on one raw material supplier.

Europe: One last remaining producer in Europe and only company who registered under REACH, monopoly supplier in Europe.

USA: One last remaining producer in USA and has not registered under REACH in Europe.

* based on our best knowledge

Page 7

Environmental situation - DNBP

Everybody is looking into "Green Chemistry"; not only for styrene production.

Companies do understand their task to safe the environment and to look for improvements.

Recent events like oil spills,... have clearly shown that the public is watching the petrochemical industry more than ever.

Reducing the overall risk in a chemical plant will also help the co-workers to reduce their individual potential to become part of an accident, contamination, ...

We have to take care!

Responsible Care®

Page 8

Technical alternatives? YES!

EVONIK INDUSTRIES

Evonik has been carrying out research for many years to devise a patented new technology who allows styrene producer to replace DNBP with a "Green" alternative.

It is called the SiYPro™ Super series.

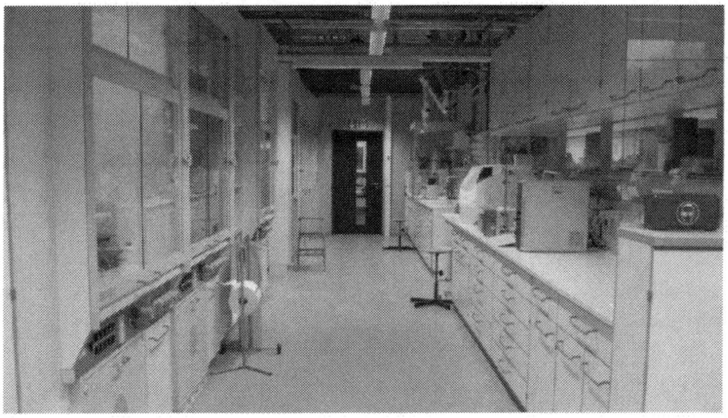

Page 9

Technical alternatives? YES!

EVONIK INDUSTRIES

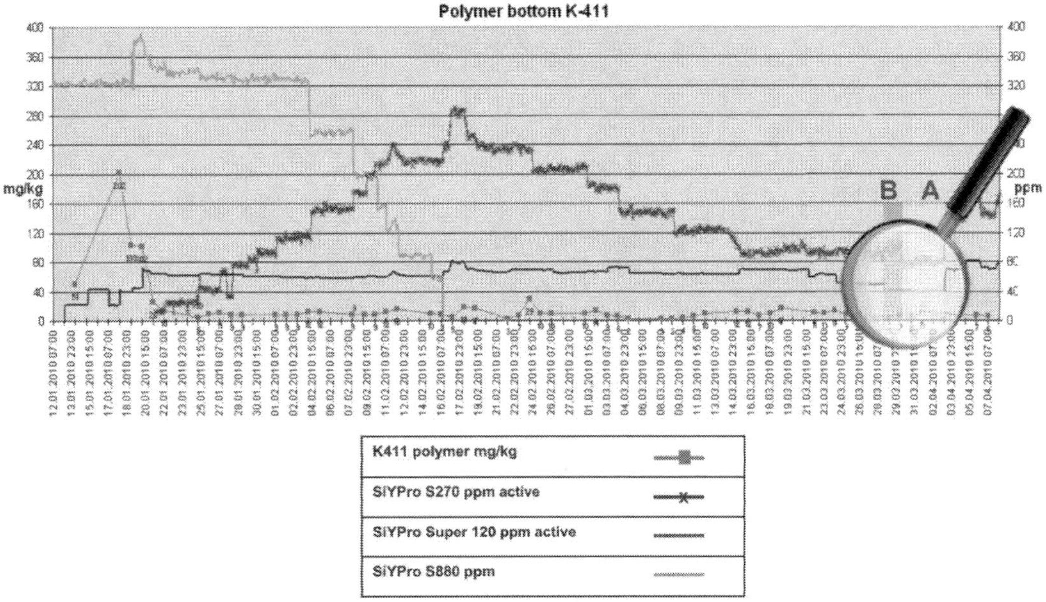

Page 10

Technical alternatives? YES!

EVONIK INDUSTRIES

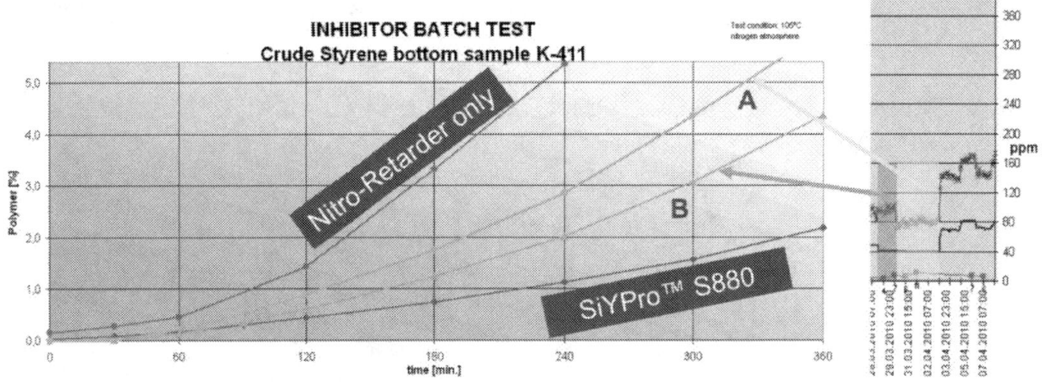

A: 80 ppm SiYPro S270 + 40 ppm SiYPro Super 120 [active]

B: 90 ppm SiYPro S270 + 40 ppm SiYPro Super 120 [active]

Page 11

Technical alternatives? YES!

EVONIK INDUSTRIES

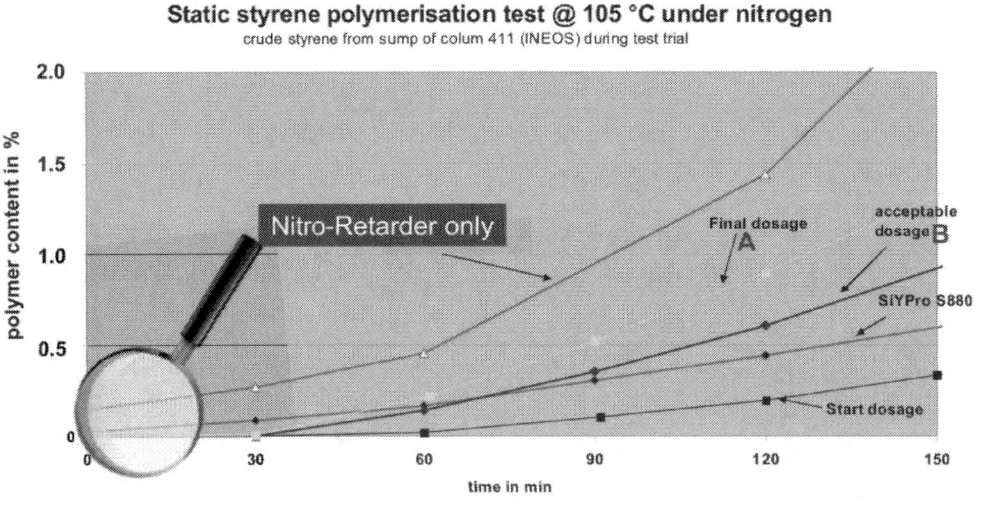

Page 12

Technical alternatives? YES!

EVONIK INDUSTRIES

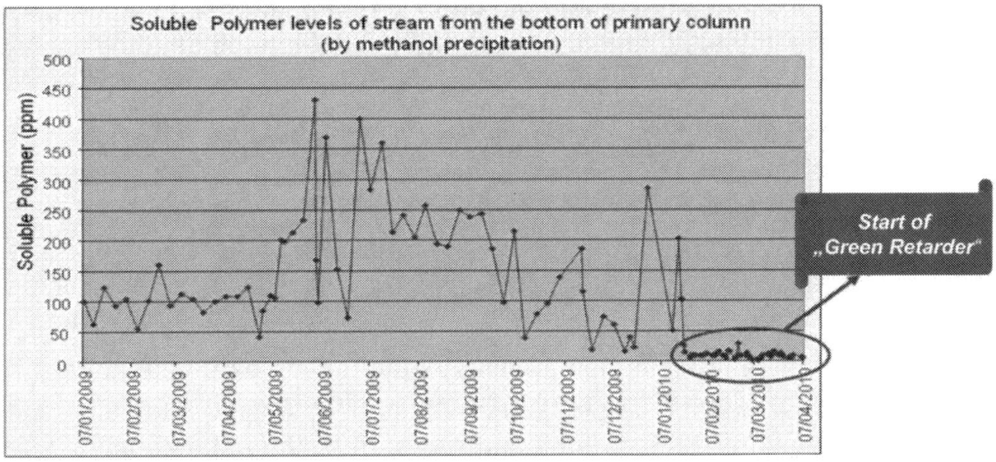

Tackling the economic challenge

EVONIK INDUSTRIES

How much would you be ready to justify from our margin of 1mt sold styrene to eliminate a product of concern in your plant?

Tackling the economic challenge

EVONIK INDUSTRIES

Which potential advantages can we see?

- Credits from lower soluble polymer level
- Tar stream is easier to handle – sale as fuel?
- Solvent credit of the additive
- No concerns of "potential corrosion"
- Alternative in case DNBP supply stops
- Reduced operational risks for co-workers
- "Green Image" for the company
- Regulatory advantages from the authorities
- …

Page 15

Tackling the economic challenge

EVONIK INDUSTRIES

Credits from lower soluble polymer level

For a reduction of 100ppm we can assume the following equation:

100ppm = 0.01% more styrene/mt styrene produced.

Sales price of styrene is about $1400 /mt
Each 100ppm reduction in soluble polymer leads to
$0.14 more margin
per 1mt styrene produced

Page 16

Tackling the economic challenge

Tar stream is easier to handle – sale as fuel?

In some areas the tar stream could be sold to power plants as fuel. With DNBP it is not that easy because burning of DNBP will lead to NOx emission. This is either not allowed or it needs to be paid for.
SiYPro™ Super series only contains the elements C, H, O.

Page 17

Tackling the economic challenge

Solvent credit of the additive

70% of the additive is solvent. This solvent is value for each styrene producer. The solvent can be either ethylbenzene, styrene or a mixture of both (depends on the SiYPro™ Super series type). Depending on the ingredients the value is about \$0.9 – 1.0 /kg additive.

No concerns of "potential corrosion"

DNBP quality can be very different. An important value is the acid number. Bad DNBP is under suspicion to cause corrosion. SiYPro™ Super series does not contain any corrosive species.

Page 18

Tackling the economic challenge

EVONIK INDUSTRIES

Alternative in case DNBP supply stops

Even if you have decided to stay with DNBP the supply may once will come into trouble. Example could be that starting material would not be available in the market. In this case a proven alternative like the SiYPro™ "green retarder" is a good alternative as fall back position.

Reduced operational risks for co-workers

All your employees are well trained experts and you are handling thousands of tons dangerous products. But this products are normally handled in a closed system. Process additives are normally delivered in remote systems and need to be connected to the plant. Also maintenance at DNBP contaminated equipment is not easy.

Page 19

Tackling the economic challenge

EVONIK INDUSTRIES

"Green Image" for the company

Being a producer- or a designing company does not matter. The image you give to the rest of the world can be important. Either in marketing your products or in the outside view to neighbours, the community, the workers union, ... With "Green Chemistry" you reduce the output of NO_x and CO_2.
Sometimes it is helpful to improve your image and gain additional benefits out of this.

Regulatory advantages from the authorities

In more and more areas in the world the authorities are concerned about dangerous goods. There is a clear trend that if possible dangerous goods are banned or the use is strictly controlled. SiYPro™ Super series is a tool to proof to the authorities that you as a company are in line with this idea and wherever possible follow that request.

Page 20

Summary

EVONIK INDUSTRIES

Green Retarder for styrene production - an economic challenge compared to DNBP?

Yes but we are able to handle it!

We have seen that technically there is an alternative. This alternative brings a number of different improvements which are either

- ✓ **directly countable in Euro, USD, GBP,...**
- ✓ **add value which is not directly countable like safety, outside image, authority communication, ...**

Page 21

Questions

EVONIK INDUSTRIES

Page 22

EVONIK
INDUSTRIES

Using *in situ* Spectroscopy for Kinetic Modeling and Scale-up

Sagar Sarsani & Ravi Dixit
The Dow Chemical Company
Freeport, TX

4th AIChE Regional Process Technology Conference

League City, TX October 4 – 5, 2012

1

Why *in situ*

- ***In situ*** analysis (v.s. ***ex situ***) is necessary when
  - Species are very labile
  - Reaction is very fast
  - Chemicals are very hazardous
- In addition, ***in situ*** analysis is desirable
  - Eliminates the need to collect samples
  - Reduces perturbation to the reaction/process
  - Enables monitoring of reaction progress in real-time
  - Does not require sample preparation and off-line analysis

2

Spectroscopy Probes

- Infrared (IR) or Fourier Transform Infrared (FTIR) is based on light transmission, while Raman is based on scattering. Any system accessible to light can be probed (no vacuum needed; no sample prep needed) by optical spectroscopic methods

- Spectroscopy (IR, Raman) yields a wealth of information
  - Functional groups
  - Concentrations

- Drawbacks & limitations
  - Spectral interference may hamper analysis of a mixture
  - Detection limit often not as low as chromatographic techniques

3

Applications of *in situ* Spectroscopy

- Process development
- Process control
- Process optimization
- To provide mechanistic insight
- Monitor product selectivity
- Rapid and reliable information in *real time*
- Generate high quality kinetic data

4

Case Study: Butylene Oxide Polymerization

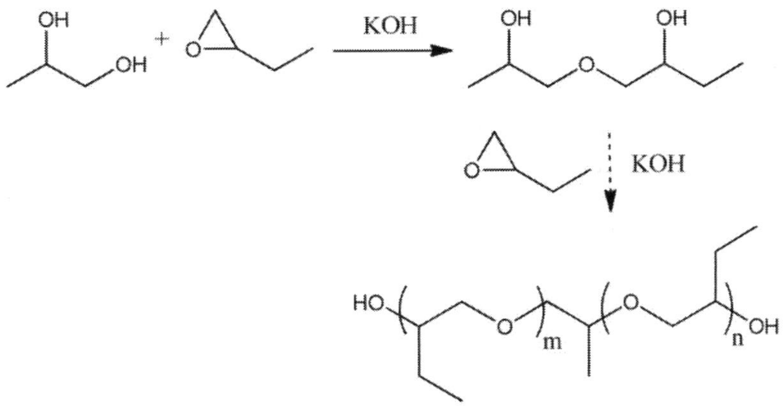

5

Conventional Methods

- Requires sampling from the reactor
- Analysis by chromatographic methods
- Challenges
  - Quenching living polymerization
  - Unreacted oxide containing sample: Hazardous
  - Sample preparation
  - Perturbation of the system

6

Experimental Setup: RC1 Reactor

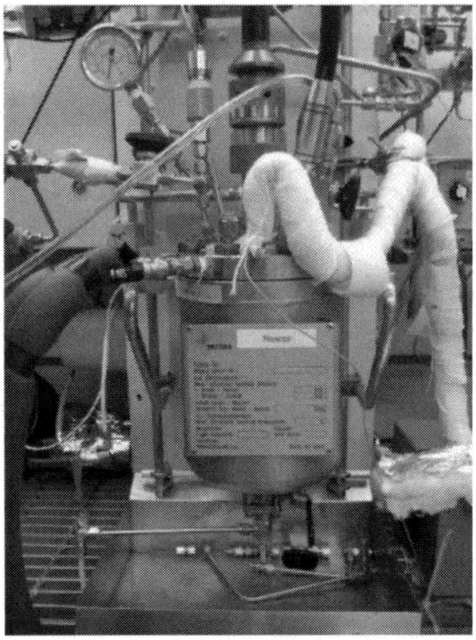

- RC1 reactor
- ATR FTIR probe (In situ for monitoring liquid [BO])
- fiber-optical cable and a diamond ATR element

7

Reactor Internals

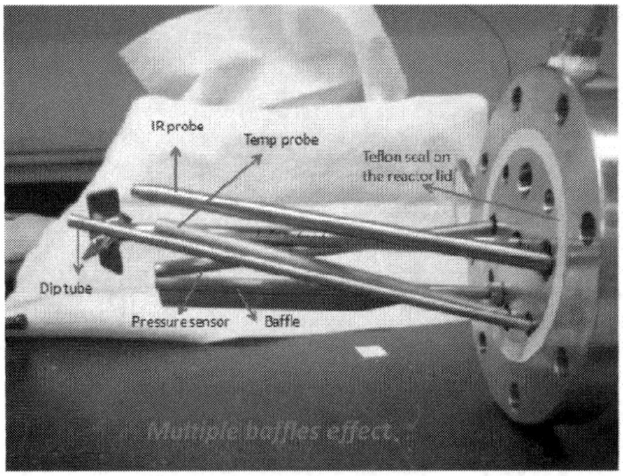

Mixing studies performed using geometrically similar glass reactor

8

Experimental Procedure

- Prepare the Catalyzed initiator mixture
- Load the Catalyzed initiator into RC1
- Nitrogen in Vapor space; Preheat to 90°C
- Increase temperature to desired set point: Agitation is started
- Load the Oxide: Addition of each shot is quick (~ few seconds)
- Analysis by IR probe (for monitoring liquid [BO])

9

Details of Experiment

#	compound added	amount added (g)	cumulative BO (g)	cumulative BO/OH molar ratio	total weight (g)	$[BO]_{liquid\ phase}$
0	PG	127.42	-	-	-	-
1	BO	41.07	41.07	0.294	168.49	22.0%
2	BO	58.39	99.46	0.712	185.81	23.8%
3	BO	120.19	219.65	1.573	247.61	33.1%
4	BO	125.74	345.39	2.474	253.16	25.5%
5	BO	245.19	590.58	4.230	372.61	33.4%
6	BO	243.73	834.31	5.976	371.15	24.9%
7	BO	183.7	1018.01	7.292	311.12	15.8%
8	BO	56.13	1074.14	7.694	183.55	4.6%

10

BO Concentration Profiles Determined by Classical Least Squares Method

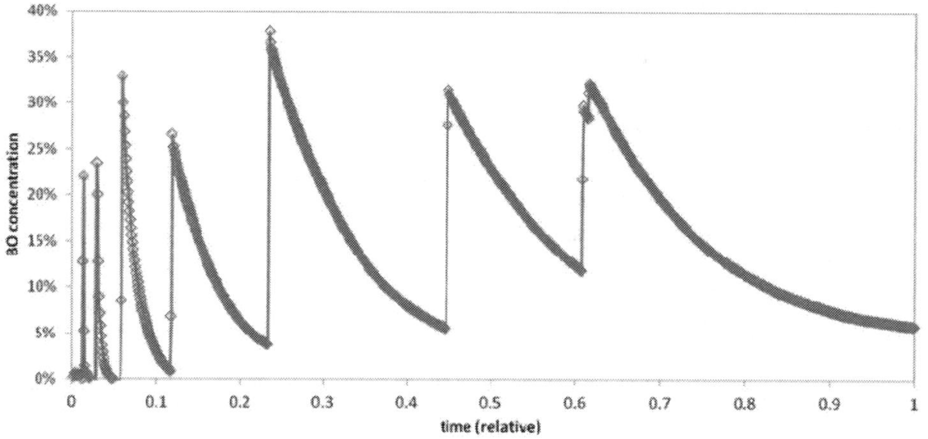

11

Comparison of Model Fit with IR Concentration Data

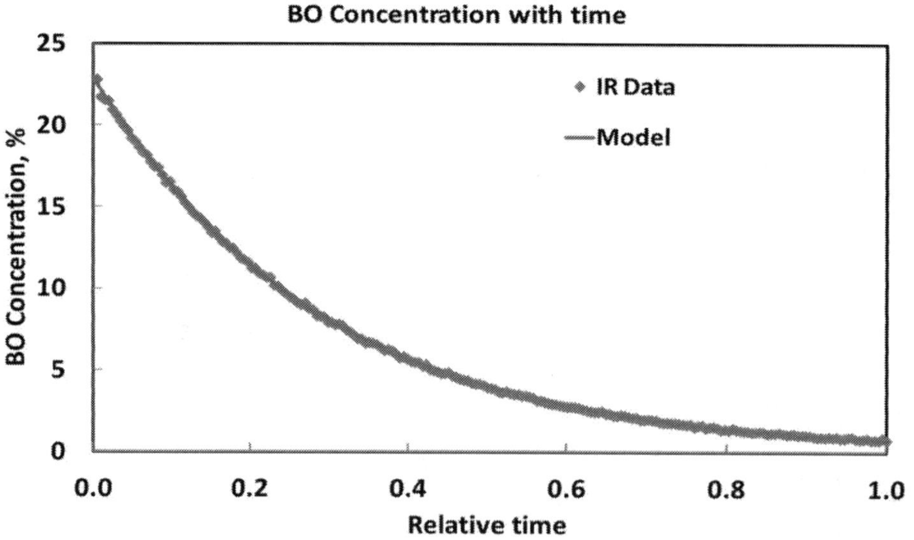

12

Determining Kinetic Parameters

$$\frac{d[BO]}{dt} = -A * e^{-\frac{E_a}{RT}} * [Catalyst]^n * [BO]^m$$

Profile of liquid phase [BO] and parameter estimation will give k.

Data at different temperatures will give Ea

Since for each feed shot, $k_i = f(M.W)$

$$k_{overall} = f(M.W.D)$$

13

Rigorous Reaction Scheme

$$\begin{array}{l} \mathbf{P_0 + M \rightarrow P_1} \\ \mathbf{P_1 + M \rightarrow P_2} \\ \mathbf{P_2 + M \rightarrow P_3} \\ \mathbf{:} \end{array}$$

$$\frac{dP_0}{dt} = -kP_0$$

$$\frac{dP_1}{dt} = kP_0 - kP_1$$

$$\vdots$$

$$\frac{dP_i}{dt} = kP_{i-1} - kP_i$$

14

Reaction Engineering

Reaction Scheme → Reactor Configuration

↓

Molecular Architecture

15

Process Development and Scale-up Factors

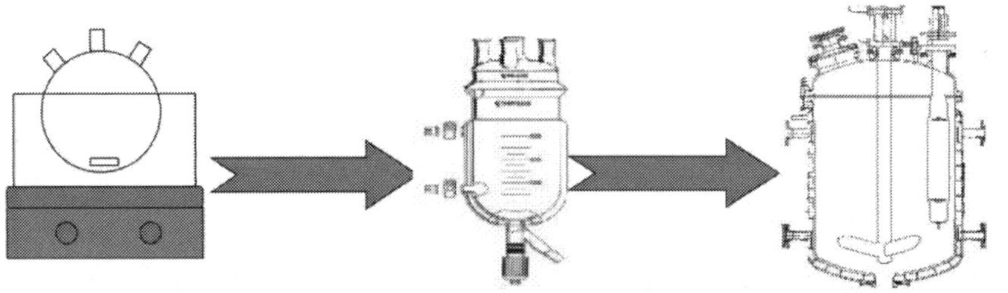

Chemistry
- Chemical equilibria
- Chemical kinetics

Physical properties
- Phases
- Densities
- Viscosities & rheology

Transport phenomena
- Hydrodynamics/Mixing
- Mass transfer
- Heat transfer

Atiemo-Obeng, 2011

Conclusions

- *In situ* spectroscopy allowed determination of kinetics with out the need for sampling or calibration of intermediate products
- The influence of mixing was evaluated in this reaction
- Kinetic parameters determined are used in reactor model for successful scale up of these processes
- We sell these products under UCON™ OSP fluids

™ Trademark of The Dow Chemical Company

17

Acknowledgements

- Carlos Villa
- Shawn Chen
- Victor Atiemo-Obeng

18

Extra Slides

19

Sucrose Hydrolysis – ReactIR™ 1000

$$\underset{\text{sucrose}}{C_{12}H_{22}O_{11}} + H_2O \xrightarrow{\text{HCl}} \underset{\text{fructose}}{C_6H_{12}O_6} + \underset{\text{glucose}}{C_6H_{12}O_6}$$

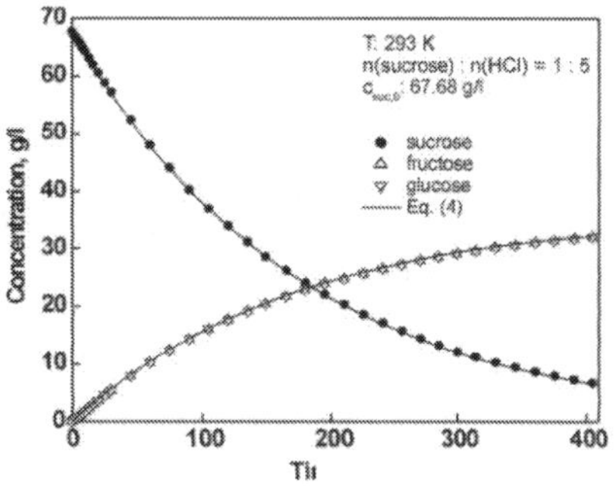

$$\frac{c_{\text{suc}}}{c_{\text{suc},0}} = e^{-k't}$$

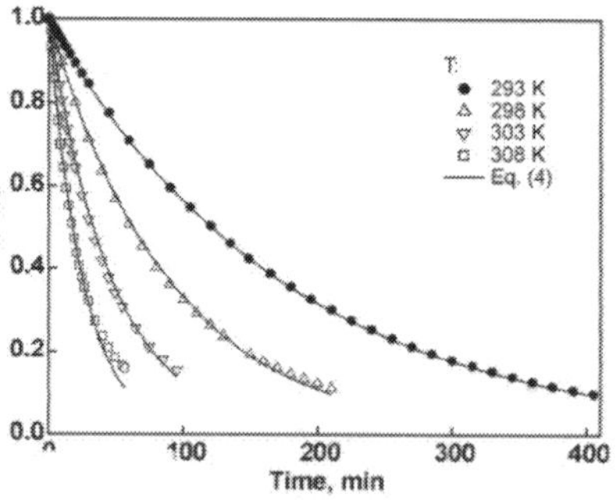

$$(-r_{\text{suc}}) = 5.075 \times 10^{14} \exp^{-\frac{95000}{RT}} c_{\text{suc}} c_{\text{H}^+}$$

Albin Pintar, Jurka Batista and Janez Levec, Analyst, 2002

20

Opportunities to Increase Process Yields through Better Mixing of Reactants – A Case Study on Dichlorobutene Production

Chilka, A., ANSYS-Fluent India Pvt. Ltd.
Kasat, G.,
Vedapuri, D.,
Originally presented on: 10/5/2012 13:15:00 - 13:40:00

In vapor phase reactors two or more vapor/gas streams are mixed together. The gases are generally fed through nozzles. A critical problem with these reactors is the incomplete mixing of the injected gas/vapor streams leading to incomplete conversions. The degree of mixing is governed by the design of the feed nozzle. This presentation focuses on improving the yield of vapor phase chlorination of butadiene to dichlorobutene. Given the large volumes in which dichlorobutene is produced there is a considerable economic incentive in improving the yields of this process or in the alternative, in increasing the capacity without a loss of yield. Mere increase of the feed rates results in poor mixing leading to loss of yield. In this study computational fluid dynamics based models are used to understand the impact of the feed nozzle design on the mixing dynamics, conversion and selectivity. The nozzle design was optimized to maximize yields. A sensitivity study of the nozzle design on reactor dimensions was also conducted.

Low Energy Solvent Separation in Solution Polymerization Processes

Guzman, J. D., The Dow Chemical Co.
Jain, P., The Dow Chemical Company
srivastava, R.,
Wang, A.,
Zogg, Jr., M. J., The Dow Chemical Company
Kiran, E., Virginia Tech
Originally presented on: 10/5/2012 13:40:00 - 14:05:00

In solution polymerization processes, after polymerization is carried out in a paraffinic solvent, the solvent is typically removed by evaporation. The evaporation step requires a considerable amount of energy, and additional energy (refrigeration work) is later required to condense the solvent vapors before they can be recirculated into the polymerization reactor.

We have developed a new polymerization process that takes advantage of liquid-liquid separation phenomena in polymer solutions. At the right combination of temperature, pressure, and composition, a fraction of the solvent in a homogenous polymer solution forms a separate liquid phase, free of polymer. The solvent separation occurs rapidly when it proceeds via spinodal decomposition. Solvent is therefore quickly separated from the reactor effluent without providing additional energy, and it can be recirculated without supplying refrigeration work for condensation.

Liquefied Natural Gas (LNG) Industry Overview
Adams, J.,
Originally presented on: 10/5/2012 12:00:00 - 12:25:00

The natural gas industry is an extremely important sector of the world's growing energy demand and LNG is a vital part of meeting that demand. The Liquefied Natural Gas (LNG) Industry Overview presentation will explore many aspects of the LNG industry. The presentation starts with the history of natural gas liquefaction and continues into the evolution of the industry while examining the reasons for the changing demand in the outlook of the US LNG market. The presentation also provides details on the major design components of an LNG liquefaction facility. Gas treating options are addressed along with a thorough description of the condensation and liquefaction step. A summary of major liquefaction equipment and types of LNG Storage Tanks will be presented to complete a comprehensive account of LNG liquefaction.

Introduction to Refining Economics

Embry, D.,

Originally presented on: 10/5/2012 12:25:00 - 12:50:00

Refineries are complex multi-product plants that are the blend of many complex trade-offs. Estimating the value of a feedstock change or process reconfiguration requires evaluation of the entire system. This talk will provide a review of the basic refining economic terms, refinery processes and the factors that drive the economics of operating a refinery. It will be demonstrated how overall system evaluation results differ from local estimates.

BAKERRISK

AIChE 4th Annual Regional Process Technology Conference

Process Safety and Risk Management

Overview of BakerRisk

- Established in 1984
- Conducted over 6,000 projects relating to explosive, fire, or toxic hazards
- Five technical sections:
  - *Process Safety and Risk Management*
  - *Blast and Explosion Effects*
  - *Protective Structures Design and Analysis*
  - *Testing, Research and Development*
  - *Incident Investigations*

BAKERRISK

Background – Process Safety

- Union Carbide. Bhopal, India disaster
- December 3, 1984. Water contamination of a methyl isocyanate tank (over 40 toms of MIC)
- Runaway reaction. Hot MIC vapor vented through the tank pressure relief system
- Due to process changes, tank refrigeration equipment and high temperature alarm were not in service
- Over 3,000 fatalities, and over 100,000 injured

BAKERRISK

Background – Process Safety

- Bhopal FOREVER changed the way that the process industry approaches process safety
- In February, 1985, 17 chemical industry leaders asked AIChE to lead an effort to eliminate catastrophic process incidents
- By the end of 1985, CCPS had been formed with 39 charter members

BAKERRISK

Background – Process Safety

- Phillips 66 Chemical Complex. Pasadena, Texas. October 23, 1989, about 1 PM
- October 22, maintenance work started on a HDPE reactor, cleaning polyethylene particles plugged in 3 of 6 reactor settling legs
- Single ball valve at the top of each leg
- Procedure: close the ball valve, disassemble the settling leg, remove the plastic material blockage/plug

BAKERRISK

Background – Process Safety

- Air hoses to the isolating ball valve were cross connected, and the valve was opened with the reactor in operation
- 85,000 pounds of highly flammable gas was released
- In < 2 minutes, the gas found an ignition source
- Series of explosions resulted in 23 fatalities, over 130 injured

BAKERRISK

Phillips 66 Chemical Complex

BAKERRISK

Background – Process Safety

- **Air hoses to the ball valve were to be disconnected prior to maintenance work. This was not done**
- **Site procedures called for isolation by a double block valve system or blind flange. This was not followed**
- **A contractor who regularly performed this maintenance operation was doing the work**

BAKERRISK

Background – Process Safety

- BP Texas City Refinery. March 23, 2005
- Start-up of the raffinate splitter tower of the ISOM unit
- Operators began filling the tower with gasoline components
- An alarm to warn about the quantity of liquid in the tower was disabled
- Operators opened a discharge valve which passed the gas components through a heat exchanger

BAKERRISK

Background – Process Safety

- A blow down drum with an atmospheric vent stack overflowed (hot flammable vapors and liquids)
- The vapor cloud explosion resulted in 15 fatalities, over 170 injured

BAKERRISK

BP Texas City

BAKERRISK

Developments – Process Safety

- 1992: the OSHA Process Safety Management Standard (29 CFR 1910.119) was published
- This standard requires most companies in the US that process chemicals and refined products to implement a PSM program
- Primary PSM objective: to prevent releases of hazardous materials that can expose employees, other personnel, and company assets to serious hazards.

BAKERRISK

Developments – Process Safety

- Process Safety: proactive identification, evaluation and mitigation or prevention of hazardous material releases that could occur as a result of failures in processes, procedures, or equipment
- PSM Standard: developed with industry participation.. Regulations are mainly performance based
- Effective PSM is a process of continual improvement, and will improve all areas of a plants operation: safety, environmental, reliability, compliance, and economic performance

BAKERRISK

PSM – 14 Elements

- Employee Participation
- Develop a written plan for employee involvement in the program
- Consult with employees in the development and implementation of the PSM program
- Provide access to PSM information (PHA's)

BAKERRISK

PSM – 14 Elements

- **Process Safety Information**
- **Company shall compile technical information on the hazardous chemicals, process and equipment prior to conducting a PHA**
- **Include hazards of chemicals and flammables in the process**
- **Include information regarding technology (design, operating limits, ...) of the process**
- **Include information pertaining to the equipment in the process**

BAKERRISK

PSM – 14 Elements

- **Process Hazards Analysis**
- **PHA focus: identify hazards and safeguards, rank (prioritize) the hazards, recommendations to mitigate or eliminate identified hazards**
- **PHA review required at least every 5 years**
- **Document the findings, recommendations, actions, and the actions schedule**

BAKERRISK

PSM – 14 Elements

- Operating Procedures
- Develop and implement written operating procedures for safely conducting the process operations
- Review and update as needed to insure the procedures are current and accurate (certify this annually)
- Develop and implement safe work practices to provide for control of hazards

BAKERRISK

PSM – 14 Elements

- Training
- Each employee operating a process shall be trained with a process overview and the operating procedures
- Training also includes safety and health hazards, emergency operations, safe work practices, and documentation of training
- Refresher training shall be provided at least every 3 years, or more often if necessary

BAKERRISK

PSM – 14 Elements

- **Contractors**
- Often involved in hazardous maintenance and construction tasks
- Company shall evaluate the contractor's safety performance and program
- Inform contractor employees of the known potential hazardous chemical release hazards related to the contractor's work
- Implement contractor safe work practices, evaluate performance, and keep an injury, illness log

BAKERRISK

PSM – 14 Elements

- **Pre-Startup Safety Review**
- Performed for new facilities, and for modified facilities with changes in process safety information
- PHA complete (new facilities) or MOC complete (modified facilities)
- Verify that construction and equipment meets design specifications, and that employee training is complete

BAKERRISK

PSM – 14 Elements

- **Mechanical Integrity**
- **Establish and implement written procedures to maintain the integrity of facility equipment, piping, instrumentation, electrical systems, and other physical items**
- **Conduct tests and inspections (T&I), and document T&I. Frequency?**
- **Correct identified deficiencies in equipment. Schedule?**
- **Inspect quality, suitability, and installation of new equipment**
- **Material control (parts, maintenance materials)**

BAKERRISK

PSM – 14 Elements

- **Hot Work**
- **Covers use of flame producing equipment, or use of equipment that can create sparks**
- **Company shall issue hot work permits for such work that is conducted on or near a covered process**
- **Permit includes fire prevention requirements, dates authorized. and the equipment, work area**

BAKERRISK

PSM – 14 Elements

- **Management of Change**
- **Establish and implement written procedures to manage changes to process chemicals, technology. equipment, and operating procedures for a covered process.**
- **Train employees prior to start up of the revised process**
- **Update PSI, PHA and operating procedures**

BAKERRISK

PSM – 14 Elements

- **Incident Investigations**
- **Investigate each incident that caused, or reasonably could have caused, a catastrophic release of hazardous chemicals (within 48 hours)**
- **Establish a qualified incident investigation team, complete an incident investigation report, and review with the affected employees**
- **Document and implement the corrective actions**

BAKERRISK

PSM – 14 Elements

- **Emergency Planning and Response**
- **Establish and implement a plant emergency action plan for response to a release of hazardous chemicals**
- **The plan and instructions should include: response team members, alarm systems, communications, external notifications, emergency equipment, steps to minimize harm to people and the environment**

BAKERRISK

PSM – 14 Elements

- **Compliance Audits**
- **Conduct a PSM audit at least once every 3 years to verify that the PSM procedures and priorities are adequate and are being followed**
- **Issue a report of the findings, and document the response to identified deficiencies and problems**
- **Internal PSM reviews and evaluations should be a regular facility activity**

BAKERRISK

PSM – 14 Elements

- **Trade Secrets**
- **Company shall make all information necessary to develop and implement an effective PSM program, without regard to possible trade secret status. Employees may be required to sign confidentiality agreements for access to trade secret information.**

BAKERRISK

Consequence and Risk

- **Consequence: The significance/magnitude of the damage should a hazardous incident occur**
- **3 Elements of Risk**
- **The hazard**
- **Consequence of the hazard**
- **Probability the hazard will occur**

- **Many companies are now evaluating risk for hazards that can result in unacceptable consequences**

BAKERRISK

Risk Assessment

- The focus of risk assessment is to better evaluate the priorities of hazards that should be addressed.
- Once the priorities are established, there are 3 options for mitigating the risk:
  1. remove the hazard
  2. reduce the consequence
  3. reduce the hazard frequency

BAKERRISK

Risk Matrix

- The risk matrix is used to evaluate the risk ranking of a specific hazard, event.
- Normally 4 or 5 squares per side of the risk matrix: 4 or 5 levels of frequency, 4 or 5 levels of consequence
- Frequencies can range from > 1 in 100 years to < 1 in 1 million years
- Consequence levels: can be based on employee safety, public safety, environmental impact, economic impact, and public relations impact

BAKERRISK

Risk Matrix

Likelihood	Consequences				
	Insignificant	Minor	Moderate	Major	Severe
Almost certain	M	H	H	E	E
Likely	M	M	H	H	E
Possible	L	M	M	H	E
Unlikely	L	M	M	M	H
Rare	L	L	M	M	H

BAKERRISK

Risk Matrix

- **Actions**

  E - immediate action required to reduce risk

  H - action required to reduce risk, schedule to be determined

  M - risk should be reduced if the measures are cost effective, no action is an option

  L - no action necessary

BAKERRISK

Risk: F-N Curve

- **Frequency- Consequence curve**
- **Vertical axis: frequency the event will occur**
- **Horizontal axis: consequence, number of people affected**
- **Acceptable risk: not defined by regulatory agencies. Must be defined by each company.**

BAKERRISK

F-N Curve

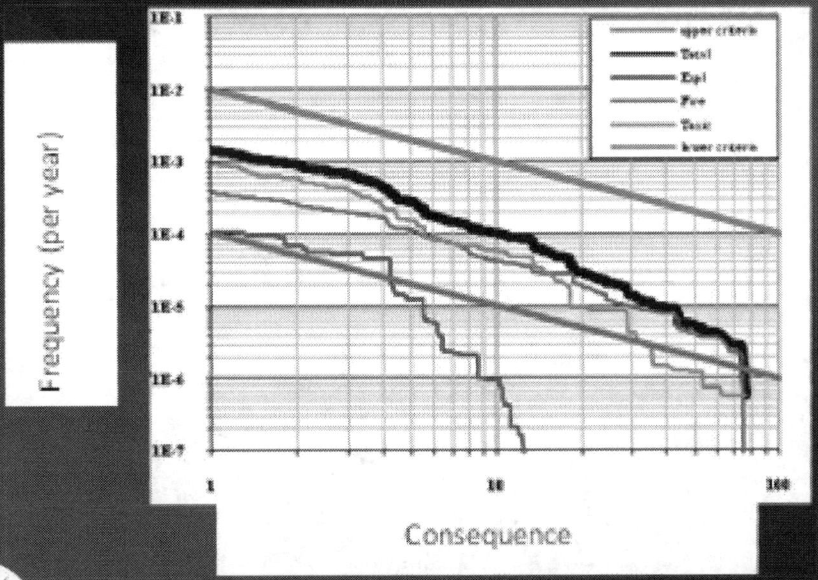

BAKERRISK

Risk Evaluations

- **Some options for risk evaluations**

- **F-N curve for risk to onsite personnel from explosion, fire and toxic hazards**
- **Building Risk Table: level of risk incurred by occupants of each building from explosion, fire and toxic hazards**

BAKERRISK

Risk Evaluations

- **Outdoor Population Area Risk Table: level of risk incurred by onsite personnel in specific plant areas from explosion, fire and toxic hazards**
- **Table of Significant Risk Sources: Ranking of release sources and locations that result in the most significant risk to onsite personnel from explosion, fire and toxic hazards.**

BAKERRISK

BAKERRISK

Structural Upgrades

BakerRisk Shock Tube

BAKERRISK

BAKERRISK

Building Component Tests

Monolithic Window Test

- Window test conducted at BakerRisk shock tube facility
- Window 50 in x 68 in, ¼" annealed glass
- Blast Load ~ 4 psi, 60 psi-msec
- Window fails in "High Hazard Mode"

BAKERRISK

Monolithic Window Test

BAKERRISK

Monolithic Window Test

- **High speed video of window test conducted at BakerRisk shock tube facility**
- **Window 50 in x 68 in, ¼" annealed glass**
- **Blast Load ~ 4 psi, 60 psi-msec**
- **Window fails in "High Hazard Mode"**
- **Note formation of glass fragments**

BAKERRISK

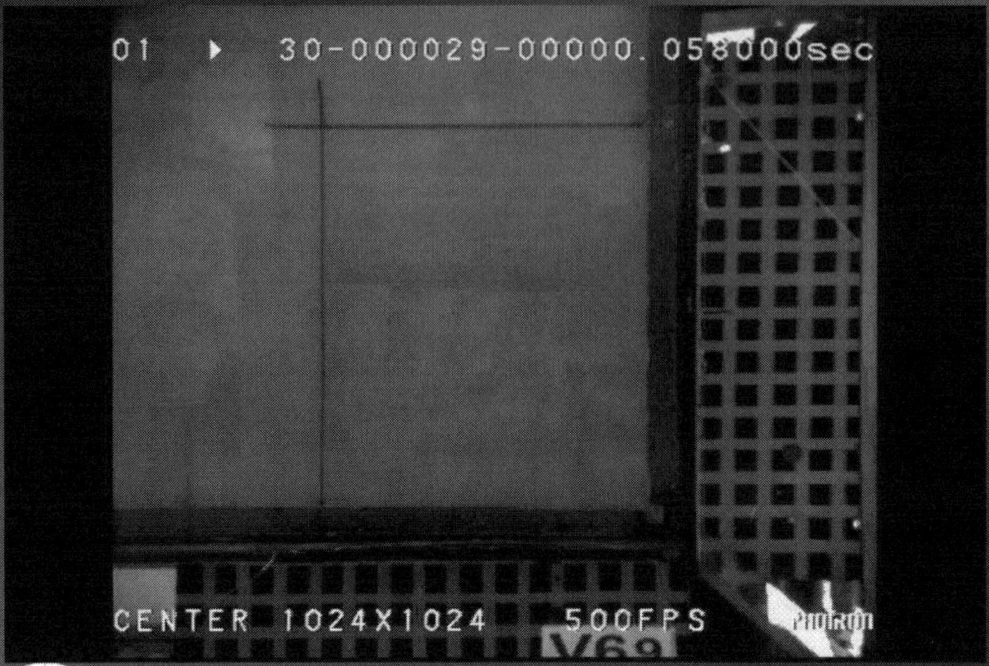

BakerRisk

Monolithic Window Test

Laminated Window

- **Window test conducted at BakerRisk shock tube facility**
- **Blast Load ~ 5 psi, 60 psi-msec**
- **Window passes in "Minimal Hazard Mode"**

BakerRisk

BAKERRISK

Laminated Window Test

Laminated Window

- High speed video of window test conducted at BakerRisk shock tube facility
- Blast Load ~ 5 psi, 60 psi-msec
- Window passes in "Minimal Hazard Mode"

BAKERRISK

BAKERRISK

Laminated Window Test

Door

- Company Demonstration
- Blast Load ~ 2 psi, 40 psi-msec

BAKERRISK

Door Blast Test

BAKERRISK

CMU Wall

- **Wall test conducted at BakerRisk shock tube facility**
- **Lightly reinforced wall**
- **Blast Load ~ 18 psi, 118 psi-msec**
- **Wall suffered severe damage, requiring replacement**

BAKERRISK

BAKERRISK

CMU Wall Test

CMU Wall

- Wall test conducted at BakerRisk shock tube facility
- Lightly Reinforced Wall
- Blast Load ~ 10 psi, 200 psi-msec
- Blowout

BAKERRISK

BAKERRISK

CMU Wall Test

Tent Testing

- **Air tent testing**
- **Blast load: ~ 4 psi, 80 psi-msec**

BAKERRISK

BAKERRISK

800 x 600 / 500 fps / 2000 us
Frame 200 / 0.400 s

BAKERRISK

Explosion Testing

Structure With Vent Panels

- **Propane (~4.0%)**
- **2D flame expansion**
- **Low congestion (<15% Area blockage ratio)**

BAKERRISK

Low Congestion Test

BAKERRISK

Vapor Cloud Explosion

- Ethylene (~5.88%)
- 3D flame expansion
- Medium congestion (15-30% Area blockage ratio)

BAKERRISK

Medium Congestion Test

Thank You

- **Questions, Discussion**

SIMPLE OPPORTUNITIES FOR IMPROVING ENERGY EFFICIENCY

Alan Rossiter,
President, Rossiter & Associates
Past Chair, STS-AIChE
alan@rossiters.org

PROGRESS IN ENERGY EFFICIENCY

- Breakthrough technologies.
- Better equipment.
- Clever design procedures.
- Advanced control & real time optimization.
- Management systems
- **Don't Forget the Basics!**

EXAMPLE 1: KEEP THE HOT STREAMS HOT

WHAT'S WRONG WITH THIS PICTURE?

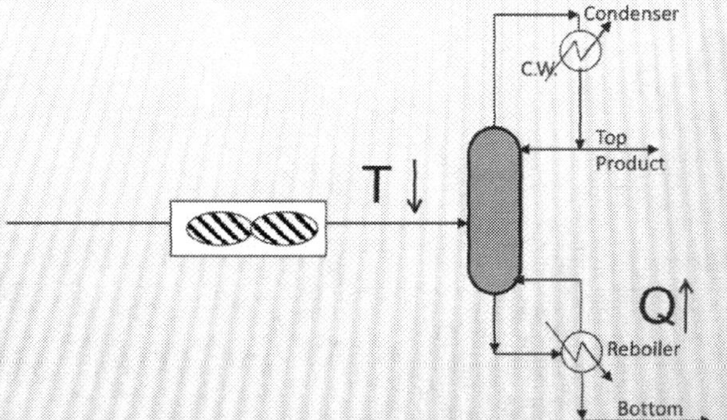

A SIMPLE PARTIAL FIX...

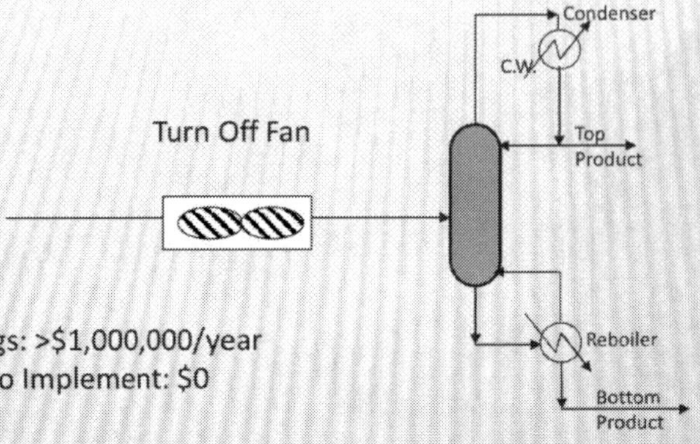

Savings: >$1,000,000/year
Cost to Implement: $0

A FURTHER CORRECTION...

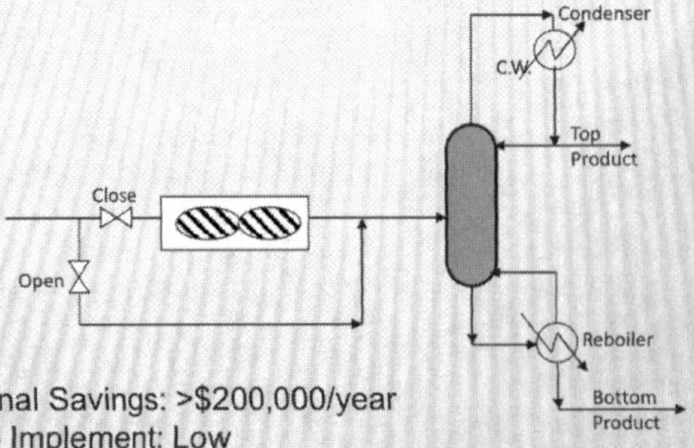

Additional Savings: >$200,000/year
Cost to Implement: Low

EXAMPLE 2:
IF YOU'VE GOT IT, USE IT!

WHY BYPASS?

Operators opened the bypass because the pumparound duty became too high in the winter. They did not close it in the summer.

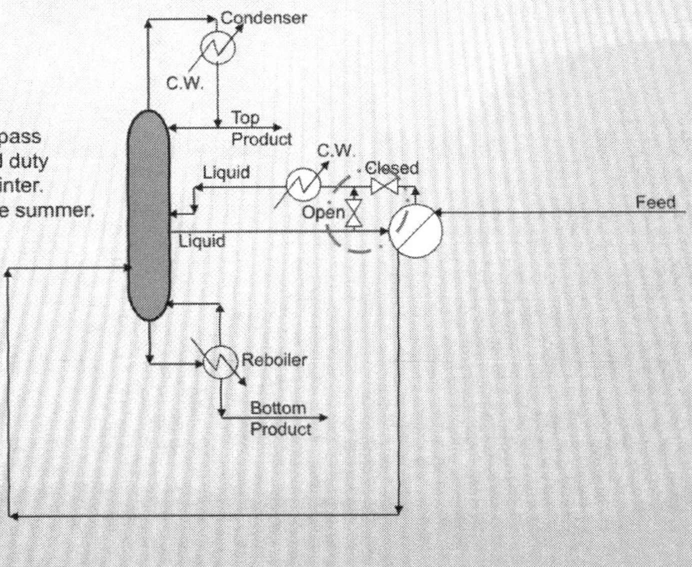

A SIMPLE FIX...

Close the bypass.
Monitor bypass valve position.
Operators may only open the bypass after obtaining authorization.

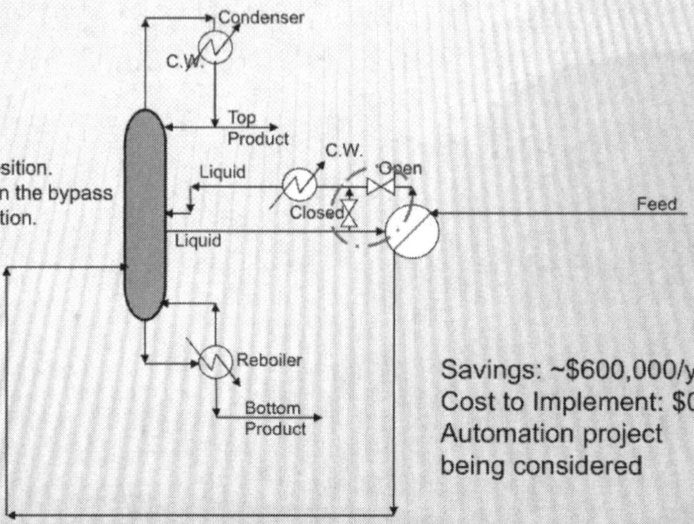

Savings: ~$600,000/year
Cost to Implement: $0
Automation project being considered

EXAMPLE 3: FINISH THE JOB

HEAT EXCHANGER BASICS

HEAT EXCHANGER BASICS

- Performance decays over time as fouling or scaling increases resistance to heat transfer
- Rate of decay: type of service and the design of the heat exchanger
- Periodic cleaning required for many heat exchangers
- Monitor performance of individual heat exchangers and trains (software available)

A Case Study in Heat Exchanger Cleaning

FAILING TO FINISH THE JOB

Cleaned, but then left out of service for 2-½ months.

Lost Opportunity ~ $100,000

Lesson Learned: Communicate!

EXAMPLE 4: DON'T PUMP MORE THAN YOU NEED

WHAT'S WRONG HERE?

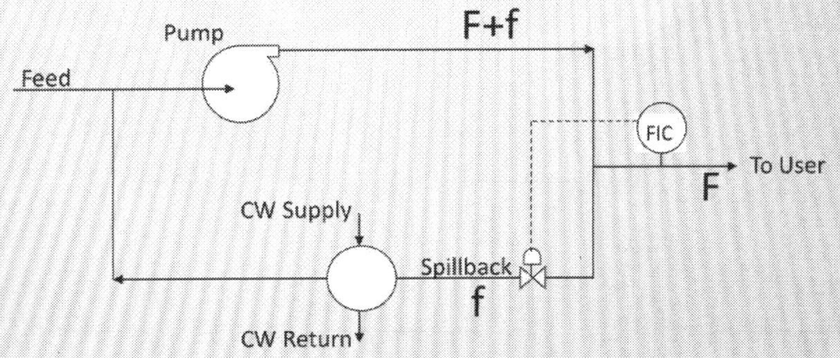

Inefficient Flow Control

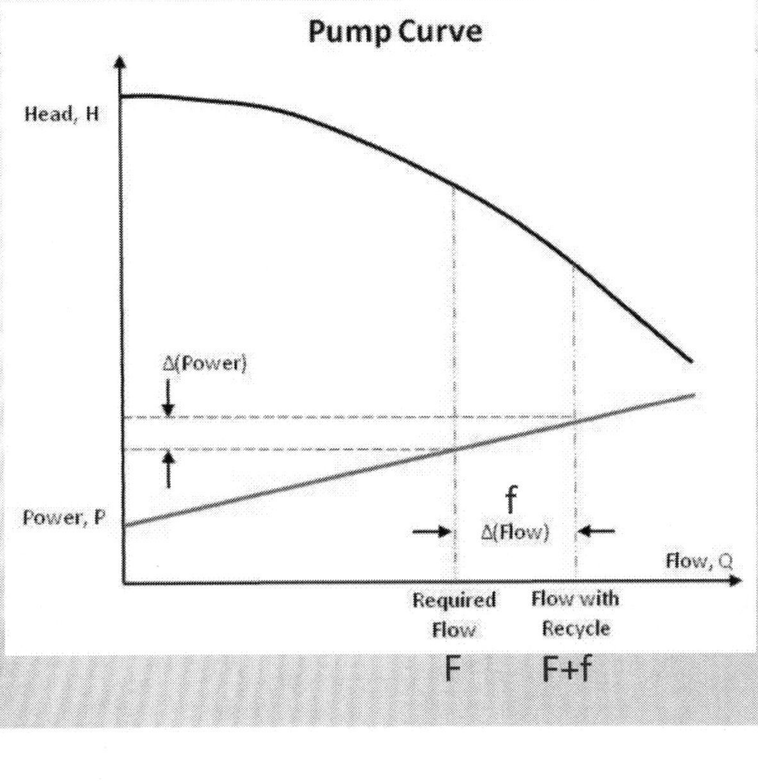

TAKE ADVANTAGE OF THE PUMP CURVE!

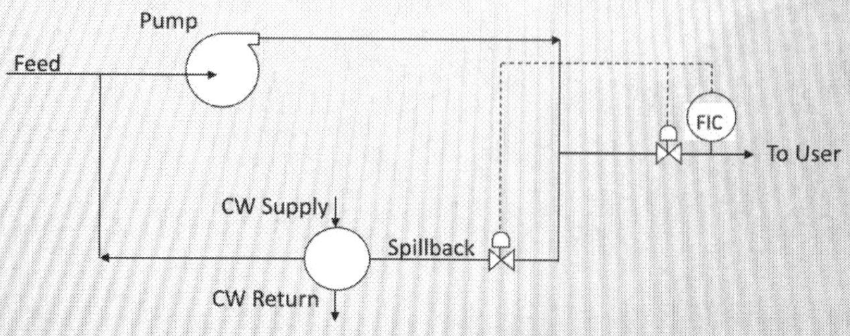

Minimize Spillback:
Saves $150,000/year

COMPARING PUMP CONTROL OPTIONS

Bypass Control:

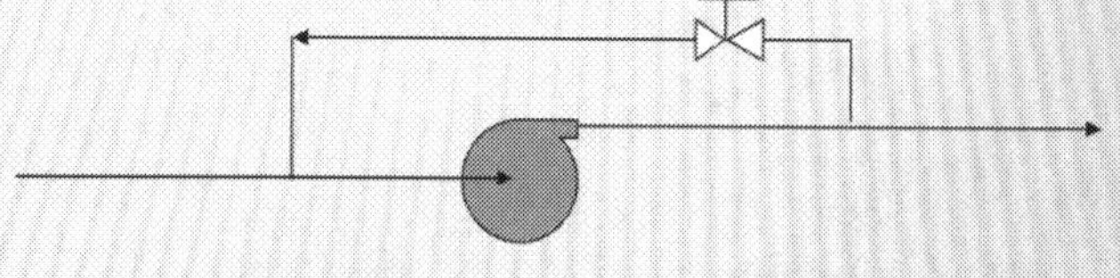

Throttle Control:

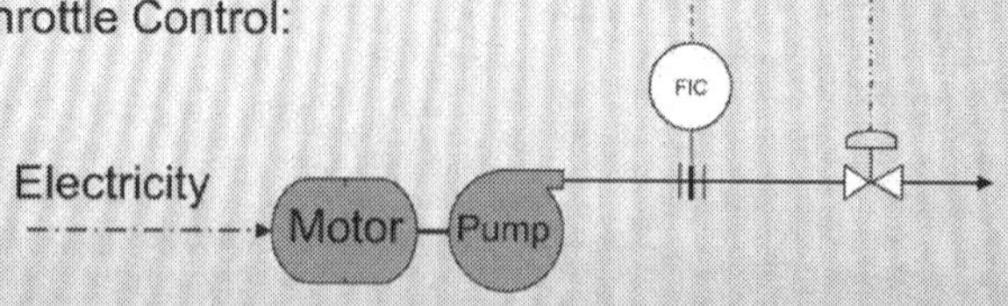

ANOTHER OPTION

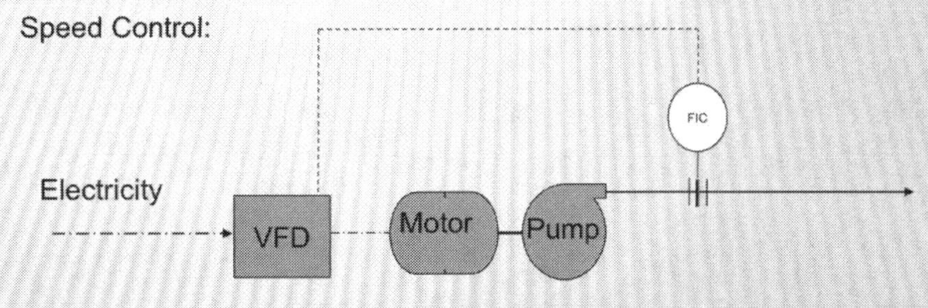

VFD = Variable Frequency Drive

POWER SAVINGS WITH VFDS

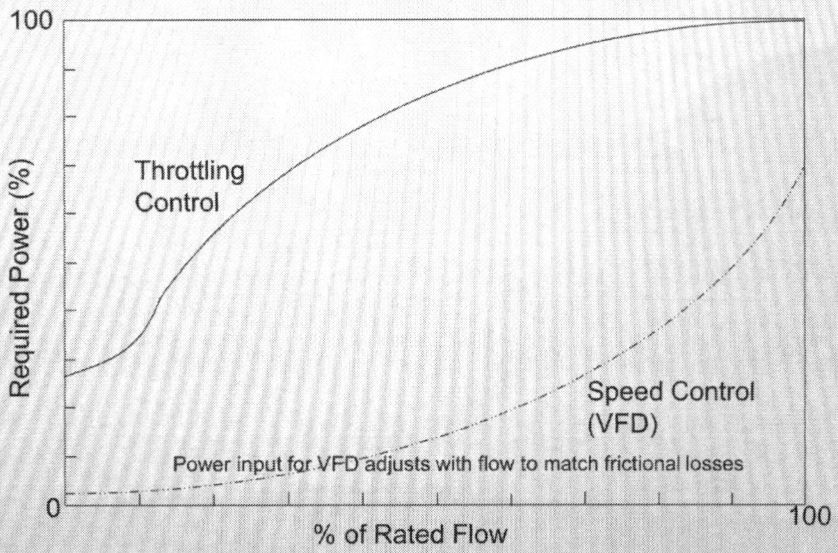

EXAMPLE 5:
BASIC PHYSICAL PROPERTIES

THERMAL DEAERATOR

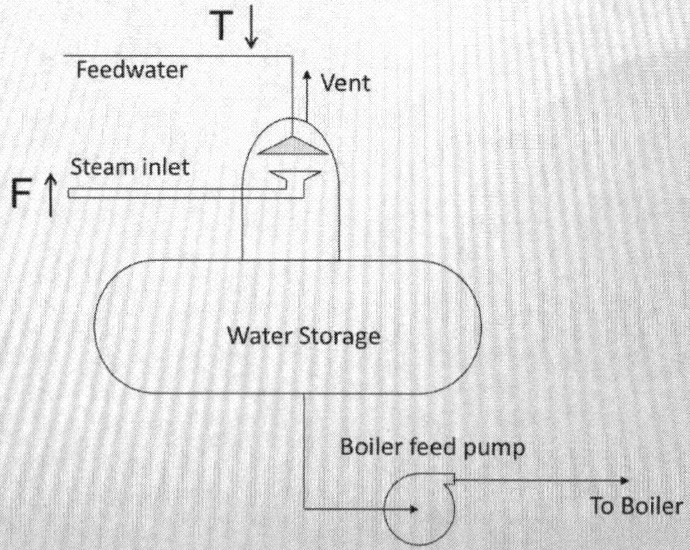

WHAT'S WRONG THIS TIME?

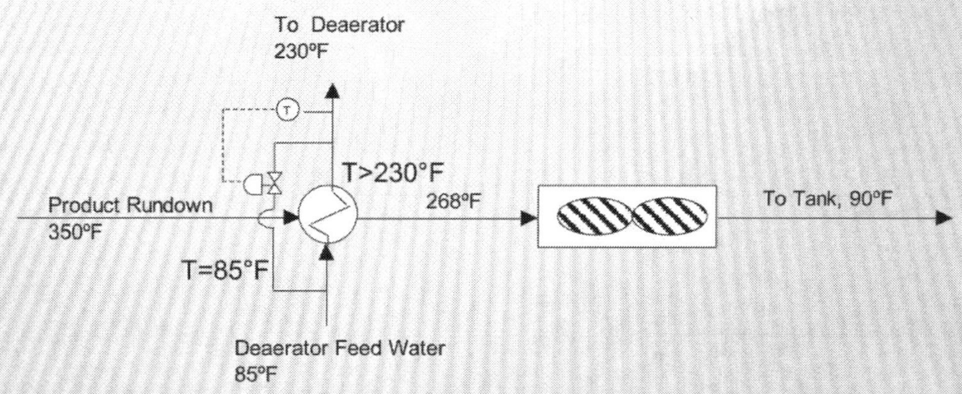

Water boils at 240°F under 10 psig pressure.
It boiled in the heat exchanger, causing vapor lock

A PARTIAL CORRECTION

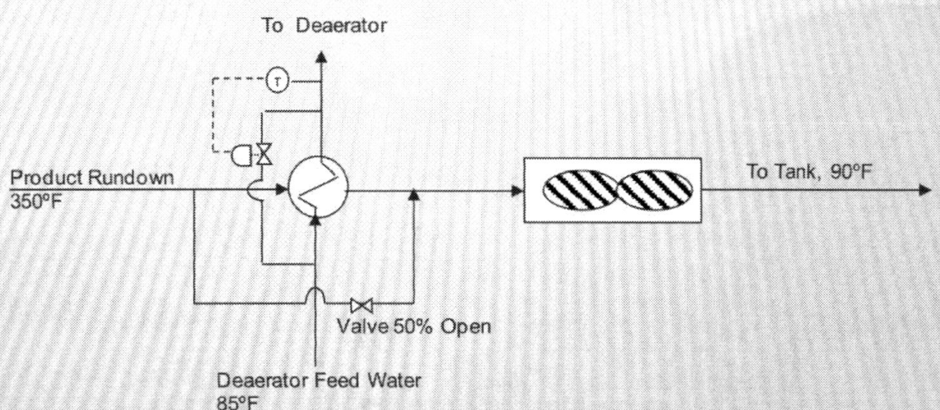

A BETTER FIX

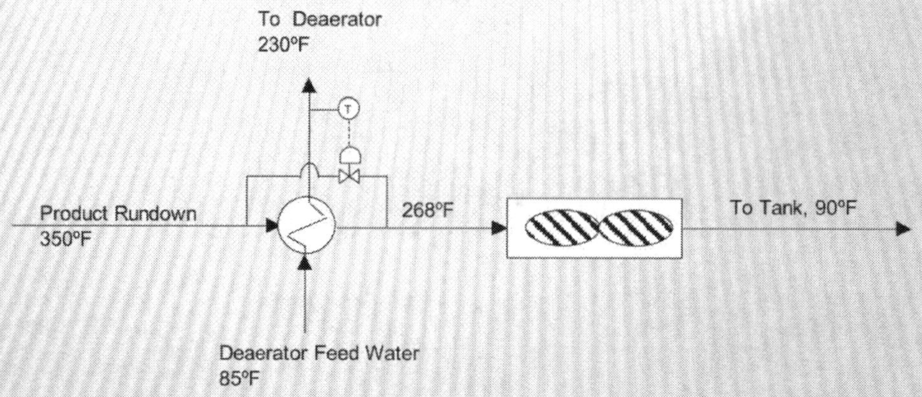

CONCLUSIONS

Use all available tools but, don't forget the basics!

- Understand the purpose and limitations of your equipment
- Exploit changes in ambient conditions
- Communicate effectively
- Consider the energy impact of your control systems
- Remember basic physical properties

AUTHOR INDEX